Microbial Fuel Cell Technology for Bioelectricity

Venkataraman Sivasankar • Prabhakaran Mylsamy
Kiyoshi Omine
Editors

Microbial Fuel Cell Technology for Bioelectricity

With a Foreword by
Kazuya Watanabe

Springer

Editors
Venkataraman Sivasankar
Post Graduate and Research
Department of Chemistry
Pachaiyappa's College
Chennai, Tamil Nadu, India

Department of Civil Engineering
Graduate School of Engineering
Nagasaki University
Nagasaki, Japan

Prabhakaran Mylsamy
Post Graduate and Research
Department of Botany
Pachaiyappa's College
Chennai, Tamil Nadu, India

Kiyoshi Omine
Department of Civil Engineering
Graduate School of Engineering
Nagasaki University
Nagasaki, Japan

ISBN 978-3-030-06542-3 ISBN 978-3-319-92904-0 (eBook)
https://doi.org/10.1007/978-3-319-92904-0

Softcover re-print of the Hardcover 1st edition 2018

Printed on acid-free paper

This Springer imprint is published by the registered company Springer Nature Switzerland AG
The registered company address is: Gewerbestrasse 11, 6330 Cham, Switzerland

Dedicated to my beloved mother Padmavathi Venkataraman and eldest brother M.V. Jothiprakash

Venkataraman Sivasankar

Dedicated to my wife Ranjith Kumari Prabhakaran and son Siddharth Prabhakaran

Prabhakaran Mylsamy

Foreword

Currently, human society faces critcal problems associated with the use of fossil fuels. The amount of fossil fuel stored in underground reservoirs has decreased rapidly due to mass consumption. Furthermore, carbon dioxide exhausted by the combustion of fossil fuels accumulates in the air, resulting in the greenhouse effect and global warming. Under such circumstances, energy sources alternative to fossil fuels are strongly needed for supporting human activity in the twenty-first century, particularly alternatives that are renewable and do not trigger global warming. Biomass has been realized as an important renewable energy source, and scientists and engineers are keen on developing technologies for utilizing biomass as an energy source. Bio-electricity generated from biomass therefore attracts many researchers as a promising option for renewable energy.

Microbial fuel cells (MFCs) are devices that use living microbes as catalysts for the conversion of organic fuels into electricity, and these devices have recently attracted extensive attention as sustainable bioenergy systems. MFC systems can be applied to the generation of electricity at water/sediment interfaces in the environment, such as bay areas, wetlands, and rice fields. In MFCs, electrochemically-active bacteria (EAB), which can transfer electrons to extracellular solid acceptors, play key roles in electricity generation. Many EAB have been isolated from MFCs and natural environments and have been characterized in the laboratory. Due to their diverse metabolic potentials and established manipulation techniques, EAB are considered to be excellent materials for biotechnology.

In this book, we will find an excellent collection of articles on the fundamental description, application, and further perspectives of MFC technology. I therefore believe that this book will exert a tremendous impact on science and technology for the research, development, and commercialization of renewable green energy. I strongly recommend this book to researchers, engineers, and students who are interested in MFC technology.

Finally, I appreciate the contributions and dedicated efforts of all the authors in this book.

Sincerely,

Kazuya Watanabe, Professor
School of Life Science
Tokyo University of Pharmacy and Life Sciences
Japan

Preface

Even the structure of the atom has been found by the mind

– Bhagvan Sri Ramana Maharshi

In eighteenth century, the birth of microbial fuel cells (MFCs) was registered by Luigi Galvani, who explored the flow of electrons in biological organisms in the legs of a frog. Later, in 1912, Michael C. Potter at University of Durham, United Kingdom, carried this research forward and made his first demonstration of MFCs. After four decades of Barnet Cohen's (1931) MFC research using multiple units, I. Karube reported further developments in 1975. The centenary of MFC research was proved with an outcome, such that the first practical application was achieved by running a small DC motor for a few seconds by H.P. Bennetto in 1985. Following this, the role of electroactive bacteria (EAB), by W. Habermann and E.-H. Pommer, and electron conduction properties of *Geobacter* species, by D.R. Bond and D.R. Lovely, were reported in the years 1990 and 2003, respectively.

With the above anecdote, the basic principle and the organization of chapters have been discussed in a nutshell as follows:

The electron transfer mechanism is significant in an electrochemical system and is sought with an absolute interaction between microbes and electrodes. Conspicuously, in bio-catalyzed systems, the transfer becomes significant in harnessing electrical energy along with the degradation of pollutants present in the system.

Microbial fuel cells (MFCs) are electrochemical reactors that function duc to the oxidation of biodegradable substrates at the expense of electroactive bacteria (EAB) to generate electric current. Microorganisms are the biocatalysts that help in the electrical energy conversion of biodegradable substrates including complex wastewater and sediments in addition to simple sugars and derivatives. The metabolic transportation of electrons from EAB to electrodes (insoluble substrates) is feasible thanks to extracellular electron transport.

On the basis of the electrogenic nature of certain bacteria, MFC technology has gained incredible attention for its production of energy from various soluble organic wastes of natural (low-carbon biomass) and synthetic sources.

The mixed cultures are preferred for inoculation in MFCs due to cost-effectiveness and practicality reasons in wastewater treatment. Studies on the choice of sewage for bacterial feeding in MFCs to drive out energy are compelled due to futuristic challenges and environmental sustainability. Although we have ample knowledge of microbiology and electro-/materials chemistry, we need to examine the technical viability in accordance with the application niche which together pave the way for commercialization.

As the possibility of drawing energy from biotic components is the need of the hour, Chap. 1 focuses on the possible sources of biomass which sustainably cater to energy needs. The progress of converting waste to worth has yet to move at a radical rate in the coming years.

The real notion behind the second chapter is to make readers understand the fundamentals, classifications, significance, and challenges of MFC technology.

Indeed, the inherent nature of the autotrophs (plant and algae) has been of spellbinding interest toward the making and breaking of electricity and organic contaminants, respectively. Intensive research on a potentially engineered rhizodeposition would facilitate an appreciable rate of electron transfer to captivate state-of-the-art applications. On the other hand, the advent of heterotrophic fungi has recently gained attention recently by utilizing the species to form biofilm around the electrode, which improves the MFC's performance. By bridging the significance of biomass and energy, Chaps. 3, 5, and 6 provide exhaustive discussions based on the previously reported findings on plant-, algae-, and fungi-based MFCs.

Above all, the quintessential microbial community is governed by the types of plants and soils upon which the output power of an MFC is relied upon. To corroborate this, Chaps. 4 and 7 discuss the significance of paddy plant and soil (with organic waste) in generating appreciable electricity by using bamboo charcoal spun with iron wire as an anode. The duality of iron wire (with bamboo charcoal) was discernible with regard to electron generation and nutrition to the soil organisms.

Exclusive chapters (8 and 10) have been devoted to energy production from animal waste; in particular, the ruminal degradation of biomass was studied by varying electrodes, pH, substrates, catholytes, and buffers.

Interestingly, electricigens are one among the meritorious factors in enhancing the efficiency of MFCs. These are microorganisms with high coulombic efficiency that completely oxidize organic fuels to carbon dioxide by actively transporting electrons to electrode. Also, these are credited with the conservation of energy which in turn equips MFCs for long-term sustainability. Electricigens are detailed in Chap. 9.

The bimodal practice of bioelectricity generation and remediation has become a dominant area of research. With profound insight, Chaps. 11 and 12 discuss the efficiency, economic feasibility, and ultimate commercialization of the process. Specifically, microbial desalination cell (MDC)–based research attempts to discuss the two unique challenges of water purification and sustainable power from a single vantage point, and it has gained great attention among interdisciplinary researchers.

Chapter 13 addresses the performance of MFCs at pilot-scale levels and the associated challenges to be overcome in large-scale applications. The final chapter of the book provides a thorough investigation of the successful implementation of MFCs in rural areas and stresses its potential to improve life conditions.

The present book will be very useful and informative as a valuable reference to those working in MFC, biofilm interactions, bio-electrochemistry, and waste management. In addition, it should gain much of appreciation among environmental chemists/biologists, bioenergy researchers, and wastewater/biofuel industrialists.

Nagasaki, Japan Venkataraman Sivasankar
Chennai, India Prabhakaran Mylsamy
Nagasaki, Japan Kiyoshi Omine

Contents

About the Editors

Venkataraman Sivasankar received his doctorate in Chemistry in 2009 from Bharathidasan University, Tiruchirappalli, Tamil Nadu, India. Presently, he is a post-doctoral fellow in the Department of Civil Engineering, Nagasaki University, Nagasaki, Japan. He has been a faculty member in the Department of Chemistry in Pachaiyappa's College, Chennai, India, since 2014. His research areas include materials synthesis and wastewater treatment. He received the prestigious JSPS fellowship in 2016. To his credit, he has more than 50 research articles in peer-reviewed journals and five book chapters with renowned publishers. For instance, he edited the book, *Surface Modified Carbons as Scavengers of Fluoride from Water* in 2016 with Springer. He collaborates and performs research with professors in universities and research laboratories in Algeria, France, Japan, Iran, and South Africa.

Prabhakaran Mylsamy is an assistant professor of Botany in Pachaiyappa's College, Chennai, Tamil Nadu, India. He completed his doctorate in 2012 at the University of Madras, Chennai, India. His research focuses on algal biotechnology including algal MFCs. He was awarded the DST – SERB Young Scientist Award in 2013. He has been accredited with both a major project (DST) and a minor project from UGC. He is credited with 15 original research papers in national and international peer-reviewed journals, one authored book, and five book chapters.

Kiyoshi Omine is a professor in the Department of Civil Engineering, Graduate School of Engineering, Nagasaki University, Nagasaki, Japan. His research areas of interest include soil microbial fuel cells for composting and power regeneration, geotechnical utilization of waste materials, and geo-environmental improvement techniques. He is a member of several technical societies of Japan. He has authored and coauthored more than 50 research papers in national and international peer-reviewed journals. He is credited with mentoring three JSPS fellows in Kyushu and Nagasaki universities.

Chapter 1
Biologically Renewable Resources of Energy: Potentials, Progress and Barriers

Vasanthy Muthunarayanan, Gueguim kana Evariste Bosco, Thamaraiselvi Chandran, Tamilselvi Duraisamy, Selvakumar Muniraj, Yeshona Sewsynker-Sukai, Preshanthan Moodley, and Zanenhlanhla Gumbi

1.1 Introduction

1.1.1 Energy

Energy revolves around various aspects of life and plays a major role in the survival of all living organisms. It may be defined as the capacity to do work and its unit of measurement is joule. Several types of energy exist, and these include chemical, mechanical, kinetic, potential and electrical, among others. Energy sources may be categorized as renewable and non-renewable (Shockey et al. 2010). Non-renewable energy resources include petroleum-based fossil fuels that are diminishing at an alarming rate as a result of the increased global energy demand. In the present time, global energy requirements are primarily met by non-renewable fossil fuels such as coal, petroleum, bitumen, natural gas and tar sand (Das and Veziroglu 2001). Dependence on conventional fossil fuels such as petroleum-based reserves has led to its depletion combined with environmental pollution (Levin et al. 2004). Thus, renewable energy production has occupied global precedence and has made the search for an efficient and sustainable energy system an imperative for sustainable socio-economic development (Barbir et al. 1990; Demirbas 2009). Renewable energy resources include solar, wind, hydro- and geothermal power (Shockey et al.

V. Muthunarayanan (✉) · T. Duraisamy · S. Muniraj
Department of Environmental Biotechnology, School of Environmental Sciences, Bharathidasan University, Tiruchirappalli, 620024, India
e-mail: vasanthy@bdu.ac.in

G. k. E. Bosco · Y. Sewsynker-Sukai · P. Moodley · Z. Gumbi
School of Life Sciences, University of KwaZulu-Natal, PMB Campus, Durban, 4041, South Africa

T. Chandran
Department of Biotechnology, Mother Teresa Woman's University, Kodaikanal, 624101, India

V. Sivasankar et al. (eds.), *Microbial Fuel Cell Technology for Bioelectricity*,
https://doi.org/10.1007/978-3-319-92904-0_1

2010) in addition to microbial-produced biofuels such as bioethanol, biogas and biodiesel (Naik et al. 2010).

1.1.2 Energy Resources and Sustainable Development

Sustainable development can be defined as living, producing and consuming in a manner that meet the needs of the present without compromising the ability of the future generations to meet their own needs. Energy needs are widely regarded as a major sustainable development challenge, with close links to the climate change and global poverty agenda. It was at the Stockholm UN conference in 1972 where the relationship between the increasing energy consumption and its relation to environmental degradation was first addressed (Petford 2004). Later in May 2002, UN secretary-general Kofi Annan identified five key areas as global critical challenges of the twenty-first century. The areas were energy, water and sanitation, health, agriculture and biodiversity. With climate change as a challenging issue to be faced, the world now focusses to tap clean and renewable energy to achieve sustainable development.

1.1.3 Current Scenario of World's Energy Usage

It will be apt to mention that the world's major energy requirement is met by fossil fuels till date. The longevity of the resources available to meet the world's energy demands is yet to be understood. The world's population is ever increasing with it presently being about 7.5 billion and is projected to reach about 9.2 billion by 2050 as per 2008 estimate (www.fao.org/docrep/014/i2280e/i2280e.pdf). With the increase in population, the energy demand is expected to increase regardless of the countries. In the International Energy Outlook (IEO) 2000, much of the growth in worldwide energy use is projected for the developing world. In particular, energy demand in developing Asia and Central and South America is projected to more than double between 1997 and 2020 (http://www.eia.doe.Gov/oiaf/ieo/tbla1_a8.html). Based on the world's energy demand, coal was estimated to last till 2080. Crude oil has been estimated to be completely exhausted by the year 2060, whereas natural gas is estimated to be exhausted by the 2050s. (www.ecotricity.co.uk). As per Beyond Petroleum (BP) statistical review in 2003, the proven reserves include 1000 million barrels of oil by 2001, of which 65% is reported to be with the Middle East, 165 trillion m^3 of natural gas and about 800 billion tons of coal. Considering both the availability and demand, it is clear that these reserves will not last long enough to support the global future demand. In addition to its exhaustion, fossil fuels have been shown to be detrimental to the environment which has further prompted the search for alternative resources.

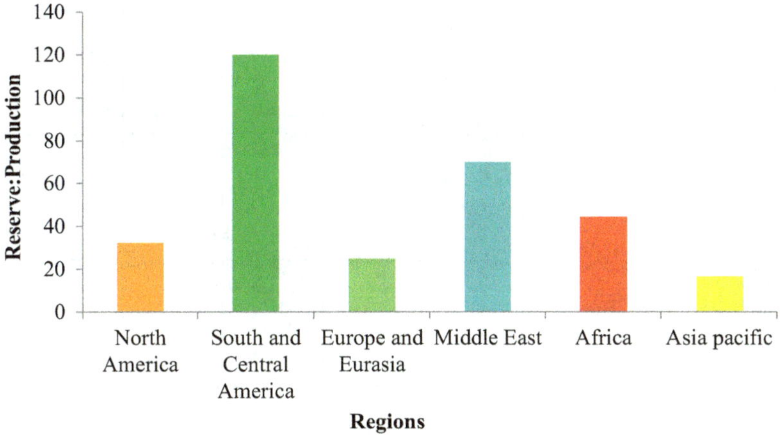

Fig. 1.1 Regional oil reserves to production ratio for 2016. (BP 2017)

Oil reserves have shown to be the most exploited energy source worldwide. The annual Beyond Petroleum (BP) statistical review of world energy 2017 reported that the total proven oil reserves at the end of 2016 increased by 0.9% (BP 2017). The Middle East is the major oil supplier contributing 47.7% of the total world oil reserves. The global oil reserve-to-production ratio (Fig. 1.1) shows that the Middle East oil reserves, according to the current production rate, are sufficient to meet nearly 50.6 years of global production (BP 2017).

The *Hubbert peak theory* says that for any given geographical area, from an individual oil-producing region to the planet as a whole, the rate of petroleum production tends to follow a bell-shaped curve (Hubbert 1956). It is one of the primary theories on peak oil. The theory is based on the observation that the amount of oil under the ground in any region is finite; therefore, the rate of discovery which initially increases quickly must reach a maximum and decline. In the USA, oil extraction followed the discovery curve after a time lag of 32–35 years (Fig. 1.2). The theory is named after American geophysicist M. King Hubbert, who created a method of modelling the production curve given an assumed ultimate recovery volume.

Human population have dwindled the natural resources of the world, be it the fossil fuels, minerals, metals, coral reefs, etc. However, as non-renewable resources are finite, different estimations were made regarding the time period these resources will last. It is found to vary between 50 and 500 years (Harris 2014). But still, certain productive ecosystems were not fully utilized by the mankind which includes the biomass.

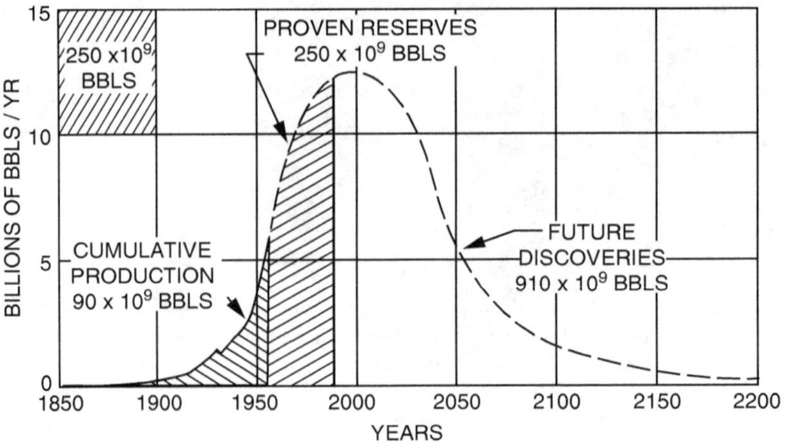

Fig. 1.2 Comparison of actual oil production rate data with the prediction made by Hubbert in 1956

1.2 Renewable Energy Resources

Renewable energy resources include wind, tidal, solar, geothermal and biomass. Though nuclear energy is categorized as renewable, it has its own demerits of disposal of the nuclear wastes and also with respect to the problems associated with the decommissioning of such nuclear power plants. Hence for the sustainability to be ensured, renewable energy resources are to be depended upon by both the current and future generations to meet their energy needs. The National Environmental Policy Act of 1969 declared the sustainability as a national policy. It is aimed "to create and maintain conditions under which humans and nature can exist in productive harmony, that permit fulfilling the social, economic and other requirements of not only the present but also the future generations". This aspect necessitates the usage of renewable resources which are nonconventional till date.

1.2.1 Potential of Biological Energy Resources

Following an oil shock experienced in the 1970s which resulted in a rise in the prices of petroleum, the attention turned towards biomass (Klass 1998). Biomass is a renewable energy resource, is carbon neutral and is found to have a good scope for future generation (http://biomasspower.gov.in/). Such biomass is the storehouse of energy as the photosynthetic fixation of CO_2 occurs in it resulting in the starches, lignocelluloses, etc. Such biomass has the possibility of being converted into fuels such as ethanol, methanol and hydrogen. Such biological origin renewable energy

Table 1.1 Energy generation potential of different biomass

Biomass type	Energy generation potential	Reference
Lignocellulose	Abundant biomass resource suitable for direct combustion for generation of heat, power Gasification and pyrolysis for energy conversion Fermentation for ethanol production but further improvements are needed for energy conversion efficiency and economic feasibility	Wang et al. (2013), Paz (2013) Bahng et al. (2009) Alvira et al. (2010), Viikari et al. (2012)
Saccharose-rich biomass, e.g. sugarcane, sugar beet, cereal grain, potatoes and cassava	On subjecting to fermentation, ethanol can be obtained	Alvira et al. (2010) Balat et al. (2008)
Algal biomass	A promising feedstock for fermentation, large-scale production under development	Wei et al. (2013)
Vegetable oil	Could be used directly or mixed with alcohol through transesterification resulting in biodiesel	Daroch et al. (2013)
High moisture or wet biomass, e.g. sludge of industrial and domestic wastewater treatment, livestock manure and food residues	Suitable for biogas production which could be used in gas turbines	Tauseef et al. (2013) Weiland (2010)

resources could be regarded as the promising sources of energy for the present and future.

Biomass refers to the wood from trees, bushes, charcoal, crop residues and animal wastes. However, it is estimated that it could provide about 7–15% of the global energy requirement (Woodward et al. 2000). It has the advantage of being grown at any place and could be utilized to meet the energy demand. Further, it could be stored and used easily for a longer time interval. The usage could be matched with the rate of growth to again ensure sustainability.

Such biomass is classified according to their origin, namely, agricultural and forest residues, residues form agro-industries and municipal waste (Paz 2013). According to the composition of the biomass, they are classified as follows (Table 1.1):

Biomass is further less site-specific than other renewable energy resources such as wind or tidal power, as some type of vegetation can be grown anywhere. From such biomass, combustion is the immediate process of energy extraction, but other processes such as fermentation, anaerobic digestion and gasification are also possible. It is important to understand that the rate of growth of such biomass must not be exceeded by the rate of harvest. However, one major consideration is the efficiency of plant leaves which could convert only about 1% of the incident solar radiation into energy (Petford 2004).

1.2.2 Potential and Progress of Biomass Utilization as Biofuel

The total primary production occurring on planet earth meets the demands of the human population. The demand is met through the agriculture, fisheries, forestry and other similar activities. The US Energy Information Administration's latest IEO (International Energy Outlook) 2017 projects that the world energy consumption would grow by about 28% between 2015 and 2040. Now the attention is focussed on the reduction of greenhouse gas emissions in accordance with the Kyoto Protocol. Thus, renewable biological energy resources have gained significant interest. Renewable biological energy resource, namely, biomass, includes the wastes and residues of plants and animals. It is organic, is carbon based, and undergoes combustion to release heat. Heat may be used to generate work and electricity. There are three different ways to utilize biomass: as mentioned above it can be burned to produce electricity and heat, it could be changed to gas-like fuels such as hydrogen, carbon monoxide and methane and it could be converted into liquid fuels also called biofuels such as ethanol and methanol (Demirdas 2010).

The heat energy available after combustion equivalent to the enthalpy or net energy density ranges from 8 MJ/kg (undried green wood) and 1.5 MJ/kg (dry wood) to 40 MJ/kg (fats and oils) and 56 MJ/kg (Methane) (Twidell and Tony Weir 2006).

Lignocellulosic biomass refers to plant material which is predominantly composed of carbohydrate polymers such as cellulose and hemicellulose bound by lignin. Cellulose is a polymer consisting of unbranched linear glucose, whereas hemicellulose is a heterogeneous polymer composed of hexose and pentose sugars (Binod and Pandey 2016). The cellulose, hemicellulose and lignin layers usually make up approximately 30–50%, 20–40% and 10–30%, respectively, depending on the type of biomass (Sindhu et al. 2016). Lignocellulosic biomass accounts for approximately 50% of the global biomass, with an annual production of 200 billion tons (Kabir et al. 2015). Biomass is a major contributor to both the agricultural and forestry industry and is considered a promising feedstock for biofuel production given its low cost and abundance, and it does not compete with food security (Cai et al. 2017). The agricultural industry generates millions of tons of lignocellulosic waste which are dumped in landfill sites or burnt in the field postharvest (Moodley and Gueguim Kana 2017a). The conventional use of lignocellulosic biomass has been mostly burning for energy, for instance, sugar-processing plants that use the sugarcane bagasse residue to power boilers (Smithers 2014). However, this process has severe negative environmental implications. In recent times, there has been a shift towards using biomass for the production of biofuels such as hydrogen, ethanol and biogas (Brethauer and Studer 2015) as shown in Table 1.2.

Prior to biofuel production, lignocellulosic biomass needs to undergo biochemical conversion to release simple sugars such as glucose and xylose either via enhanced enzymatic or chemical hydrolysis (Taherzadeh and Karimi 2008). Various pretreatment strategies exist that can achieve efficient biochemical conversion by degrading the recalcitrant lignin layer thereby allowing either microorganisms or

Table 1.2 Biofuels produced from lignocellulosic biomass

Biomass	Biofuel	Yield	Reference
Sugarcane leaves	Biohydrogen	248 ml/g sugar	Moodley and Gueguim kana (2015)
Sorghum leaves	Biohydrogen	213.14 ml/g sugar	Rorke and Gueguim kana (2016)
Sorghum leaves	Bioethanol	17.15 g/L	Rorke and Gueguim kana (2017)
Seaweed	Bioethanol	14.89 g/L	Adams et al. (2011)
Corn cobs	Bioethanol	24 g/L	Li et al. (2016)
Duckweed	Biogas	11,620 ml	Yadav et al. (2017)
Pine	Biogas	17 L/kg	Brown et al. (2012)

enzymes access to the cellulose and hemicellulose fractions (Moodley and Gueguim Kana 2017b). These pretreatments are detailed in Table 1.3. Currently, acid and alkali pretreatment are being explored at a pilot scale. Kapoor et al. (2017) examined pilot-scale dilute acid pretreatment of rice straw, whereas Skiba et al. (2017) examined pilot-scale alkali pretreatment of oat hulls. New pretreatment technologies have significantly enhanced enzymatic saccharification and sugar recovery from lignocellulosic biomass. Inorganic salt has emerged as an effective and low-cost chemical for biomass pretreatment (Sewsynker-Sukai and Gueguim Kana 2017). A previous report by Sewsynker-Sukai and Gueguim Kana (2017) optimized a sequential alkalic and metal salt pretreatment using corn cobs and gave a high reducing sugar yield of 1.10 g/g with 14.02% $Na_3PO_4.12H_2O$, 3.65% $ZnCl_2$ and 5% solid loading. The optimized pretreatment gave a tenfold increase in sugar yield compared to previous studies on corn cobs (Sewsynker and Gueguim Kana 2017). Likewise, Moodley and Gueguim Kana (2017c) reported on a sequential microwave-assisted salt-alkali pretreatment of sugarcane leaves that yielded 1.17 g/g reducing sugar using 1.67 M $ZnCl_2$ and 1.52 M NaOH at 400 W for 10 min. This pretreatment technique showed a 62% sugar yield improvement compared to previous reports (Moodley and Gueguim Kana 2017c).

1.2.3 Production of Ethanol from Biomass

Present-day energy security coupled with environmental concerns has skewed research towards alternative energy sources that are renewable and sustainable, low-cost and environmentally friendly (Zabed et al. 2016). Ethanol has gained significant interest as a potential replacement for fossil fuel-derived sources (Aguilar-Reynosa et al. 2017). This is largely due to its several advantages over fossil fuels such as gasoline which include its renewable and sustainable nature, ease of storage, higher oxygen content and higher octane number, among others (Zabed et al. 2016). In the present time, countries such as the United States of America (USA), Brazil, China, Canada and several European Union (EU) member states have publicized their commitment to bioethanol development programmes in an attempt to reduce the reliance on conventional fossil fuels. The annual global

Table 1.3 Commonly employed pretreatment technologies

Pretreatment	Mode of action	Advantage (s)	Disadvantage (s)	Reference
Steam explosion	Mechanical shearing and defibrillation of fibres Amorphous cellulose potentially depolymerized	Cost effective No toxic chemicals required	Partial degradation of lignocellulosic matrix Produces toxic inhibitor compounds	Wang et al. (2015)
Irradiation	Cellulose is degraded into fragile fibres and oligosaccharides	Improves enzymatic hydrolysis	High cost Challenges with scale-up	Akhtar et al. (2015)
Pyrolysis	High temperature causes cellulose to decompose	Less volatile compounds are produced	Process is slow	Akhtar et al. (2015)
Alkaline	Cleaves linkages in lignin and glycosidic bonds of polysaccharides	Requires low temperature and pressure Low inhibitors generated Produces highly digestible substrate	High cost Generation of irrecoverable salts	Sindhu et al. (2015)
Acid	Hydrolyses hemicellulose to xylose Modifies lignin structure	Simple method. Thermal energy not required	High cost Produces toxic inhibitor compounds	Jung and Kim (2015)
Microwave chemical	Dipolar polarization achieves heating Rapid oscillation causes molecules to vibrate	Uniform heating Improves pretreatment speed Decreased energy input	Dependent on properties of the material Formation of hot spots Challenges with scale-up	Xu (2015)
Inorganic salt	Act as Lewis acids Dissociate into complex ions and rupture glycosidic linkages	Low cost Low toxicity	Partial degradation of lignocellulosic matrix	Sewsykner-Sukai and Gueguim kana (2017)

bioethanol production has stealthily increased over the last 10 years and is depicted in Fig. 1.3. The USA contributed to the highest ethanol production and has been projected to be more than 50% of the total global ethanol produced in 2016 (RFA 2017). More specifically, second-generation lignocellulosic bioethanol production has received much attention as a suitable process for bioethanol production without raising food security concerns (Aguilar-Reynosa et al. 2017; Zabed et al. 2016). The annual global production of lignocellulosic biomass has been estimated at 200 billion t/year, of which nearly 8–20 billion t can be used for biofuel production (Saini

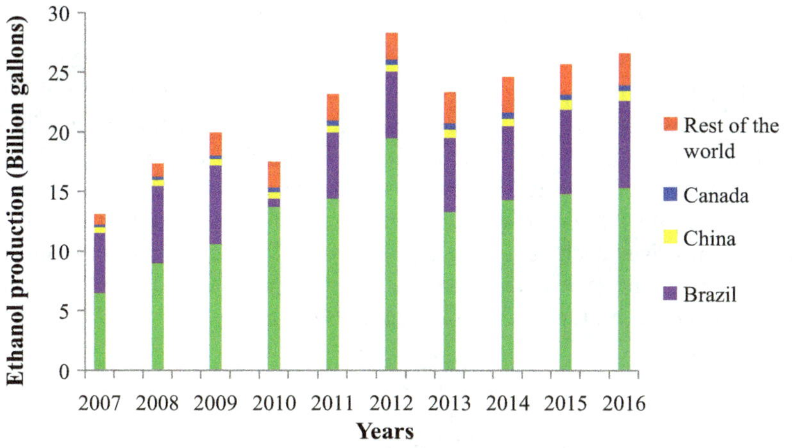

Fig. 1.3 Global ethanol production by country in the last 10 years. (RFA 2016)

et al. 2014). The majority of agricultural waste residues are derived from corn, sugarcane, rice and wheat. Corn stover has shown to be one of the most promising agro-residues for lignocellulosic bioethanol production and consists of stalks, leaves, cobs and husks. The global average annual production rate of corn stover is estimated at 4 t/acre (Kim and Dale. 2004). Another major agro-residue for bioethanol production is sugarcane bagasse which consists of stalks, leaves and tops and is the leftover waste from sugarcane processing at approximately 317–380 million t/year (Sánchez 2009). Other biomasses that show bioethanol potential include wheat straw (1–3 t/acre annually) and rice straw (731 million t/year) (Saini et al. 2014). Both rice and wheat straws are majorly produced in Asian countries, whereas corn stover and sugarcane bagasse are generated in large quantities in North and South America, respectively. Bioconversion of these agro-residues to ethanol involves three crucial steps: biomass pretreatment, enzymatic hydrolysis of sugar polymers to fermentable sugar monomers and fermentation of sugar monomers to bioethanol (Zhao et al. 2015). The biomass pretreatment step is necessary for the improvement of enzymatic accessibility to the structural matrix. A previous report on corn stover, corn cobs and wheat straw gave yields of 450, 510 and 490 L/t, respectively (Kreith and Krumdieck 2013). Presently, corn, wheat, rice and sugarcane lignocellulosic wastes display significant potential feedstocks for bioethanol production and require further exploration to maximize several key technological issues. These include cost-effective biomass pretreatment regimes that result in high sugar yields, efficient utilization of feedstock and fermentation processes that result in high ethanol yields with shorter fermentation times (Aguilar-Reynosa et al. 2017; Zhao et al. 2015; www. ⟨http://ethanolrfa.org/resources/industry/statistics/#1454099788442-e48b2782-ea53⟩; 2016.

1.2.4 *Production of Biodiesel from Biomass*

The biodiesel is scientifically defined as alkyl esters of especially long-chain fatty acids. Such acids are derived from either animal fats or vegetable oils. This biodiesel has higher energy density, facilitates favourable combustion and has an enhanced lubricating property. Biomass is an energy resource which could facilitate the production of such biodiesel. Various reports explain the possibility of utilizing agricultural residues such as tea waste, bagasse of sugarcane and hazelnut, cotton waste, corn cobs and residues of olive seeds for the production of different valuable products including biofuels (Conesa et al. 2009; Blanco et al. 2002; Demiral and Sensoz 2008).

1.2.4.1 Production of Biodiesel from Microalgae

Microalgae are a large group of unicellular autotrophic microorganisms that carry out photosynthesis. These microorganisms can convert solar energy into chemical energy with a greater efficiency than plants due to their unicellular structure (Harun et al. 2010). This group of microorganisms has received significant attention as a promising biodiesel feedstock due to their widespread availability, ability to grow on nonarable land and being cultivable on wastewater or saltwater (Brennan and Owende 2010; Mata et al. 2010). Microalgae have higher growth rates and oil yields compared to crop plants with a productivity unit per area of 13,69 l/m^2 oil yield which is higher than the 0,6 l/m2 oil yield obtained from feedstocks such as soybean, coconut and palm oil (Chisti 2007). Microalgae are also a source of different value-added products such as carotenoids, β-carotene and chlorophyll which all have uses in food and pharmaceutical industries thereby increasing the economic potential of microalgal bio-refineries (Mata et al. 2010; Harun et al. 2010).

As seen in Table 1.4, microalgae can produce much higher amounts of biodiesel as compared to plant crops. Under optimal growth conditions, microalgae are capable of producing and accumulating hydrocarbons of between 30% and 70% of their dry weight (Kong et al. 2007); most commonly reported oil yields are between 20% and 50% of the biomass dry weight (Spolaore et al. 2006). Parameters that are important for microalgal cultivation include light, CO_2, temperature and pH (Faried et al. 2017).

Microalgal fatty acids can be divided in two groups based on the polarity of the molecular headgroup: polar and neutral lipids. Polar fatty acids consist of phospholipids and glycolipids. The neutral fatty acids consist of free fatty acids and acylglycerols, which are of interest in terms of biodiesel production (Cuellar-Bermudez et al. 2015). Acylglycerols are fatty acid esters that have been bonded onto a glycerol backbone and according to the number of fatty acid chains can be classified as triacylglycerols, diacylglycerols or monoacylglycerols (Halim et al. 2011). The content and lipid profile of microalgae are species dependent but can also be affected by culture conditions such as light intensity periods, nitrogen depletion, salinity

Table 1.4 Comparison of some sources of biodiesel

Crop	Oil yield (L/ha)	Land area needed (M ha)[a]	Percent of existing US cropping area
Corn	172	1540	846
Soybean	446	594	326
Canola	1190	223	122
Jatropha	1892	140	77
Coconut	2689	99	54
Oil palm	5950	45	24
Microalgae[b]	136,900	2	1,1
Microalgae[c]	58,700	4,5	2,5

Chisti (2007)
[a]For meeting 50% of all transport fuel needs of the USA
[b]70% oil (by wt) in biomass
[c]30% oil (by wt) in biomass

stress, temperature change and pH (Richmond. 2008; Guschina & Harwood. 2006). Nitrogen limitation is a commonly used strategy in the increase of lipid and triacylglycerols in microalgae to produce biodiesel (Cuellar-Bermudez et al. 2015; Xin et al. 2010).

The ASTM definition of biodiesel is a fuel comprised of mono-alkyl esters of long-chain fatty acids derived from vegetable oils and animal fats (Hoekman et al. 2012). It can serve as alternative to diesel fuel that could be used in diesel engines, only if its physical and chemical properties conform to the international standard specification. The relevant standard in the USA is the ASTM Biodiesel Standard D 6571 (Knothe 2006). The European Union uses separate standards for biodiesel used in vehicles (standard EN 14214) and biodiesel used as heating oil (standard EN 14213) (Knothe 2006). In South Africa, the SANS 342:2016 is used to specify automotive diesel fuel blended with 5% of biodiesel.

Microalgae are rich in polyunsaturated fatty acids with four or more double bonds; for example, eicosapentaenoic acid which has five double bonds and docosahexaenoic acid which has six double bonds occur very commonly in microalgal oils; unfortunately such fatty acids and fatty acid methyl esters (FAMEs) are susceptible to oxidation during storage which reduces their acceptability for use as biodiesel (Chisti 2007). Triglycerides consist of three chains of fatty acids joined to a glycerol backbone (Halim et al. 2011); the process of transesterification replaces the glycerol molecule with methanol forming fatty acid methyl esters (Harun et al. 2010). The fatty acid profiles of microalgae are influenced by specific growth conditions such as nutrient levels, temperatures and light intensities; this can make it difficult to define a compositional profile for algal biodiesel (Hoekman et al. 2012). Ashokkumar et al. (2014) showed that the major fatty acids found in *Botryococcus braunii* were methyl palmitate and methyl oleate and the acid number of 0.49 mg KOH/g and a cetane number of 55.4 were both within the ASTM standards, whereas a study done by De Alva et al. (2013) using *Scenedesmus acutus* showed that the biodiesel produced from this microalgal species did not meet ASTM standards and

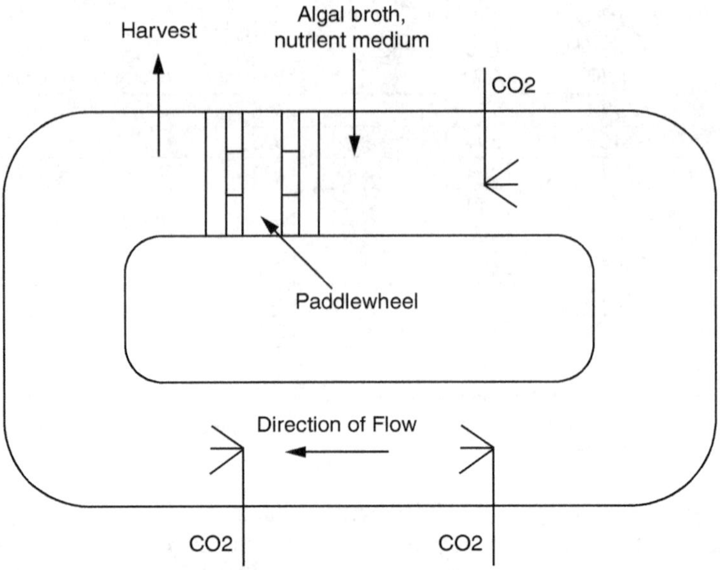

Fig. 1.4 Aerial view of an open raceway pond. (Adapted from: Brennan and Owende 2010)

the dominant fatty acids being palmitic acid, hexadecadienoic acid and linoleic acid were over the ASTM limit and the 1.08 mg KOH/g acid value did not comply with both ASTM D6751 and EN14214 standards (de Alva et al. 2013), highlighting the different fatty acid compositions found in different microalgal species.

The production of microalgal biomass is generally more expensive than crop cultivation since microalgal cultivation requires light, carbon dioxide, water and inorganic salts. Therefore, to minimize costs, cultivation should rely solely on freely available sunlight (Chisti 2007). Cultivation of algae is most commonly carried out in raceway ponds or photo bioreactors.

Raceway ponds are open circular ponds and can be in the form of natural waters such as lakes and lagoons or artificial ponds and containers. The configuration of raceway ponds is a closed loop oval recirculation channel that is typically 0.2–0.5 m deep, the depth is limited due to the penetration limit of light and an increase in depth would result in a decrease in the efficiency of the pond (Brennan and Owende 2010). A paddlewheel provides mixing and circulation in the pond, and cooling is achieved only by evaporation which often results in significant water losses; also there is no temperature control as temperature fluctuates seasonally (Chisti 2007) (Fig. 1.4). Carbon dioxide can be sparged at the bottom of the pond as a carbon source for autotrophic cultivation as the atmosphere only contains about 0.03–0.06% CO_2; therefore, it is expected that the mass transfer limitation could slow down the growth of the microalgae (Mata et al. 2010). Ashokkumar et al. (2014) used a $25m^2$ open raceway pond in semi-continuous mode for the cultivation of *B. braunii*-TN101 resulting in a biomass productivity of 33.8 g m^{-3} day^{-1} (Ashokkumar et al. 2014). Raceway ponds are less expensive cultivation method but

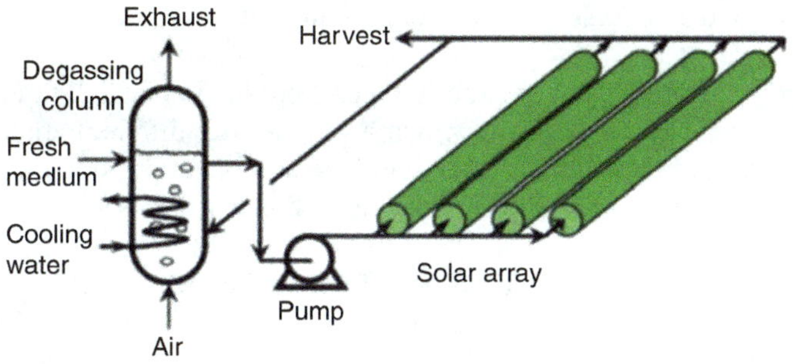

Fig. 1.5 Photo bioreactor schematic with horizontal tubular solar array. (Adapted from: Chisti 2007)

have several limitations such as high risks of contamination that cannot be completely prevented especially in open outdoor raceway ponds, and this can greatly affect the productivity of the algae. Poor mixing in the pond can also affect biomass concentration (Chisti 2007). Raceway ponds are currently used in research and in industry in the form of shallow big ponds, circular pond tanks and closed ponds; these are usually operated in continuous mode to prevent sedimentation (Harun et al. 2010).

The algal culture is introduced into the pond at a point after the paddlewheels and follows the shape of the pond with mechanical aeration from CO_2 spargers; culture is harvested before the paddlewheel point.

There are several types of photo bioreactors such as airlift, tubular, flat plate and vertical column photo bioreactors. The main advantage of photo bioreactors is that it allows for the maintenance of optimum growth conditions allowing for consistency in biomass and lipid productivity (Lam and Lee 2012). Sunlight or artificial light is captured in an array of transparent tubes that are made of plastic or glass, less than 0.1 m in diameter (Chisti 2007). These tubular arrays can be aligned horizontally, vertically, inclined or as a helix (Brennan and Owende 2010) (Fig. 1.5). The tubing configurations can influence a number of parameters in energy usage; horizontal tubing is more scalable but requires large areas of land (Halim et al. 2011). A degassing column functions in circulating the culture medium to the tubes and back (Chisti 2007). Zhu et al. (2010) cultivated *Chlorella zofingiensis* on piggery wastewater in tubular bubble column photo bioreactor resulting in a net biomass productivity of 1.314 g l^{-1} day^{-1}. *Scenedesmus acutus* was cultivated in a tubular photo bioreactor with six vertical cylinders housed in a greenhouse illuminated by solar light. A biomass of 113.7 g dry weight was obtained from 123.1 l of wastewater (de Alva et al. 2013). Feng et al. (2011) used four 2.2 L column aeration photo bioreactors for the cultivation of *Chlorella vulgaris* in artificial wastewater resulting in a cell concentration of 0.28 g/l (Feng et al. 2011).

Microalgal broth is circulated from the degassing column into the solar array of horizontal tubes and back to the degassing column.

1.2.4.2 Current Progress in Biodiesel Production

The most pressing and urgent need for biodiesel production is high lipid accumulation in microalgal strains. Naturally, microalgae produce different amounts of proteins and lipids, and these amounts vary strain to strain, but physiological stress has been used as a technique for driving metabolic fluxes towards biosynthesis of the target products.

Nitrogen limitation/deficiency has been found to be the most efficient stimulant for high lipid production in several different microalgal species (Mallick et al. 2016). A lipid content increase of 33% dry cell weight (dcw) in *Choricystis minor* was achieved under simultaneous nitrate and phosphate deficiencies (Sobczuk and Chisti 2010). Mallick et al. (2016) observed an increase in lipid content from 8% to 57% (dcw) under simultaneous nitrate, phosphate and iron limitations in the microalgae *Chlorella vulgaris*. A high light intensity together with nitrogen-depleted conditions has also been found to increase lipid contents to 54% (dcw) in *Nannochloropsis oceanica* IMET1 (Xiao et al. 2015). Nitrogen limitations can also be applied to municipal wastewater mediums; Robles-Heredia et al. (2015) achieved a 63% (dcw) lipid content in *Chlorella vulgaris* (Robles-Heredia et al. 2015).

Other methods that may be used to increase the accumulation of lipids in algae include metabolic engineering. In order to achieve optimal metabolic engineering for optimal lipid production, an in-depth knowledge and understanding of the microalgal lipid biosynthesis is required. This has not been extensively examined till this date (Mallick et al. 2016). Mass cultivation of microalgae in raceway ponds or enclosed photo bioreactors is another technique to increase the feasibility of microalgal biodiesel by producing algae biomass at a large scale for various other products including biodiesel (Mallick et al. 2016).

The cultivation of microalgae at a laboratory scale differs when compared to common cultivation methods at large scale. This poses a problem with scale-up studies concerning microalgae production. A miniature parallel raceway pond was developed in our laboratory to bridge the gap between laboratory-scale and commercial-scale production. The similar cultivation methods will be similar in the important configurational structures and therefore allow for easy translation of laboratory results to commercial results.

1.2.4.3 Challenges with the Commercialization of Biodiesel

The cost of microalgal biodiesel production at a commercial scale is one of the major drawbacks to the commercialization of this product. Factors that contribute to these costs are harvesting, drying and oil extraction and transesterification.

Harvesting

The solid-liquid separation is a critical step in the production of microalgal biodiesel. Though, it is most commonly achieved by centrifugation, this is an extremely cost-intensive step. Research has shifted to finding less energy and cost-intensive methods to separate microalgae from their growth liquid. The use of chemical flocculants such as $FeCl_3$ and $AlCl_3$ increases the speed of the sedimentation process but is not environmentally friendly; therefore, less toxic and cheaper flocculants need to be investigated (Mallick et al. 2016). In a study carried out by Knuckey et al. (2006), chitosan was found to be an excellent flocculant for freshwater microalgae such as *Chlorella* sp. and the marine alga *Isochrysis galbana*. Electro-coagulation-flocculation (ECF) is an effective method for microalgal flocculation resulting in faster flocculation at higher current densities; when using an aluminium node, the ECF method is effective at laboratory scale, but at large scale the use of much more power due to the higher current densities required is not feasible (Vandamme et al. 2011). Magnetic separation is a simple, easy, low energy consuming and low-cost separation technique that can be applied to separation of microalgae from growth medium (Yavuz et al. 2009). The separation is based on the movement of magnetically tagged particles in response to a magnetic field (Yavuz et al. 2009; Borlido et al. 2013). Fe_3O_4 particles have been successful in the separation of *Botryococcus braunii*, *Chlorella ellipsoidea* and *Nannochloropsis maritima* (Hu et al. 2013; Xu et al. 2011). Iron oxides are preferred due to their biocompatibility, strong paramagnetic behaviour, low toxicity and easy synthesis (Reddy and Lee 2013).

Drying

The drying of algal cells after harvesting is necessary for the storage of the feedstock as well as for downstream processing. The drying step can account for up to 30% of the total production costs of algal fuel therefore making this a big hurdle in the commercialization of algal biofuel (Grima et al. 2003). Generally, heat is used to dry algal biomass, but due to the high moisture content of microalgae, more heat is required to dry larger quantities of biomass relating to higher energy costs, and the high moisture content of microalgae makes sun drying very low in efficiency. Several artificial drying methods have been used in food industries such as drum, freeze, spray, oven and vacuum drying to name a few (Mallick et al. 2016). Solar drying is the most feasible drying method that can be employed at large scale but poses problems relating to large areas of land required for the drying of large amounts of biomass. There is a pressing need for the development of solar dryers that can overcome the issues of feasibility at a commercial scale.

1.2.5 Production of Biogas from Biomass

Currently, approximately 80% of the world's energy is supplied from fossil fuels which will eventually become exhausted, thus leading to rising fuel prices (Zabed et al. 2016). The combustion of these finite fuel sources also contributes to the emission of greenhouse gases (GHGs) which have negative effects on the environment (Taherzadeh and Karimi 2008). For this reason, renewable and sustainable sources of energy are being investigated. Biogas through anaerobic digestion has been identified as a biofuel to meet global demand; thus, efforts are being channelled towards developing technologies to enable the scale-up of biogas production (Kabir et al. 2015). Anaerobic digestion allows for the significant reduction of waste and can generate bio-fertilizer or soil conditioner from undigested material (Lettinga 2005).

Biogas is generated during the multistep anaerobic digestion process which involves various diverse groups of microorganisms. In the first stage, hydrolytic bacteria degrade insoluble complex molecules such as carbohydrates, proteins and fats into simpler molecules such as sugars, fatty acids and amino acids. This is followed by acidogenesis where fermentative bacteria convert sugars and other compounds into acids, alcohols, carbon dioxide, hydrogen and ammonia. The third stage involves hydrogen-producing microorganisms synergistically converting volatile fatty acids to acetate. The final stage involves methanogens using the products of acidogenesis (Jha and Schmidt 2017).

These microorganisms are able to degrade a wide range of organic matter such as food waste, municipal waste and animal manure (Kabir et al. 2015). Lignocellulosic biomass, however, is gaining much attention as a feedstock for biogas production given its renewable and sustainable nature (Brown et al. 2012). Since lignocellulosic material is high in lignin, a pretreatment is required prior to anaerobic digestion. For instance, untreated wheat straw generated 0.189 m^3/kg VS methane, whereas steam-exploded wheat straw gave 0.275 m^3/kg VS methane, a 45% improvement (Bauer et al. 2009). Similarly, untreated straw produced 0.165 m^3/kg VS methane compared to extruded straw which produced 0.281 m^3/kg VS methane (Hjorth et al. 2011).

1.3 Barriers of Utilization of Renewable Biological Energy Resources for Fuel Production

When compared with the cost invested for refinement of available fossil fuels facilitating their commercialization, the production cost of bioethanol and biodiesel from renewable biological energy resources is expected to be more expensive. Maybe governments could encourage the usage of such bio-based fuels with a lesser tax levied for the same in order to promote their usage. Another method of encouraging the usage of fuels of biological origin is to provide subsidies for the same. However, the usage of biofuels needs certain changes in the design of the engines employed for the same. Presently, the major drawback for lignocellulosic biogas is the relatively low-methane yield; therefore, research is focussing on investigating key parameters affecting anaerobic digestion (Seppala et al. 2007).

1.4 Future Possibilities of Utilization of Renewable Biological Energy Resources for Fuel Production

Based on the current energy reserves, it becomes mandatory to depend on the biological energy resources in the future. However, the release of CO_2, CH_4 and H_2S from such energy resources should be investigated to understand the possibility of gaseous emissions during the biofuel production. Much scope exists for the usage of biofuels as the greenhouse gas emissions are expected to be comparatively lesser than that is obtained from the combustion of the fossil fuels. The probable reason is that the combustion of the biofuels is complete which guarantees lesser GHG emissions including SO_2 and even particulates (Twidell and Tonyweir 2006). The authors suggest the possible reason for the same to be the lesser amount of sulphur present in the biofuels when compared with the fossil fuels. Shifting the focus from the fossil fuels to fuels of renewable ethanol, hydrogen and biodiesel would be a boon to the mankind. However, the technologies must be framed to meet the demand of the population in the days to come.

1.5 Concluding Remarks

Environmentally, substituting biofuels for fossil fuels would reduce greenhouse gas emissions thereby reducing urban air pollution. However complete combustion of such biofuels in engines, turbines, etc. needs much sophistication. Similarly, when biomass is either burnt off or completely processed, there may be a net loss of nitrogen as the crops or the plantation does not reach nature, hence necessitating the addition of artificial manure. However, a more sustainable energy system has to evolve to meet the ever-increasing energy demand of the exploding population. Of all the renewable energy resources, biological renewable energy resources occupy a prime position to meet the global energy demand in a clean way in time to come.

As biomass is a widely available resource unlike the area-specific nature of wind, hydro- and solar power, it has the largest potential. New concept of biorefineries is being developed to optimize the utilization of biomass after being converted into energy, food and feed (Caldeira-Pires et al. 2013, Cherubini 2010a, b).

However, life cycle analysis has highlighted that some bioenergy systems could have relatively more carbon emissions than the fossil fuels which are to be replaced. Hence, it is essential to perform accurate analysis before choosing the system (Gnansoumou et al. 2009, Lapola et al. 2010). The positive aspects to be considered include the employability, utilization of resources, sustainability and above all meeting the energy demand in the coming centuries.

For the future, a tie-up between the researchers, educational institutions, government, NGOs and stakeholder partnership could definitely play an important role for the sustenance of life.

Acknowledgement The authors from Bharathidasan University acknowledge the UGC-NON SAP, DST-FIST and UGC SAP for providing the necessary facilities in the department.

References

Adams JM, Toop TA, Gallagher JM, Donnison IS (2011) Seasonal variation in Laminaria digitata and its impact on the biochemical conversion routes to biofuels. Bioresour Technol 102:9976–9984

Aguilar-Reynosa A, Romaní A, Rodríguez-Jasso RM, Aguilar CN, Garrote G, Ruiz HA (2017) Comparison of microwave and conduction-convection heating autohydrolysis pretreatment for bioethanol production. Bioresour Technol 243:273–283

Akhtar N, Gupta K, Goyal D, Goyal A (2015) Recent advances in pretreatment technologies for efficient hydrolysis of lignocellulosic biomass. AICHE J 35:489–511

Alvira P, Toma's-Pejo E, Ballesteros M, Negro MJ (2010) Pretreatment technologies for an efficient bioethanol production process based on enzymatic hydrolysis: a review. Bioresour Technol 101(13):4851–4861

Ashokkumar V, Agila E, Sivakumar P, Salam Z, Rengasamy R, Ani FN (2014) Optimization and characterization of biodiesel production from microalgae Botryococcus grown at semi-continuous system. Energy Convers Manag 88:936–946

Bahng M, Mukarakate C, Robichaud D, Nimlos MR (2009) Current technologies for analysis of biomass thermochemical processing: a review. Anal Chim Acta 651:117–138

Balat M, Balat H, O z C (2008) Progress in bioethanol processing. Prog Energy Combust Sci 34(5):551–573

Barbir F, Veziroglu TN, Plass HJ (1990) Environmental damage due to fossil fuels use. Int J Hydrog Energy 10:739

Bauer A, Bösch P, Friedl A, Amon T (2009) Analysis of methane potentials of steam-exploded wheat straw and estimation of energy yields of combined ethanol and methane production. J Biotechnol 142:50–55

Binod P, Pandey A (2016) Introduction. In: Pandey A, Negi S, Binod P, Larroche C (eds) Pretreatment of biomass, processes and technologies. Elsevier, Amsterdam, p 3

Blanco López MC, Blanco CG, Martínez-Alonso A, Tascón JMD (2002) Composition of gases released during olive stones pyrolysis. J Anal Appl Pyrolysis 65(2):313–322

Borlido L, Azevedo AM, Roque ACA, Aires-Barros MR (2013) Magnetic separations in biotechnology. Biotechnol Adv 31(8):1374–1385

BP (2003) BP statistical review of world energy. BP plc, p 40

BP (2017) https://www.bp.com/content/dam/bp/en/corporate/pdf/energy-economics/statistical-review-2017/bp-statistical-review-of-world-energy-2017-full-report.pdf

Brennan L, Owende P (2010) Biofuels from microalgae—a review of technologies for production, processing, and extractions of biofuels and co-products. Renew Sust Energ Rev 14(2):557–577

Brethauer S, Studer MH (2015) Biochemical conversion processes of lignocellulosic biomass to fuels and chemicals – a review. Chimia 69:572–581

Brown D, Shi J, Li Y (2012) Comparison of solid-state to liquid anaerobic digestion of lignocellulosic feedstocks for biogas production. Bioresour Technol 124:379–386

Cai J, He Y, Yu X, Banks SW, Yang Y, Zhang X, Yu Y, Liu R, Bridgwater AV (2017) Review of physicochemical properties and analytical characterization of lignocellulosic biomass. Renew Sust Energ Rev 76:309–322
Caldeira-pires A, da Luz SM, Palma-Rojas S, Rodrigues TO, Silverio VC, Vilela F, Barbosa PC, Alves AM (2013) Sustainability of the biorefinery industry for fuel production. Energies 6:329–350
Cherubini F (2010a) GHG balances of bioenergy systems – overview of key steps in the production chain and methodological concerns. Renew Energy 35:1565–1573
Cherubini F (2010b) The biorefinery concept: using biomass instead of oil for producing energy and chemicals. Energy Convers Manag 51:1412–1421
Chisti Y (2007) Biodiesel from microalgae. Biotechnol Adv 25(3):294–306
Conesa JA, Font R, Fullana A (2009) Comparison between emissions from the pyrolysis and combustion of different wastes. J Anal Appl Pyrolysis 84(1):95–102
Cuellar-Bermudez SP, Aguilar-Hernandez I, Cardenas-Chavez DL, Ornelas-Soto N, Romero-Ogawa MA, Parra-Saldivar R (2015) Extraction and purification of high-value metabolites from microalgae: essential lipids, astaxanthin and phycobiliproteins. Microb Biotechnol 8(2):190–209
Daroch M, Geng S, Wang G (2013) Recent advances in liquid biofuel production from algal feedstocks. Appl Energy 102:1371–1381
Das D, Veziroglu TN (2001) Hydrogen production by biological processes: a survey of literature. Int J Hydrog Energy 26:13–28
de Alva MS, Luna-Pabello VM, Cadena E, Ortíz E (2013) Green microalga Scenedesmus acutus grown on municipal wastewater to couple nutrient removal with lipid accumulation for biodiesel production. Bioresour Technol 146:744–748
Demiral I, Şensöz S (2008) The effects of different catalysts on the pyrolysis of industrial wastes (olive and hazelnut bagasse). Bioresour Technol 99(17):8002–8007
Demirbas A (2009) Political, economic and environmental impacts of biofuels: a review. Appl Energy 86:S108–S117
Demirdas A (ed) (2010) Biorefineries-for biomass upgrading facilities. Green energy and Technology series. Springer Publications. New York, p 240
Faried M, Samer M, Abdelsalam E, Yousef RS, Attia YA, Ali AS (2017) Biodiesel production from microalgae: processes, technologies and recent advancements. Renew Sust Energ Rev 79:893–913
Feng Y, Li C, Zhang D (2011) Lipid production of Chlorella vulgaris cultured in artificial wastewater medium. Bioresour Technol 102(1):101–105
Gnansounou E, Dauriat A, Villegas J, Panichelli L (2009) Life cycle assessment of biofuels: energy and greenhouse gas balances. Bioresour Technol 100:4919–4930
Grima EM, Belarhi EH, Fernandez FA, Medina AR, Ahisti Y (2003) Recovery of microalgal biomass and metabolites: Process options and economics. Biotechnol Adv. 20(7):491–515
Guschina IA, Harwood JL (2006) Lipids and lipid metabolism in eukaryotic algae. Prog Lipid Res 45(2):160–186
Halim R, Gladman B, Danquah MK, Webley PA (2011) Oil extraction from microalgae for biodiesel production. Bioresour Technol 102(1):178–185
Harris F (ed) (2014) Global environmental issues. Wiley and sons. UK
Harun R, Singh M, Forde GM, Danquah MK (2010) Bioprocess engineering of microalgae to produce a variety of consumer products. Renew Sust Energ Rev 14(3):1037–1047
Hjorth M, Gränitz K, Adamsen APS, Moller HB (2011) Extrusion as a pretreatment to increase biogas production. Bioresour Technol 102:4989–4994
Hoekman SK, Broch A, Robbins C, Ceniceros E, Natarajan M (2012) Review of biodiesel composition, properties, and specifications. Renew Sust Energ Rev 16(1):143–169
Hu YR, Wang F, Wang SK, Liu CZ, Guo C (2013) Efficient harvesting of marine microalgae Nannochloropsis maritima using magnetic nanoparticles. Bioresour Technol 138:387–390

Huang G, Chen F, Wei D, Zhang X, Chen G (2010) Biodiesel production by microalgal biotechnology. Appl Energy 87:38–46

Hubbert MK (1956) Nuclear energy and fossil fuel. Spring Meeting of the Southern District, American Petroleum Institute, San Antonio, TX, March 7–9, Publication No. 95, Shell Development Co

Jha P, Schmidt S (2017) Reappraisal of chemical interference in anaerobic digestion processes. Renew Sust Energ Rev 75:954–971

Jung YH, Kim KH (2015) Acidic pretreatment. In: Pandey A, Negi S, Binod P, Larroche C (eds) Pretreatment of biomass, Processes and Technologies. Elsevier, Amsterdam, p 27

Kabir MM, Forgacs G, Horvath S (2015) Biogas from lignocellulosic materials. In: Karimi K (ed) Lignocellulose-Based Bioproducts. Springer, Cham, p 207

Kapoor M, Soam S, Agrawal R, Gupta RP, Tuli DK, Kumar R (2017) Pilot scale dilute acid pretreatment of rice straw and fermentable sugar recovery at high solid loadings. Bioresour Technol 224:688–693

Kim S, Dale BE (2004) Global potential bioethanol production from wasted crops and crop residues. Biomass Bioenergy 26:361–375

Klass D (1998) Biomass for renewable energy, fuels and chemicals. Academic Press, San Diego

Knothe G (2006) Analyzing biodiesel: standards and other methods. J Am Oil Chem Soc 83(10):823–833

Knuckey RM, Brown MR, Robert R, Frampton DM (2006) Production of microalgal concentrates by flocculation and their assessment as aquaculture feeds. Aquac Eng 35(3):300–313

Kong Q, Chen P, Ruan R (2007) High oil content microalgae selection for biodiesel production. In: 2007 ASAE annual meeting. American Society of Agricultural and Biological Engineers, p 1

Kreith F, Krumdieck S (2013) Principles of sustainable energy systems. CRC Press. Taylor and Francis group. London

Lam MK, Lee KT (2012) Microalgae biofuels: a critical review of issues, problems and the way forward. Biotechnol Adv 30(3):673–690

Lapola D, Schaldach R, Alcamo J, Bondeau A, Koch J, Koelking C, Priess J (2010) Indirect land-use changes can overcome carbon savings from biofuels in Brazil. Proc Natl Acad Sci 8:3388–3393

Lettinga G (2005) The anaerobic treatment approach towards a more sustainable and robust environmental protection. Water Sci Technol 52:1–11

Levin DB, Pitt L, Love M (2004) Biohydrogen production: prospects and limitations to practical application. Int J Hydrog Energy 29:173–185

Li P, Cai D, Luo Z, Qin P, Chen C, Wang Y, Zhang C, Wang Z, Tan T (2016) Effect of acid pretreatment on different parts of the corn stalk for second generation ethanol production. Bioresour Technol 206:86–92

Mallick N, Bagchi SK, Koley S, Singh AK (2016) Progress and challenges in microalgal biodiesel production. Front Microbiol 7:1019

Mata TM, Martins AA, Caetano NS (2010) Microalgae for biodiesel production and other applications: a review. Renew Sust Energ Rev 14(1):217–232

Moodley P, Gueguim Kana EB (2015) Optimization of xylose and glucose production from sugarcane leaves (Saccharum officinarum) using hybrid pretreatment techniques and assessment for hydrogen generation at semi-pilot scale. Int J Hydrog Energy 40:3859–3867

Moodley P, Gueguim Kana EB (2017a) Microwave-assisted inorganic salt pretreatment of sugarcane leaf waste: effect on physiochemical structure and enzymatic structure. Bioresour Technol 235:35–42

Moodley P, Gueguim Kana EB (2017b) Comparison of a two-stage and a combined single stage salt-acid based lignocellulosic pretreatment for enhancing enzymatic hydrolysis. Ind Crop Prod 108:219–224

Moodley P, Gueguim Kana EB (2017c) Development of a steam or microwave-assisted sequential salt-alkali pretreatment for lignocellulosic waste: effect on delignification and enzymatic hydrolysis. Energy Convers Manag 148:801–808

Naik SN, Goud VV, Rout PK, Dalai AK (2010) Production of first and second generation biofuels: a comprehensive review. Renew Sust Energ Rev 14:578–597

Paz AM (2013) Biological resources for energy. In: Earth Systems and Environmental Sciences. https://doi.org/10.1016/B978-0-12-409548-9.05881-4

Petford N (2004) Meeting society's demand for energy. In: Harries F (ed) Global environmental issues. John Wiley and sons ltd. USA

Reddy DHK, Lee SM (2013) Application of magnetic chitosan composites for the removal of toxic metal and dyes from aqueous solutions. Adv Colloid Interf Sci 201:68–93

RFA (2016) Industry Statistics, Renewable Fuels Association, Washington, DC

RFA (2017) Industry Statistics, Renewable Fuels Association, Washington, DC

Richmond A (ed) (2008) Handbook of microalgal culture: biotechnology and applied phycology. John Wiley & Sons. UK

Robles-Heredia JC, Sacramento-Rivero JC, Canedo-López Y, Ruiz-Marín A, Vilchiz-Bravo LE (2015) A multistage gradual nitrogen reduction strategy for increased lipid productivity and nitrogen removal in wastewater using Chlorella vulgaris and Scenedesmus obliquus. Braz J Chem Eng 32(2):335–345

Rorke DCS, Gueguim Kana EB (2016) Biohydrogen process development on waste sorghum (Sorghum bicolor) leaves: optimization of saccharification, hydrogen production and preliminary scale up. Int J Hydrog Energy 41(30):12941–12952

Rorke DCS, Gueguim Kana EB (2017) Kinetics of bioethanol production from waste sorghum leaves using Saccharomyces cerevisiae BY4743. Fermentation 3:19

Saini JK, Saini R, Tewari L (2014) Lignocellulosic agriculture wastes as biomass feedstocks for second-generation bioethanol production: concepts and recent developments. 3 Biotech 2014:1–17

Sánchez C (2009) Lignocellulosic residues: biodegradation and bioconversion by fungi. Biotechnol Adv 27:185–194

Seppälä M, Paavola T, Rintala J (2007) Methane yields of different grass species on the second and third harvest in boreal conditions. In: 11th IWA world congress on anaerobic digestion, Brisbane, pp 23–27

Sewsynker-Sukai Y, Gueguim Kana EB (2017) Optimization of a novel sequential alkalic and metal salt pretreatment for enhanced delignification and enzymatic saccharification of corn cobs. Bioresour Technol 243:785–792

Shockey I, Kinicki R, Delorey J, Arruda K, Baldwin S, Quinn A (2010) Feasibility study of alternative energy sources and conservation techniques for implementation in El Yunque National Forest. Available at http://web.cs.wpi.edu/~rek/Projects/USF_Proposal. Accessed 2 Dec 2015

Sindhu R, Pandey A, Binod P (2015) Alkaline pretreatment. In: Pandey A, Negi S, Binod P, Larroche C (eds) Pretreatment of biomass, Processes and Technologies. Elsevier, Amsterdam, p 51

Sindhu R, Binod P, Pandey A (2016) A novel sono-assisted acid pretreatment of chilli post-harvest residue for bioethanol production. Bioresour Technol 213:58–63

Skiba EA, Baibakova OV, Budaeva VV, Pavlov IN, Vasilishin MS, Makarova EI, Sakovich GV, Ovchinnikova EV, Banzaraktsaeva SP, Vernikovskaya NV, Chumachenko VA (2017) Pilot technology of ethanol production from oat hulls for subsequent conversion to ethylene. Chem Eng J 329:178. https://doi.org/10.1016/j.cej.2017.05.182

Smithers J (2014) Review of sugarcane trash recovery systems for energy cogeneration in South Africa. Renew Sust Energ Rev 32:915–925

Sobczuk TM, Chisti Y (2010) Potential fuel oils from the microalga Choricystis minor. J Chem Technol Biotechnol 85(1):100–108

Spolaore P, Joannis-Cassan C, Duran E, Isambert A (2006) Commercial applications of microalgae. J Biosci Bioeng 101(2):87–96

Taherzadeh MJ, Karimi K (2008) Pretreatment of lignocellulosic wastes to improve ethanol and biogas production: a review. Int J Mol Sci 9:1621–1651

Tauseef SM, Premalatha M, Abbasi T, Abbasi SA (2013) Methane capture from livestock manure. J Environ Manag 117:187–207

Twidell J, Tonyweir (eds) (2006) Renewable energy resources, 2nd edn. Taylor and Francis. USA

Vandamme D, Pontes SCV, Goiris K, Foubert I, Pinoy LJJ, Muylaert K (2011) Evaluation of electro-coagulation–flocculation for harvesting marine and freshwater microalgae. Biotechnol Bioeng 108(10):2320–2329

Viikari L, Vehmaanpera J, Koivula A (2012) Lignocellulosic ethanol: from science to industry. Biomass Bioenergy 46:13–24

Wang M, Liu K, Dai L, Zhang J, Fang X (2013) The structural and biochemical basis for cellulose biodegradation. J Chem Technol Biotechnol 88:491–500

Wang K, Chen J, Sun S-N, Sun RC (2015) Steam explosion. In: Pandey A, Negi S, Binod P, Larroche C (eds) Pretreatment of biomass, Processes and Technologies. Elsevier, Amsterdam, p 75

Wei N, Quarterman J, Jin Y (2013) Marine macroalgae: an untapped resource for producing fuels and chemicals. Trends Biotechnol 31:70–77

Weiland P (2010) Biogas production: current state and perspectives. Appl Microbiol Biotechnol 85:849–860

Woodward J, Place C, Arbeit K (2000) Energy resources and the environment. In: Ernst WG (ed) Earth Systems: Processes and Issues. Cambridge University Press, Cambridge, UK

www. ⟨http://ethanolrfa.org/resources/industry/statistics/#1454099788442-e48b2782-ea53⟩; 2016

Xiao Y, Zhand J, Cui J, Cui Q (2015) Simultaneous accumulation of neutral lipids and biomass in Nannochloropsis oceanica IMET1 under high light intensity and nitrogen replete conditions. Algal Res 11:55–62

Xin L, Hong-ying H, Ke G, Ying-xue S (2010) Effects of different nitrogen and phosphorus concentrations on the growth, nutrient uptake, and lipid accumulation of a freshwater microalga Scenedesmus sp. Bioresour Technol 101(14):5494–5500

Xu J (2015) Microwave pretreatment. In: Pandey A, Negi S, Binod P, Larroche C (eds) Pretreatment of biomass, processes and technologies. Elsevier, Amsterdam, p 157

Xu L, Guo C, Wang F, Zheng S, Liu CZ (2011) A simple and rapid harvesting method for microalgae by in situ magnetic separation. Bioresour Technol 102(21):10047–10051

Yadav D, Barbora L, Bora D, Mitra S, Rangan L, Mahanta P (2017) An assessment of duckweed as a potential lignocellulosic feedstock for biogas production. Int Biodeterior Biodegrad 119:253–259

Yavuz CT, Prakash A, Mayo JT, Colvin VL (2009) Magnetic separations: from steel plants to biotechnology. Chem Eng Sci 64(10):2510–2521

Zabed H, Sahu JN, Boyce AN, Faruq G (2016) Fuel ethanol production from lignocellulosic biomass: an overview on feedstocks and technological approaches. Renew Sust Energ Rev 66:751–774

Zhao X, Moates GK, Elliston A, Wilson DR, Coleman MJ, Waldron KW (2015) Simultaneous saccharification and fermentation of steam exploded duckweed: improvement of the ethanol yield by increasing yeast titre. Bioresour Technol 194(2015):263–269

Zhu L, Cheung CS, Zhang WG, Huang Z (2010) Emissions characteristics of a diesel engine operating on biodiesel and biodiesel blended with ethanol and methanol. Sci Total Environ 408:914–921

Chapter 2
Microbial Fuel Cells: Fundamentals, Types, Significance and Limitations

L. Benedict Bruno, Deepika Jothinathan, and M. Rajkumar

2.1 Introduction

The tremendous increase in world energy requirement and shortage in energy supply pose a severe threat to energy resources. The overconsumption of these non-renewable energy resources creates a global crisis owing to liberation of CO_2 and various toxic gases to the environment (IPCC 2007; Sharma et al. 2009). Various strategies are taken to reduce the energy requirement and issues related to energy. However few challenges are still there to overcome the crisis; for example, reduction of carbon footprint is one of the major problems. In the past few decades, scientists are focusing on renewable energy resources as an alternative to alleviate the environmentally related issues. The major alternative sources are hydrogen from biomass, solar power, wind energy, tidal energy, hydropower energy, etc. Among the renewable resources, bioenergy showed a promising alternative way. Microbial fuel cell (MFC) has been considered as one of the efficient alternative renewable bioenergy technologies, since the demonstration of MFC in 1911 (Potter 1911) and generation of electricity (Kim et al. 1999). Numerous studies have been conducted in recent decades for increasing power output and achieved substantial advancements and developments in MFC. MFC has various applications in addition to electricity production; it could be exploited for industrial wastewater treatment (Du et al. 2007), dairy wastewater treatment (Mohan et al. 2010), leachate treatment (Choi and Ahn 2015), agricultural wastewater treatment, treatment of toxic gases,

L. B. Bruno (✉) · M. Rajkumar
Department of Environmental Sciences, Bharathiar University, Coimbatore, 641046, India

D. Jothinathan
Department of Life Sciences, Central University of Tamil Nadu, Thiruvarur, 610005, India

V. Sivasankar et al. (eds.), *Microbial Fuel Cell Technology for Bioelectricity*,
https://doi.org/10.1007/978-3-319-92904-0_2

degradation of petroleum components, production of hydrogen and methane, etc. (Greenman et al. 2009; Yi et al. 2009; Evelyn et al. 2014).

MFCs are microorganism-mediated electrochemical devices that oxidize organic substrate by metabolic processes of microorganisms to produce electrical energy with high efficiencies for long periods of time. The reactions are as follows where acetate is used as a substrate.

Anodic compartment:

$$\begin{array}{l} \mathrm{H3C-C=O+2(O-H-O)} \xrightarrow{\text{Microbe}} \mathrm{O=C=O+H^{+}+e^{-}} \\ \quad\quad\;\; | \\ \quad\quad\;\; \mathrm{O} \end{array}$$

Cathodic compartment:

$$2(\mathrm{O=O})+8\mathrm{H}^{+}+8\mathrm{e}^{-} \rightarrow 4(\mathrm{O-H-O})$$

Typical MFC consists of two chambers—anode chamber and cathode chamber—which are separated by proton/cation exchange membrane (PEM/CEM). The basic steps of MFC are as follows: • As a first step, the conversion of organic substrate into electrons and protons takes place in anode compartment. The decaying of biomass/substrate occurs in the anode compartment by the activity of microbes and delivers H+ ions and electrons. • The electrons are transferred to electrodes from the bacterial cell either directly or by the help of mediators. • Further the transferred electrons pass through external circuit, from anode to cathode. Simultaneous protons permeation occurs from anode to cathode. The migration of protons in facilitated by PEM/CEM, whereas the electrons from anode chamber to cathode chamber flows through electrodes. While travelling through the electrical circuit, an electrical current is produced. • The diffusion of H+ ions from anode to cathode creates a high electro-chemical gradient close to the anode. • The reduction of oxygen into water, acceptance of electrons and protons takes place in cathode chamber which facilitate more diffusion of H + ions from anode to cathode (Logan et al. 2005; Du et al. 2007; Kumar et al. 2016). The overall process and types of anode chamber classifications and cathode chamber classifications are represented in Fig. 2.1. Research on MFC was intensively increasing in recent decades. According to Web of Science ™ database, a total of 5324 papers were published in the area of MFC as of 02.11.2017. The progress of research publication is increased every year. In 2016 a total of 746 publications were published in the area of MFC. However most of the researches were conducted at bench scale due to difficulties (operational cost, maintenance) in the large-scale operation.

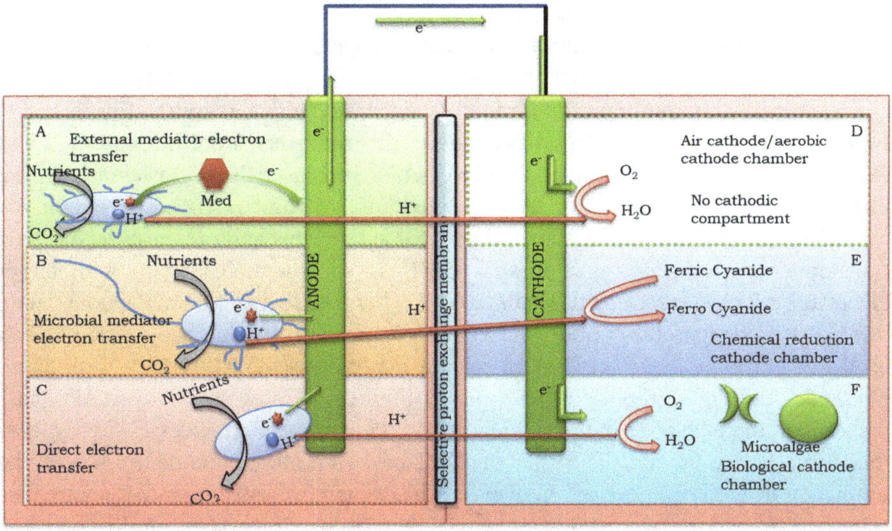

Fig. 2.1 Schematic representation of various types of microbial fuel cell. (A) Mediator-dependent anode chamber, (B) bacterial mediator-dependent anode chamber, (C) mediator-less/direct electron transfer anode chamber, (D) air cathode chamber, (E) chemical electron-accepting cathode chamber, (F) biological electron-accepting cathode chamber

2.2 Basic Configuration and Mechanism of MFC

2.2.1 Anode Chamber

In general MFC is a half-biological system, where the biological activity occurs in the anode chamber. The anode chamber could be considered as the heart of the configuration because the electricity driven by microbes, reduction of organic compounds, reduction of chemical oxygen demand (COD), various wastewater treatments, biofilm formation and hydrogen production occur in this half. The potential of the MFC lies in the anode chamber. In particular, the microbial consortia, substrate used, mediators, pH and anode material used in the anode chamber are considered as important factors of MFC efficiency. Extensive studies on anode material, arrangement, compatibility and stability have been developed to improve the efficiency of the MFC. The selection of anode material is important, and it should exhibit several characteristics such as high physical stability, high chemical stability, high conductance, low resistance, etc. The frequently used nonmetallic anode materials for MFCs include carbon paper, graphite sheet, carbon mesh, carbon cloth, carbon felt, carbon foam, carbon brush, graphite rod, graphite plates and reticulated vitreous carbon (RVC) (Wei et al. 2011; Zhou et al. 2011). However the bottleneck in the traditional carbon electrodes is meagre interaction with microbes. To circumvent this barrier, modified anode material is being used to improve the efficiency of the MFC. Paul et al. (2017) used graphene oxide zeolite anode as modification of carbon felt anode which increases the efficiency of the MFC by 3.6 times higher than the carbon felt anode. Some of the studies were conducted and revealed that the interaction of microbes with anode and electrochemical oxidation was improved by decorated anode material. For instance, Jiang et al. (2017) used a modified anode macroporous graphitic carbon foam (MGCF) coated with polydopamine (PDA) which increased the bacterial interaction, loading capacity and electron transfer mediated by flavin. To some extent the modification tends to provide larger electrical active sites and increase the electron transfer efficiency. Carbon nanotube (CNT) is such a modification with larger electron active sites (Delord et al. 2017). There have been many studies with different carbon nanotube anode materials for increasing efficiency of the MFC. Fan et al. (2017) exploited multiwalled carbon nanotubes (MWCNTs) as a modified electrode, and they reported that the biofilm development of exoelectrogenic bacteria on the electrode increased leading to an increase in power density by 49% than unmodified anode. The use of CNT-based modified anode improves the efficiency of electron transfer. However it has several disadvantages. Recently the use of natural anode material has been increased due to its degradability and low cost. The natural anode material is prepared from the natural resources by carbonization method. Natural materials such as recycled paper, coconut shell, plant trunk, bamboo, etc. are carbonized and used as electrode in the anode chamber. These natural anodes are inexpensive, which increase the attachment of biofilm on the surface by providing 3D structure (Chen et al. 2012; Zhang et al. 2014; Karthikeyan et al. 2015). Similarly metal anode has also been reported to produce high power density. Several metals are being used in MFC. The metals like gold, tungsten, platinum, aluminium, stainless steel, titanium, nickel, etc. are used to produce increased power density (Schroder et al. 2003; Pocaznoi et al. 2012; Zhou et al. 2016). However the corrosive nature of certain metals and cost of the metals prevent from the wide application of these metal anodes. In addition some studies demonstrated the uses of carbon-metal composite materials for improved efficiency. For example, graphite linked with Mn^{4+}/Ni^{4+}/polytetrafluoroethane/ceramic paste increases the power density thereby increased the efficiency of the MFC (Lowy et al. 2006).

Besides anode material, the substrate also plays an important role in electricity generation in the anode chamber. The catalytic reaction of the microorganisms converts the chemical energy into electrical energy where the various nutrients and carbon sources for the microbial metabolism are provided by the substrate. So far various substrates have been utilized as a carbon source in anode chamber from pure compounds to waste sludges. In several studies, acetate is used as a substrate for electrogenic bacteria in anode chamber (You et al. 2015). Ieropoulos et al. (2017) used gelatine as feedstock for MFC operation at starvation condition which provided better longevity. Apart from acetate, a diverse use of various substrates has been reported for effective electricity production, e.g. arabitol (Catal et al. 2008), glucose (Chae et al. 2009), ethanol (Kim et al. 2007), propionate (Chae et al. 2009), butyrate (Chae et al. 2009), cysteine (Logan et al. 2005) and furfural (Luo et al. 2010). Meanwhile the reliability of the microbial population and electricity production were tested with different combined substrates. For instance, You et al. (2015)

investigated the biofilm formation ability and reliability with acetate and casein. Zhang et al. (2013) fed synthetic wastewater with methanol and sodium nitrate as a carbon and nitrogen source. They found that increased concentration of nitrate increases the electricity generation up to 1999.95 ± 2.86 mg/L. Further increase of the substrate concentration to 3560 ± 36.80 mg/L decreased the voltage output. Colombo et al. (2017) compared the electricity generation and organic removal with mixed kitchen waste (fibres), whey from dairy industries (sugar), fisheries residues previously processed to recover oils (proteins) and pulp waste from citrus juice production (acidic). Sevda et al. (2014) evaluated the feasibility of phosphate-buffered solution (PBS) with acetate as carbon source for electricity generation as well as with glucose-rich synthetic wastewater and enriched fermented molasses in sewage from wastewater treatment plant. Further, several studies used meat processing wastewater (Heilmann and Logan 2006), brewery wastewater (Zhuang et al. 2010), starch processing wastewater (Lu et al. 2009), swine wastewater (Kim et al. 2008), dairy industry wastewater (Mardanpour et al. 2012), paper recycling wastewater (Huang and Logan 2008), chocolate industry wastewater (Patil et al. 2009) and landfill leachate wastewater (Sonawane et al. 2017) as a substrate in anode compartment.

The microbial community in anode compartment is an important biological factor for assessing the electron transfer efficiency (Chae et al. 2009). In general organic substrate provides the carbon source and acts as an electron donor. Different mechanisms are involved in transferring of electrons. The substrates enter into glycolysis pathway and undergo a series of various reactions and enter into tri-carboxylic acid cycle and electron transport cycle to synthesize adenosine triphosphate (ATP). In this synthesis process, a series of compounds are involved in electron shuttle (NADH coenzyme Q reductase, nicotinamide adenine dinucleotide (NADH), ubiquinone, succinate dehydrogenase, cytochrome bc_1, cytochrome c, cytochrome c oxidase) from microbial to cell membrane to electron acceptor via redox reaction (Park and Zeikus 2000; Bond and Lovley 2003). Hence potential bacteria which could be able to transport externally are being exploited for MFC (Aelterman et al. 2006). Pure cultures were used mostly in the starting era of MFC. However the power density of the pure culture is low and certain microbes could not be able to produce more protons and electrons. Due to this limitation, several researchers used mixed culture to enhance biofilm formation and power density. The findings also suggested that the mixed culture increases the power density compared with pure culture. In addition microbes could be able to develop biofilm on the surface of the anode. The efficiency of the MFC improved due to biofilm formation on the electrode surface. The exopolysaccharides, proteins secreted by microbes compactly, adhere individual cells and form biofilm (Chae et al. 2009). The excreted electrons are transferred to electrodes directly through pili, c-Cyts or some electron transport complexes produced by microorganisms (Park and Zeikus 2000; Bond and Lovley 2003). Microbes that are not able to transport electrons outside from the membrane need external mediators such as methylene blue, thionin, resazurin, humic acid, methyl orange, neutral red or ferricyanide, etc. for shuttling the electrons from membrane to electrodes (Luu and Ramsay 2003; McKinlay and Zeikus 2004; Watanabe et al. 2009; Rahimnejad et al. 2011).

2.2.2 Cathode Chamber

In conventional two-chambered MFC, cathode chamber is considered as electron and proton acceptor, whereas in single-chambered MFC, air cathode is used as an electron acceptor. The electrons reach the cathode compartment through external circuit, where they get reduced by the electron-accepting molecules and complete the electron cycle. The cathode chamber is a limiting factor where it could affect the MFC performance in huge degree (Rismani-yazdi et al. 2008). In cathode chamber different electron acceptors are being employed to cathodic reduction such as chlorate, perchlorate, nitrite, nitrate, dichloroethane, hexavalent chromium, tetrachloroethane, ferricyanide, potassium ferricyanide, ferric ions, fumarate, sodium bromated, etc. (Logan et al. 2006; Jadhav et al. 2014; Yang et al. 2014; Cui et al. 2016: Lian et al. 2017; Kim et al. 2017). In contrast, oxygen acts as an electron acceptor molecule in single-chambered MFC. Besides the electron acceptor molecules, cathode (i.e. electrode) is also a limiting factor in the cathode chamber. The general cathodic reaction is as follows.

Most of the anode materials are being used as cathode, but a catalyst and a binder are necessary for the air cathode. Several materials such as carbon mesh, carbon cloth, double-sided cloth, graphite fibre brush, carbon felt, carbon fibre brush, graphite felt and stainless steel mesh were used as cathode. However, if these metals are used to prepare air cathode, it is important to incorporate catalyst such as activated carbon, co-tetra-methyl phenylporphyrin, platinum, N-doped carbon nanotubes, etc. and binder such as polytetrafluoroethylene (PTFE), poly(−dimethylsiloxane), Nafion, etc. (Liu and Logan 2004; Cheng et al. 2006; Li et al. 2011; Zhang et al. 2011a). Whereas, catalyst is not required for some cathode material when the electron acceptor such as potassium ferricyanide and ferricyanide is used in the cathode chamber. For instance, Santoro et al. (2013) compared the cathode performance with different platinum loading for increased power production. The incorporation of air cathode with 0.5–1.0 mg of pt./cm^2 showed higher power density. Luo and He (2016) investigated the efficiency of power density with activated carbon-coated nickel carbon fibre, where it increases the power density and reduced the low charge transfer resistance. In another study, Cheng and Wu (2013) reported that the low-cost cathode prepared with nickel foam with activated carbon (Catalyst) and PTFE (binder) produces comparable current density than carbon cloth platinum electrode. Hence numerous studies are undergoing to reduce the coat of the air cathode and increase the efficiency of the current production.

2.2.3 Separator Membrane

Separator is placed in between anode and cathode chamber and also transfers protons from anode compartment to cathode compartment. In general, separator must possess the quality of high proton transfer rate, low gas permeability, good thermal stability and resistance against biofouling (Liu and Logan, 2004; Zhao et al.

2009; Li et al. 2011; Hernandez-Flores et al. 2015). Several studies investigated the influence of various separators for reducing internal resistance, for low gas diffusion and for cost-effective membrane. According to material and exchange mechanism, the separator has been classified into ion exchange membrane and size selective separators. Despite materials, the membranes are majorly classified into five major divisions: proton exchange membrane (PEM), anion exchange membrane (AEM), cation exchange membrane (CEM), bipolar membrane (BPM) and polymer composite porous membrane (PCPM).

Among the various membranes Nafion™ was intensively investigated as a separator for most of the studies. Regarding the power generation, Nafion™ is considered as a good separator, because of its high proton transfer capacity, stability and low resistance. However it has several disadvantages: the first and foremost is that cost of the membrane is high, second is high permeability of oxygen across the membrane, and third is the exchange of cation across the membrane (Wang et al. 2010; Leong et al. 2013; Ercelik et al. 2017). Hence, many alternative separators were investigated to overcome these issues. Tiwari et al. (2016) developed composite membrane with polyvinyl alcohol borosilicate and polyvinyl Nafion borosilicate as a separator which reduced the cost of the membrane by 11-fold compared with traditional Nafion™. In another study, Daud et al. (2017) discovered that the use of ceramic membrane could increase the efficiency of proton transfer leading to high power density output compared to Nafion™.

Several other studies reported for reducing the cost of the technique by using various alternatives to Nafion™, for example, sulphonated polyether ether ketone (SPEEK) membrane (Ayyaru and Dharmalingam 2011), polysulphone-based anion exchange membrane (Elangovan and Dharmalingam 2016), sulphonated polybenzimidazole as a proton exchange membrane (Singha et al. 2016), sulphonated poly(ether imide)s with fluorenyl and trifluoromethyl for membrane fabrication (Kumar et al. 2016), quartz sand chamber used as alternate to PEM (Gao et al. 2017), glass wool (Ghangrekar and Shinde 2007), unglazed wall ceramic and unglazed floor ceramic (Khalili et al. 2017), electrospun microtube array membranes (Chew et al. 2016), salt bridge (Min et al. 2005), glass fibre membrane (Zhang et al. 2011), etc. Recently, these advancements in the membrane fabrication reduced the cost of the technology. However, challenges still persist for fabricating membrane for the application of MFC in large scale.

2.3 Mechanism of Pre-Treatment for Increased Power Output

MFC power output is generally influenced by several factors which limit the increased electricity generation by increasing the activation over potential, ohmic over potential and concentration over potential leading to decrease in Coulombic efficiency and power output. Pre-treatment of electrodes, pre-treatment of substrate and modifying the pH of the substrate were employed to reduce the effect of the above factors. The necessity of these treatments and modifications will differ based on the type of electrode used in the experiment, type of substrate used and type of microorganism employed.

2.3.1 *Pre-Treatment of Electrode for Increased Power Output*

The physical and chemical properties of electrode differ according to the electrode material. The electrode material and surface interfere with the attachment of the microbial communities thereby impeding the transfer of electrons. Therefore, in order to promote the biofilm attachment and good electron transfer, electrode is treated before being employed in MFC experiment. The electrode is treated either with acid or with base. The pre-treatment of electrode with acid/base is carried out with HNO_3, H_2SO_4, HCl, NH_4Cl, NaOH, etc. The pre-treatment of the electrode increases the surface roughness of the electrode thereby facilitating increased attachment of microbes. In addition, pre-treatment increases the catalytic activity and decreases the ohmic loss which increases the current generation by increasing the oxidation of the carbon sources and reducing the resistance of the electrode. Some electrode materials contain nickel oxide which reduces the conductivity of electrons, and therefore the reduction of nickel oxide to nickel is accomplished by pre-treatment. Beside these pre-treatment changes, the pore structure facilitates extra adhesion of biofilm and changes the composition of microbial community (Jung et al. 2014).

2.3.2 *Pre-Treatment of Substrate for Increased Power Output*

Apart from the pre-treatment of electrode, MFC performance could be increased by substrate pre-treatment. Various substrates are been employed in MFC: anaerobic sludge, industrial waste water, dairy waste water, biological substrates, etc. The complex substrates consume a little energy for disintegration/degradation into simpler organic compounds. This phenomenon will attribute to increase in activation loss and/or concentration over potential and leads to decrease in Coulombic efficiency and current production. The pre-treatment of substrate such as ultrasonic wave treatment, thermal exposure, heating, alkaline digestion, acidic digestion, microwave digestion, biological fermentation, thermo-chemical treatment etc. converts the complex form of the substrate into simple soluble organic compound thereby reducing the consumption of energy and improving the electricity output.

2.3.2.1 Physical/Chemical Pre-Treatment

The release of soluble organic matters from the sludge increases the metabolic rate of the microbes thereby increasing the power output. In certain studies, the combination of treatments is been used, for example, ultrasonic wave with alkaline treatment increases the soluble organic compounds which facilitate to reduce the activation loss and increase Coulombic transfer. The pre-treatment is important when biological substrate is used in MFC. For example, algal cells could be used as a substrate but the cells are highly resistant to hydrolysis. Therefore, pre-treatment is necessary to provide easily degraded organic compounds to the exo-electrogens. In certain cases, pre-treatment alone could not improve the electricity production. Therefore, addition of fermenting bacteria in addition to exoelectrogens is necessary to accomplish improved power output.

2.3.2.2 Biological Treatment

In certain cases, when a mixed consortium is used in MFC the electron liberation rate and activity of electrogens could be suppressed by other non-electrogenic consortia (e.g. methanogenic archaea). Under this condition the substrate may be totally utilized by non-electrogenic microbe which decreases the availability of substrate to electrogens and suppresses the growth and metabolism of the electrogens. Persistence of these non-electrogenic microbes increases the concentration over potential or decreases the Coulombic efficiency. The incorporation of specific predatory microbes suppresses and destroys the non-electrogenic bacterial growth thereby facilitating the electrogens to survive in the consortium. For example, inoculation of marine algae inhibits the growth of methanogenic archaea and increases the electrogenic community in MFC contained mixed consortia of microbes thereby increasing the power output (Rajesh et al. 2015).

2.4 Classification

The MFC could be classified based on the configuration, microbe used, mode of nutrition, mediators and usage of membrane. The different types of MFC are represented in Fig. 2.2. The types and design of the MFC are not permanent; new designs and types will emerge to overcome the barriers of the existing type. Hence here a few types based on the above-said factors have been discussed. The various types of MFC configurations, electrodes used, substrate utilized, power output and its applications are summarized in Table 2.1.

2.4.1 Based on Mediator

Electrons can shuttle to anode in three different ways:

(i) The soluble exogenous mediators accept the electron from microbe and transfer it to anode (e.g. methylene blue, neutral red, etc.)

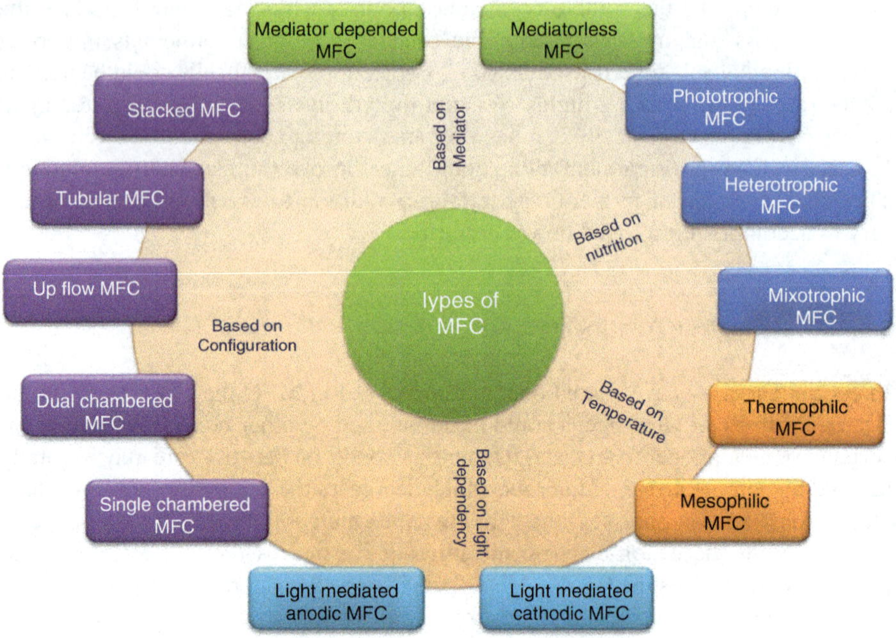

Fig. 2.2 Schematic representations of various proposed types of MFC

(ii) The soluble mediators produced by the microbe shuttle the electron from microbe membrane to anode (e.g. flavin, pyocyanin, etc.)
(iii) Direct transfer of electron from microbe to electrode by nanowires (e.g. c-Cyts, pili)

The mechanism of electron transfer in the anode compartment tends to divide MFC into mediator-dependent MFC and mediator-less MFC. Certain microbes (*Desulfobulbus* spp., *Geobacter* spp., *Pseudomonas* spp., *Rhodopseudomonas* spp., *Shewanella* spp.) are capable of transferring electrons extracellularly through pilin-related proteins (*type IV pil, pil S, pil R*), membrane-bound c-type cytochrome and cell-secreted compounds like flavin and pyocyanin. For instance, microbes like *Geobacter sulfurreducens* and *Shewanella oneidensis* transfer the electrons by formation of biofilm on the anode surface. The potential of the electron transfer is mediated by genes like gene *pilA*, long *PilA*, gene *pilC*, gene *pilR* and *OmcZ*, which regulates the biogenesis of type IV pili, the biofilm attachment on the electrode surface and the synthesis of structural proteins and outer membrane c-cytochrome Z (Xing et al. 2008; Rollefson et al. 2011; Luo et al. 2013; Smith et al. 2014). Zhang et al. (2018) studied the acceleration of electron transfer by addition of rahmnolipids. The metagenomic analysis of the anode surface bacteria revealed that the addition of rhamnolipid increased the activity of the genes cluster COG0642 and COG2204 present in the electroconductivity pili and the

Table 2.1 Types of MFC, configuration and applications

Type of MFC construction	Microbes utilized	Anode material	Cathode material	Substrate in anode	Catalyst used	Separator material	Power output	Application apart from electricity	Reference
Single chambered (super capacitive microbial fuel cells)	Electro active bacteria	Carbon brush	Air-breathing cathode	Potassium phosphate buffered saline Potassium chloride Activated sludge	Fe-Aminoantipyrine		8.1 mW	Hydrogen production	Santoro et al. (2016)
Dual chambered H type	Phototrophic consortium, from a wastewater treatment plant	Graphite felt	Graphite felt	Synthetic media	–	Ultrex membrane	2650m Wm^{-2},	–	Cao et al. (2008)
Single chamber	Mixed photosynthetic consortia	Plain graphite plates	Plain graphite plates	Designed synthetic wastewater	–	Nafion-117	3.2 mA	Volatile fatty acids removal, treatment of acid-rich effluent, wastewater treatment	Chandra et al. (2012)
Dual chamberd MFC	Intestinal microbes	Carbon fiber cloth	Carbon paper electrode	Stimulated intestinal fluid with fresh faecal samples	Pt catalyst	Carbon paper	73.3 mW/m^2	Power for implantable medical devices	Han et al. (2010)
Stacked MFC	Sludge microbial community	Carbon cloth	Carbon cloth	Activated sludge	–	Nafion 117	22.2 mW	Power production	Zhang et al. (2017)

(continued)

Table 2.1 (continued)

Type of MFC construction	Microbes utilized	Anode material	Cathode material	Substrate in anode	Catalyst used	Separator material	Power output	Application apart from electricity	Reference
Phototrophic MFC	*Rhodo pseudomonas* spp.	Carbon felts	Carbon felts	Wetland sediment Anaerobic wastewater	–	Cation exchange membranes	202.9 ± 18.1 mW/m^2	Remove pollutant	Zheng et al. (2017)
Mixotrophic MFC	*Synechocystis* sp. PCC 6803 *Shewanella oneidensis* MR-1	Carbon-based material	Air cathode	LB-broth media	Pt/C	–	8 μA/cm^2	Miniature solar-driven MFCs	Liu and Choi (2017).
Thermophilic MFC	Thermophilic bacteria from anaerobic reactor	Carbon fiber felt	Carbon fiber felt	Synthetic medium	–	Cation exchange membrane CMI-7000	437 mWm^{-2}	–	Dai et al. (2017)
Thermophilic MFC	Municipal solid waste as innoculum	Carbon felt	Graphite electrode	Thermophilic anaerobic sludge		Cation-exchange membrane	10 mA	Biogas production Anaerobic digestion	Liu et al. (2017)
Tubular MFC	Microbial consortia	Poly vinyl alcohol-coke	Air-cathode	Benzene	–	Poly vinyl alcohol hydrogel	30 mWm^{-2}	Benzene removal	Chang et al. (2017)

transfer rate of electrons to the anode (Zhang et al. 2018). In addition certain microbes synthesize the redox-active compounds such as flavin and pyocyanin which create an electron gradient in the outer membrane that facilitates the transfer of the electrons (Rollefson et al. 2011).

However, some bacteria and yeast cells require external addition of mediator for electron transfer from the outer membrane. In addition exoelectrogens require addition of mediators to overcome the slow transfer kinetics in the later stage. It has been reported that external mediators such as ferricyanide, humic acid, methylene blue (MB), methyl viologenare, neutral red (NR) and thionin or the secondary substances from the wastewater treatment such as intermediates of azo dyes and malachite green are utilized and involved in electron transport from membrane to anode (Chen et al. 2014; Xu et al. 2014). The mediator should possess characteristics such as easy diffusion into the cell membrane, high solubility rate, accepting electrons from membrane and being nontoxic to microbes (Park et al. 2000; Park and Zeikus 2000; Chen et al. 2014; Zhang et al. 2018). Ferricyanide and methylene blue possess the above characteristic features, however these mediators are expensive and increase the cost of the technique; in addition the use of these mediators reduces the viability of the microbe. Hence numerous studies are focused to improve the efficiency through mediator-less MFC and secondary metabolite-mediated electron transfer mechanisms. For instance, Xu et al. (2014) proposed that the secondary compounds from dye decolourization could be used as electron mediators. The intermediates of azo dye orange I and orange II from wastewater act as a decolourizer and electron mediator (Xu et al. 2014). To support this further, Chen et al. (2014) studied the secondary compound thionin and malachite green (a non-azo dye) for the electron-mediating and decolourization mechanism. The intemediates/secondary compounds mediated MFC is an appealing research area and promising technology for electron shuttling and reducing expenses.

2.4.2 *Based on Dependency of Microbial Nutrition*

The type and mode of nutrition tend to classify the MFC into three following classifications.

2.4.2.1 Phototrophic MFC

The phototrophic MFC majorly depended on light for the metabolism of the microorganism. The photosynthetic microbes utilize the light energy for the production chemical energy rich compound where the chemical compound is converted into electric energy. Hence, in phototrophic MFC light play an important role and may have strong effect on microbial process and metabolism. Recently the attention on phototrophic microalgae for the generation of electricity has been increased due to their simultaneous applications such as high electricity poduction, waste water treatment, ability to produce high-potential valuable products, etc. The microalgae utilize carbon dioxide to produce electrons in the presence of light. The simultaneous advantages of these phototrophic algal MFC in wastewater treatment, biofuel production and electricity generation facilitate cost-effective wide application. Several microalgae (e.g. *Arthrospira maxima*, *Chlorella vulgaris*, *Dunaliella tertiolecta*, *Scenedesmus obliquus*) have been investigated for the simultaneous generation of electricity and wastewater treatment (Wang et al. 2010a; Lakaniemi et al. 2012; Inglesby and Fisher 2013; Kakarla and Min 2014).

2.4.2.2 Heterotrophic MFC

The second category in nutritional-dependent MFC is utilizing heterotrophic microbes for the production of electricity. In general most of the MFC studies are supported by heterotrophic bacteria where it consumes various organic carbon sources for the growth. For example, in algal MFC under light condition the growth and metabolism is supported by photosynthesis, whereas in dark condition the algal cells uptake the organic compounds for its metabolism from the medium/substrate (Sanders et al. 2001; Liang et al. 2009; Wang et al. 2015). The advantages of employment of heterotrophic microbes in MFC, it could achieve faster removal of organic content in the wastewater, electricity production, nitrogen removal and phosphorus removal (Kessel et al. 2015; Wang et al. 2015).

2.4.2.3 Mixotrophic MFC

The third category in nutritional-dependent MFC is employment of algae with bacteria referred to as mixotrophic MFC. The mixotrophic MFCs have several advantages over phototrophic and heterotrophic MFCs. In mixotrophic mode of operation, the requirement of additional substrate to the bacteria (electrogens) is substituted by the dead algae. The addition of substrate increases the performance of the MFC. In certain studies, the algal cells were incorporated in the anode chamber as a substrate. Several microalgae (e.g. *Chlorella*, *Scenedesmus*, *Dunaliella tertiolecta*, *Microcystis aeruginosa*, *Arthrospira maxima*, etc.) were utilized in MFC for providing organic substrate to the bacteria (Lakaniemi et al. 2012; Inglesby et al. 2012; Inglesby and Fisher 2013; Nishio et al. 2013; Cui et al. 2014; Wang et al. 2015). The use of algae with bacteria has simultaneous benefits of energy supply, oxygen release, COD reduction and CO2 utilization.

2.4.3 Based on Dependency of Light

Since the integration of phototrophic microorganism in MFC required light for the metabolism, the light source has been provided either to the anode side or to the cathode side according to the nature of the microorganism used in the chamber. The phototrophic organism either could be used as an electron donor or electron acceptor. The microalgae grown on the anode compartment act as an electron donor, whereas if they are grown in the cathodic compartment, then they act as an electron acceptor. This scenario leads to classify MFC further into two types:

- Light-mediated anodic MFC
- Light-mediated cathodic MFC

In light-mediated anodic MFC, either phototrophic or mixotrophic microorganism is used in the anode compartment. In this type the light source is placed on the anode side. The construction of MFC varies to single-chamber to two-chamber configuration. For example, single-chamber MFC has constructed with air cathode for producing electricity using phototrophic microalgae *Chlamydomonas reinhardtii* in the Anode chamber (Pisciotta et al. 2011; Nishio et al. 2013a; Li and Zhen 2014), whereas two-chambered MFC has been used for the microalgae *Rhodobacter sphaeroides*. The microalga was placed in the anode compartment for power generation, where it produced 7.3 W/m^3 of power density (Rosenbaum et al. 2005). Similarly in light-mediated cathodic MFC, the phototrophic microbes placed in the cathode compartment are facilitated with light source (Powell et al. 2009; Wu et al. 2013; El-Mekawy et al. 2014; Mohan et al. 2014).

2.4.4 Based on Dependency of Temperature

In recent years several studies integrated thermophiles for the production of electricity and wastewater treatment using MFC. Some unique characteristics of thermophiles make them advantageous over mesophiles. Therefore the dependency of the temperature for growing microorganism in MFC leads to classify into thermophilic MFC and mesophilic MFC. The mesophilic MFC operated in ambient temperature for bioenergy and wastewater treatment, whereas thermophilic microbe requires high temperature for metabolism; therefore the MFC operated at elevated temperature. Carver et al. (2011) conducted a study with thermophilic consortium at 57 °C in anaerobic condition with glucose. The batch mode operation produced 375 mWm^{-2} power density (Carver et al. 2011). In another study Dai et al. (2017) reported a power density of 437 mWm^{-2} by using thermophilic bacterium with ethanol as a substrate.

2.4.5 *Based on Configuration*

Despite mediators, microbes utilized, requirement of light and nutrition mode, the MFC can be divided into several basic types based on the configuration. Till date the following major five types are used hugely for bioenergy production and wastewater treatment:

- Single-chambered MFC
- Dual-chambered MFC
- Upflow MFC
- Tubular MFC
- Stacked MFC

However the configurations of the MFC vary, and the development of new configuration is still emerging based on the barriers that exist in MFC. Several reports described the configuration of the above basic types; hence the overall advantages and disadvantages of these types are discussed in this section. The single-chambered MFC has advantages such as low cost for construction, being membrane-less, low internal resistance, high power output and being easy to construct (Liu and Logan 2004; Zhang et al. 2011; Logroño et al. 2017). However it has several limitations such as high oxygen diffusion which leads to the reduction in coulombic efficiency (Nimje et al. 2012; Yao et al. 2014). The dual-chambered MFC has advantages over the single-chambered one such as eliminating diffusion of oxygen by employing alternating membrane between anode and cathode. In addition the adjustment of distance between the cathode improved the efficiency of power output by reducing the internal resistance (Park and Zeikus 2003; Rozendal et al. 2006). The simultaneous advantage of bioenergy and wastewater treatment is enhanced by upflow and tubular MFC. However several studies suggested that the electricity production is reduced in upflow and tubular MFC. In addition to that, external assistance is required in upflow MFC which increases the cost of the technique and requires high volume space (Zhou et al. 2013; Tee et al. 2016). The major drawback in continuous flow mode of wastewater treatment MFC is increased internal resistance this could be reduced by utilizing the stacked MFC instead of using single large anode and/or cathode chamber. In stacked MFC, a series of individual models are stacked in parallel or in vertical position. This will reduce the internal resistance due to using of small set-up and increase the overall power output (Aelterman et al. 2006; Ledezma et al. 2013; Zhou et al. 2013).

2.5 Proposed Application of MFC

The MFC has a wide range of applications. The major applications of MFCs are electricity production, wastewater treatment, degradation of chemical compounds, biohydrogen production, biosensor fabrication, etc. Since bioelectricity production is the major goal in MFC, many studies focused on enhancement of power density. Various modifications, various microorganisms, alteration in substrate, alteration in pH, alteration in anode and cathode, alteration in membrane constituents, etc. were carried out to increase the power output (Kim et al. 2007; Li et al. 2011a; Luo et al. 2013; Santoro et al. 2013; Cui et al. 2014; Fan et al. 2017; Ercelik et al. 2017; Gao et al. 2017). Besides electricity production, simultaneous wastewater treatment was also studied significantly with MFC. Different types of wastewater have been utilized in MFC for treatment and bioelectricity production. Li et al. (2016) employed coupled system MFC for azo dye decolouration studies with acetate as substrate. The performance of the model was evaluated and reported that the system increased 36.52–75.28% of azo dye decolourization than single MFC (Li et al. 2016). Karra et al. (2013) investigated wastewater treatment potential of MFC with different flow pattern (plug flow and complete mixing) and multiple anodes and multiple cathodes in a continuous mode. The results indicated that the continuous mode with mixing flow increased the efficiency of wastewater treatment, whereas the plug flow created a gradient which increases the power density than mixing flow (Karra et al. 2013). In another approach, Gajda et al. (2014) utilized MFC for simultaneous biodegradation and bioenergy study. The study proved that the formation of the catholyte increased the performance of the electricity production and assists in wastewater treatment and biodegradation of organic wastes (Gajda et al. 2014). In addition MFCs also have the potential to degrade chemical compounds. *Cloacibacterium* sp. was used to degrade phenol and production of electricity in MFC. The microbe potentially enhances the phenol degradation (41%) with a power density of 156 mA/m^2 (Hassan et al. 2018). Certain studies coupled MFC with carbon sequestration. The potential integration of microalgae in MFC has added advantages of electricity production, wastewater treatment and carbon capture. MFC integrated with *Chlorella vulgaris* showed high phosphorus and nitrogen removal. In addition the higher biomass yield suggested the sequestration of inorganic carbon dioxide. The lipid content of several microalgae is around 40–60%, where it could be utilized for biodiesel production (Bruno et al. 2013a, b; Bruno and Sandhya 2014; Ramakrishnan et al. 2014; Huang et al. 2017). Srikanth et al. (2016) studied MFC for removal of oil grease, phenol and sulphide from refinery wastewater.

Furthermore MFC can be modified for hydrogen production. The potential production of hydrogen was 0.132 L day^{-1} in modified supercapacitive microbial fuel cells where platinum electrode is utilized without any external power sources (Santoro et al. 2016). Liu et al. (2005) reported that conventional fermentation produced 8–9 moles of hydrogen for each mole of glucose in MFC. Apart from the electricity generation, wastewater treatment and hydrogen production MFC have potential application in biosensor area (Han et al. 2010). The water alert system based on pH makes possible for MFC to be designed as water alert sensor. The water alert MFC sensors assure safe water, and according to requirement they could be changeable (Jiang et al. 2017). The flow configuration MFC-based toxicity sensor overcomes the obstacle of limited sensitivity towards toxicity (Jiang et al. 2015). Schievano et al. (2017) described a MFC-based sensor for monitoring volatile organic compounds (VOC) concentration. The biosensor was developed to monitor the increased concentration of VOC and provided a basic knowledge by early warning of the parameter increase. The overall application and barriers are represented in Fig. 2.3.

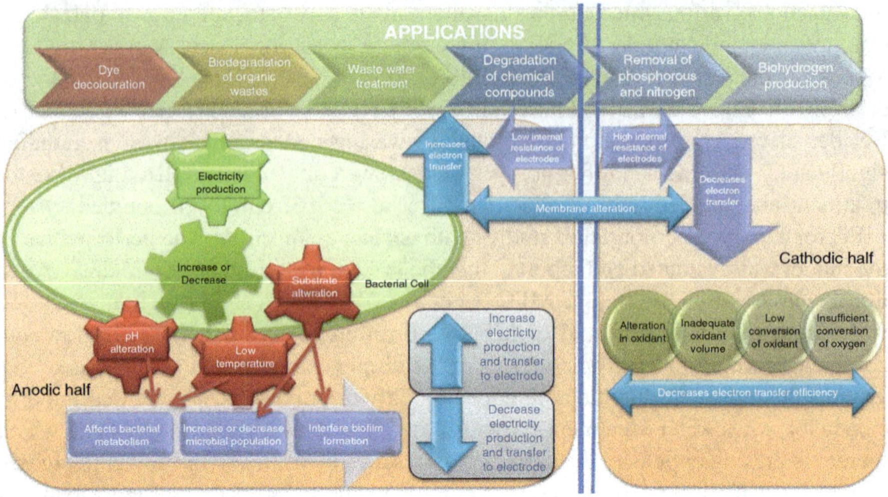

Fig. 2.3 Schematic representation of applications and barriers of microbial fuel cell

2.6 Barriers and Challenges in MFC

The development of MFC offers various advantages as discussed in the previous section. However, several limitations are persisting in MFC scale-up. Some of the challenges have to be addressed to overcome for successful development in this area. The major limitation of this technique is voltage instability. A stable voltage is required to run a system, but voltage production in MFC is comparatively low, and stability of the production is also quite uncertain. This will lead to power inadequate to run a system like biosensor, etc. (Shantaram et al. 2005; Paitier et al. 2017). Another important limitation is high internal resistance; some electrodes require voltage for efficient transfer of electrons through the circuit due to internal resistance (Aelterman et al. 2008: Fan et al. 2008; Paitier et al. 2017). In addition oxidant activation loss is also accompanied in MFC during the reduction in conversion of oxidant to reduced form (Gil et al. 2003). The most common limitation in continuous voltage production is mass transport loss. The slow rate flow or inadequate volume of oxidant in cathodic chamber leads to reduce the electron-accepting process due to the saturation of oxidant. This will attribute to potential loss in stable voltage production (Kim et al. 2002; Oh et al. 2004). Besides, seasonal variations also influence the MFC performance. This is mainly due to limiting factor that affects the growth of microbe. In particular low temperature affects the growth of microbe by suppressing the metabolism that tends to decrease in electron excretion (Shantaram et al. 2005). In addition the biofouling of membrane decreases the efficiency of proton transfer rate. Therefore a periodical change of membrane is necessary to operate in continuous mode. This will increases the cost of the process and tend to be a major barrier in MFC technology. Some of the possible solution for existing barrier is summarized in Table 2.2

Table 2.2 Barriers and solutions for MFC

MFC barriers	Possible solutions
Voltage instability	Periodical adjustment of operating parameter, microbial load, oxidant concentration
High internal resistance	Employing stacked type with less distance between electrodes
Low electron transfer	Identifying good mediator without toxicity
Low power density	Identifying potential microbe. Mutant strains and DNA recombinant strains. Using microbial consortia
Biofouling of membrane	Using ceramic membrane. Synthesis of antifouling membranes

2.7 Conclusion

MFC is a potential and emerging technology that has various simultaneous advantages. The growing concern on the future environmental energy issue and energy demand could be reduced by this kind of alternative technology. Most interestingly the MFCs provide opportunity for the selection of microbes either bacteria or algae. In addition the recent progress in MFC such as bacterio-algal fuel cell provides additional advantages such as wastewater treatment, toxicity removal, bioenergy production, biofuel production, VOC removal, etc. This extended application of MFC reduces the cost of fabrication and makes the technology feasible. However the major limitations reduce the power output. Therefore a particular development and management should be taken care of before starting up. MFC will be a vital technology for electricity production if these limitations and cost-related issues are eliminated. A critical research is necessary for any sustainable and promising technology. Hence further research and development still require for enhancing the power generation and for large-scale operations.

Acknowledgement Dr. L. Benedict Bruno thankfully acknowledges the Science and Engineering Research Board, National Postdoctoral fellowship Scheme (Grant No. PDF/2017/001074), for financial support.

References

Aelterman P, Rabaey K, Pham HT, Boon N, Verstraete W (2006) Continuous electricity generation at high voltages and currents using stacked microbial fuel cells. Environ Sci Technol 40:3388–3394

Aelterman P, Versichele M, Marzorati M, Boon N, Verstraete W (2008) Loading rate and external resistance control the electricity generation of microbial fuel cells with different three-dimensional anodes. Bioresour Technol 99:8895–8902

Ayyaru S, Dharmalingam S (2011) Development of MFC using sulphonated polyether ether ketone (SPEEK) membrane for electricity generation from waste water. Bioresour Technol 102:11167–11171

Bond DR, Lovley DR (2003) Electricity production by *Geobacter sulfurreducens* attached to electrodes. Appl Environ Microbiol 69:1548–1555

Bruno LB, Sandhya S (2014) Systematic investigation of biomass fatty acid productivity and CO_2 sequestration from generator gases by fresh water microalgae in photobioreactor for biodiesel application. Global J Sci Front Res G Bio-Tech Genet 14(1):1.0
Bruno BL, Udhaya R, Sandhya S (2013a) Biomass and lipid productivity by four fresh water microalgae in photoreactor. J Mod Biotechnol 2(3):82–88
Bruno BL, Udhaya R, Sandhya S (2013b) Operation of photobioreactor on refinery gases for CO_2 sequestration and fatty acid production using algae. J Mod Biotechnol 2(6):98–103
Cao X, Huang X, Boon N, Liang P, Fan M (2008) Electricity generation by an enriched phototrophic consortium in a microbial fuel cell. Electrochem Commun 10(9):1392–1395
Carver SM, Vuoriranta P, Tuovinen OH (2011) A thermophilic microbial fuel cell design. J Power Sources 196:3757–3760
Catal T, Li K, Bermek H, Liu H (2008) Electricity production from twelve monosaccharides using microbial fuel cells. J Power Sources 175:196–200
Chae KJ, Choi MJ, Lee JW, Kim KY, Kim S (2009) Effect of different substrates on the performance, bacterial diversity, and bacterial viability in microbial fuel cells. Bioresour Technol 100:3518–3525
Chandra R, Subash GV, Mohan SV (2012) Mixotrophic operation of photo-bioelectrocatalytic fuel cell under anoxygenic microenvironment enhances the light dependent bioelectrogenic activity. Bioresour Technol 109:46–56
Chang SH, Wu CH, Wang RC, Lin CW (2017) Electricity production and benzene removal from groundwater using low-cost mini tubular microbial fuel cells in a monitoring well. J Environ Manag 193:551–557
Chen S, He G, Hu X, Xie M, Wang S, Zeng D, Hou H, Schroder U (2012) A three-dimensionally ordered macroporous carbon derived from a natural resource as anode for microbial bioelectrochemical systems. ChemSusChem 5:1059–1063
Chen BY, Xu B, Qin LJ, Lan JCW, Hsueh CC (2014) Exploring redox-mediating characteristics of textile dye-bearing microbial fuel cells: thionin and malachite green. Bioresour Technol 169:277–283
Cheng S, Wu J (2013) Air-cathode preparation with activated carbon as catalyst, PTFE as binder and nickel foam as current collector for microbial fuel cells. Bioelectrochemistry 92:22–26
Cheng S, Liu H, Logan BE (2006) Increased performance of single chamber microbial fuel cells using an improved cathode structure. Electrochem Commun 8:489–494
Chew CH, Wub CC, Chen CC (2016) A novel electrospun Microtube Array Membrane (MTAM) based low cost conceptual tubular Microbial Fuel Cell (MFC). Eur Polym J 83:138–147
Choi J, Ahn Y (2015) Enhanced bioelectricity harvesting in microbial fuel cells treating food waste leachate produced from biohydrogen fermentation. Bioresour Technol 183:53–60
Colombo A, Schievano A, Trassati SP, Morrone R, D'Antona N, Cristiani P (2017) Signal trends of microbial fuel cells fed with different food-industry residues. Int J Hydrog Energy 42:1841–1852
Cui Y, Rashid N, Hu N, Rehman MSU, Han JI (2014) Electricity generation and microalgae cultivation in microbial fuel cell using microalgae-enriched anode and bio-cathode. Energy Convers Manag 79:674–680
Cui H, Yang H, Liu X, Zhao Y, Liang Z (2016) Performance of sodium bromate as cathodic electron acceptor in microbial fuel cell. Bioresour Technol 202:220–225
Dai K, Wena JL, Zhang F, Maa XW, Cui XY, Zhang Q, Zhao TJ, Zeng RJ (2017) Electricity production and microbial characterization of thermophilic microbial fuel cells. Bioresour Technol 243:512–519
Daud SM, Daud WRW, Kim BH, Somalu MR, Bakar MHA, Muchtar A, Jahim JM, Lim SS, Chang IS (2017) Comparison of performance and ionic concentration gradient of two chamber microbial fuel cell using ceramic membrane (CM) and cation exchange membrane (CEM) as separators. Electrochim Acta In Press
Delord B, Neri W, Bertaux K, Derre A, Ly I, Mano N, Poulin P (2017) Carbon nanotube fiber mats for microbial fuel cell electrodes. Bioresour Technol 243:1227–1231

Du H, Li H, Gu T (2007) A state of the art review on microbial fuel cells: a promising technology for wastewater treatment and bioenergy. Biotechnol Adv 25:464–482

Elangovan M, Dharmalingam S (2016) Preparation and performance evaluation of poly (ether-imide) based anion exchange polymer membrane electrolyte for microbial fuel cell. Int J Hydrog Energy 41:8595–8605

El-Mekawy A, Hegab HM, Vanbroekhoven K, Pant D (2014) Techno-productive potential of photosynthetic microbial fuel cells through different configurations. Renew Sust Energ Rev 39:617–627

Ercelik M, Ozden A, Devrim Y, Colpan CO (2017) Investigation of Nafion based composite membranes on the performance of DMFCs. Int J Hydrog Energy 42:2658–2668

Evelyn LY, Marshall A, Gostomski PA (2014) Gaseous pollutant treatment and electricity generation in microbial fuel cells (MFCs) utilising redox mediators. Rev Environ Sci Biotechnol 13:35–51

Fan Y, Sharbrough E, Liu H (2008) Quantification of the internal resistance distribution of microbial fuel cells. Environ Sci Technol 42:8101–8107

Fan M, Zhang W, Sun J, Chen L, Li P, Chen Y, Zhu S, Shen S (2017) Different modified multi-walled carbon nanotube–based anodes to improve the performance of microbial fuel cells. Int J Hydrog Energy 42:22786–22795

Gajda I, Greenman J, Melhuish C, Santoro C, Li B, Criatiani P, Ieropoulos I (2014) Water formation at the cathode and sodium recovery using microbial fuel cells (MFCs). Sustain Energy Technol Assess 7:187–194

Gao C, Liu L, Yang F (2017) Development of a novel proton exchange membrane-free integrated MFC system with electric membrane bioreactor and air contact oxidation bed for efficient and energy-saving wastewater treatment. Bioresour Technol 238:472–483

Ghangrekar MM, Shinde VB (2007) Performance of membrane-less microbial fuel cell treating wastewater and effect of electrode distance and area on electricity production. Bioresour Technol 98:2879–2885

Gil GC, Chang IS, Kim BH, KimM JJK, Park HS, Kim HJ (2003) Operational parameters affecting the performannce of a mediator-less microbial fuel cell. Biosens Bioelectron 18(4):327–334

Greenman J, Galvez A, Giusti L, Ieropoulos I (2009) Electricity from landfill leachate using microbial fuel cells: comparison with a biological aerated filter. Enzym Microb Technol 44:112–119

Han Y, Yu C, Liu H (2010) A microbial fuel cell as power supply for implantable medical devices. Biosens Bioelectron 25:2156–2160

Hassan PH, Jin B, Donner E, Vasileiadis S, Saint C, Dai S (2018) Microbial community and bioelectrochemical activities in MFC for degrading phenol and producing electricity: microbial consortia could make differences. Chem Eng J 332:647–657

Heilmann J, Logan BE (2006) Production of electricity from proteins using a microbial fuel cell. Water Environ Res 78:531–537

Hernandez-Flores G, Poggi-Varaldo HM, Solorza-Feria O, Romero-Castanon T, Rios-Leal E, Galindez-Mayer J, Esparza-Garcia F (2015) Batch operation of a microbial fuel cell equipped with alternative proton exchange membrane. Int J Hydrog Energy 40(48):17323–17331

Huang LP, Logan BE (2008) Electricity generation and treatment of paper recycling wastewater using a microbial fuel cell. Appl Microbiol Biotechnol 80:349–355

Huang Y, Huang Y, Liao Q, Fu Q, Xia A, Zhu X (2017) Improving phosphorus removal efficiency and *Chlorella vulgaris* growth in high-phosphate MFC wastewater by frequent addition of small amounts of nitrate. Int J Hydrog Energy 42:27749–27758

Ieropoulos I, Theodosiou P, Taylor B, Greenman J, Melhuish C (2017) Gelatin as a promising printable feedstock for microbial fuel cells (MFC). Int J Hydrog Energy 42:1783–1790

Inglesby AE, Fisher AC (2013) Downstream application of a microbial fuel cell for energy recovery from an *Arthrospira maxima* fed anaerobic digester effluent. RSC Adv 3:17387–17394

Inglesby AE, Beatty DA, Fisher AC (2012) *Rhodopseudomonas palustris* purple bacteria fed *Arthrospira maxima* cyanobacteria: demonstration of application in microbial fuel cells. RSC Adv 2:4829–4838

IPCC (2007) Climate change 2007: synthesis report, in Contribution of Working Groups I, II and III to the Fourth Assessment. Report of the Intergovernmental Panel on Climate Change, R. K. Pachauri and A. Reisinger, Eds. p 104, IPCC, Geneva, Switzerland

Jadhav DA, Ghadge AN, Mondal D, Ghangrekar MM (2014) Comparison of oxygen and hypochlorite as cathodic electron acceptor in microbial fuel cells. Bioresour Technol 154:330–335

Jiang Y, Liang P, Zhang C, Bian Y, Yang X, Huang X, Girguis PR (2015) Enhancing the response of microbial fuel cell based toxicity sensors to cu(II) with the applying of flow-through electrodes and controlled anode potentials. Bioresour Technol 190:367–372

Jiang Y, Liang P, Liu P, Yan X, Bian Y, Huang X (2017) A cathode-shared microbial fuel cell sensor array for water alert system. Int J Hydrog Energy 42:4342–4348

Jung GB, Fang LH, Chiou MJ, Nguyen XV, Su A, Lee WT, Chang SW, Kao IC, Yu JW (2014) Effects of pretreatment methods on electrodes and SOFC performance. Energies 7:3922–3933

Kakarla R, Min B (2014) Photoautotrophic microalgae *Scenedesmus obliquus* attached on a cathode as oxygen producers for microbial fuel cell (MFC) operation. Int J Hydrog Energy 39:10275–10283

Karra U, Troop E, Curtis M, Scheible K, Tenaglier C, Patel N, Li B (2013) Performance of plug flow microbial fuel cell (PF-MFC) and complete mixing microbial fuel cell (CM-MFC) for wastewater treatment and power generation. Int J Hydrog Energy 38(2013):5383–5388

Karthikeyan R, Wang B, Xuan J, Wong JW, Lee PKH, Leung MKH (2015) Interfacial electron transfer and bioelectrocatalysis of carbonized plant material as effective anode of microbial fuel cell. Electrochim Acta 157:314–323

Kessel MAHJV, Speth DR, Albertsen M, Nielsen PH, Camp HJMOD, Kartal B, Jetten MSM, Lücker S (2015) Complete nitrification by a single microorganism. Nature 528:555–559

Khalili HB, Mohebbi-Kalhori D, Afarani MS (2017) Microbial fuel cell (MFC) using commercially available unglazed ceramic wares: low-cost ceramic separators suitable for scale-up. Int J Hydrog Energy 42:8233–8241

Kim BH, Kim HJ, Hyun MS, Park DH (1999) Direct electrode reaction of Fe(III)-reducing bacterium, *Shewanella putrefaciens*. J Microbiol Biotechnol 9(2):127–131

Kim HJ, Park HS, Hyun MS, Chang IS, Kim M, Kim BH (2002) A mediator-less microbial fuel cell using a metal reducing bacterium, *Shewanella putrefaciens*. Enzym Microb Technol 30:145–152

Kim JR, Jung SH, Regan JM, Logan BE (2007) Electricity generation and microbial community analysis of alcohol powered microbial fuel cells. Bioresour Technol 98:2568–2577

Kim JR, Dec J, Bruns MA, Logan BE (2008) Removal of odors from swine wastewater by using microbial fuel cells. Appl Environ Microbiol 74:2540–2543

Kim C, Lee CR, Song YE, Heo J, Choi SM, Lim DH, Cho J, Park C, Jang M, Kim JR (2017) Hexavalent chromium as a cathodic electron acceptor in a bipolar membrane microbial fuel cell with the simultaneous treatment of electroplating wastewater. Chem Eng J 328:703–707

Kumar R, Singh L, Wahid ZA (2016) Exoelectrogens: recent advances in molecular drivers involved in extracellular electron transfer and strategies used to improve it for microbial fuel cell applications. Renew Sust Energ Rev 56:1322–1336

Lakaniemi AM, Tuovinen OH, Puhakka JA (2012) Production of electricity and butanol from microalgal biomass in microbial fuel cells. Bioenergy Res 5:481–491

Ledezma P, Greenman J, Ieropoulos I (2013) MFC-cascade stacks maximise COD reduction and avoid voltage reversal under adverse conditions. Bioresour Technol 134:158–165

Leong JX, Daud WRW, Ghasemi M, Liew KB, Ismail M (2013) Ion exchange membranes as separators in microbial fuel cells for bioenergy conversion: a comprehensive review. Renew Sust Energ Rev 28:575–587

Li X, Zhen H (2014) Applications and perspectives of phototrophic micro-organisms for electricity generation from organic compounds in microbial fuel cells. Renew Sust Energ Rev 37:550–559

Li WW, Sheng GP, Liu XW, Yu HQ (2011) Recent advances in the separators in microbial fuel cells. Bioresour Technol 102(1):244–252

Li X, Hu B, Suib S, Lei Y, Li B (2011a) Electricity generation in continuous flow microbial fuel cells (MFCs) with manganese dioxide (MnO_2) cathodes. Biochem Eng J 54(1):10–15
Li Y, Yang HY, Shen JY, Mu Y, Yu HQ (2016) Enhancement of azo dye decolourization in a MFC–MEC coupled system. Bioresour Technol 202:93–100
Lian J, Tian X, Li Z, Guo J, Guo Y, Yue L, Ping J, Duan L (2017) The effects of different electron donors and electron acceptors on perchlorate reduction and bioelectricity generation in a microbial fuel cell. Int J Hydrog Energy 42:544–552
Liang Y, Sarkany N, Cui Y (2009) Biomass and lipid productivities of *Chlorella vulgaris* under autotrophic, heterotrophic and mixotrophic growth conditions. Biotechnol Lett 31:1043–1049
Liu H, Logan BE (2004) Electricity generation using an air-cathode single chamber microbial fuel cell in the presence and absence of a proton exchange membrane. Environ Sci Technol 38:4040–4046
Liu H, Grot S, Logan BE (2005) Electrochemically assisted microbial production of hydrogen from acetate. Environ Sci Tchnol 39(11):4317–4320
Liu L, Choi S (2017) Self-sustaining, solar-driven bioelectricity generation in micro-sized microbial fuel cell using co-culture of heterotrophic and photosynthetic bacteria. J Power Sources 348:138–144
Liu SY, Charles W, Ho G, Cord-Ruwisch R, Cheng KY (2017) Bioelectrochemical enhancement of anaerobic digestion: comparing single- and two-chamber reactor configurations at thermophilic conditions. Bioresour Technol 245:1168–1175
Logan BE, Murano C, Scott K, Gray ND, Head IM (2005) Electricity generation from cysteine in a microbial fuel cell. Water Res 39:942–952
Logan BE, Hamelers B, Rozendal R, Schroder U, Keller J, Freguia S, Aelterman P, Verstraete W, Rabaey K (2006) Microbial fuel cells: methodology and technology. Environ Sci Technol 40:5181–5192
Logroño W, Pérez M, Urquizo G, Kadier A, Echeverría M, Recalde C, Rákhely G (2017) Single chamber microbial fuel cell (SCMFC) with a cathodic microalgal biofilm: a preliminary assessment of the generation of bioelectricity and biodegradation of real dye textile wastewater. Chemosphere 176:378–388
Lowy DA, Tender LM, Zeikus JG, Park DH, Lovley DR (2006) Harvesting energy from the marine sediment–water interface II—kinetic activity of anode materials. Biosens Bioelectron 21:2058–2063
Lu N, Zhou SG, Zhuang L, Zhnag JT, Ni JR (2009) Electricity generation from starch processing wastewater using microbial fuel cell technology. Biochem Eng J 43:246–251
Luo S, He Z (2016) Ni-coated carbon fiber as an alternative cathode electrode material to improve cost efficiency of microbial fuel cells. Electrochim Acta 222:338–346
Luo Y, Liu G, Zhang R, Zhang C (2010) Power generation from furfural using the microbial fuel cell. J Power Sources 195:190–194
Luo J, Yang J, He H, Jin T, Zhou L, Wang M, Zhou M (2013) A new electrochemically active bacterium phylogenetically related to *Tolumonas onensis* and power performance in MFCs. Bioresour Technol 139:141–148
Luu YS, Ramsay JA (2003) Microbial mechanisms of accessing insoluble Fe(III) as an energy source. World J Microbiol Biotechnol 19:215–225
Mardanpour MM, Esfahany MN, Behzad T, Sedaqatvand R (2012) Single chamber microbial fuel cell with spiral anode for dairy wastewater treatment. Biosens Bioelectron 38(1):264–269
McKinlay JB, Zeikus JG (2004) Extracellular iron reduction is mediated in part by neutral red and hydrogenase in Escherichia coli. Appl Environ Microb 70:3467–3474
Min B, Cheng S, Logan BE (2005) Electricity generation using membranes and salt bridge microbial fuel cells. Water Res 39:1675–1686
Mohan SV, Mohanakrishnan G, Velvizhi G, Babu VL Sarma PN (2010) Bio-catalyzed electrochemical treatment of real field dairy wastewater with simultaneous power generation. Biochem Eng J 51(1–2):32–39

Mohan SV, Srikanth S, Chiranjeevi P, Arora S, Chandra R (2014) Algal biocathode for in situ terminal electron acceptor (TEA) production: synergetic association of bacteria–microalgae metabolism for the functioning of biofuel cell. Bioresour Technol 166:566–574

Nimje VR, Chen CC, Chen HR, Chen CY, Tseng MJ, Cheng KC, Shih RC, Chang YF (2012) A single-chamber microbial fuel cell without an air cathode. Int J Mol Sci 13(3):3933–3948

Nishio K, Hashimoto K, Watanabe K (2013) Digestion of algal biomass for electricity generation in microbial fuel cells. Biosci Biotechnol Biochem 77:670–672

Nishio K, Hashimoto K, Watanabe K (2013a) Light/electricity conversion by defined cocultures of *Chlamydomonas* and *Geobacter*. J Biosci Bioeng 115:412–417

Oh S, Min B, Logan BE (2004) Cathode performance as a factor in electricity generation in microbial fuel cells. Environ Sci Technol 38(18):4900–4904

Paitier A, Godaina A, Lyona D, Haddoura N, Vogela TM, Monierb JM (2017) Microbial fuel cell anodic microbial population dynamics during MFC startup. Biosens Bioelectron 92:357–363

Park DH, Zeikus JG (2000) Electricity generation in microbial fuel cells using neutral red as an electronophore. Appl Environ Microbiol 66:1292–1297

Park DH, Zeikus JG (2003) Improved fuel cell and electrode designs for producing electricity from microbial degradation. Biotechnol Bioeng 81:348–355

Park AH, Kim SK, Shin IH, Jeong YJ (2000) Electricity production in biofuel cell using modified graphite felt electrode with neutral red. Biotechnol Lett 22:1301–1304

Patil SA, Surakasi VP, Koul S, Ijmulwar S, Vivek A, Shouche YS, Kapadnis BP (2009) Electricity generation using chocolate industry wastewater and its treatment in activated sludge based microbial fuel cell and analysis of developed microbial community in the anode chamber. Bioresour Technol 100:5132–5139

Paul D, Noori T, Rajesh PP, Ghangrekar MM, Mitra A (2017) Modification of carbon felt anode with graphene oxide-zeolite composite for enhancing the performance of microbial fuel cell. Sustainable Energy Technol. Assess (in press)

Pisciotta JM, Zou Y, Baskakov IV (2011) Role of the photosynthetic electron transfer chain in electrogenic activity of cyanobacteria. Appl Microbiol Biotechnol 91:377–385

Pocaznoi D, Calmet A, Etcheverry L, Erable B, Bergel A (2012) Stainless steel is a promising electrode material for anodes of microbial fuel cells. Energy Environ Sci 5:9645–9652

Potter MC (1911) Electrical effects accompanying the decomposition of organic compounds. Proc R Soc B 84:260–276

Powell EE, Mapiour ML, Evitts RW, Hill GA (2009) Growth kinetics of *Chlorella vulgaris* and its use as a cathodic half-cell. Bioresour Technol 100:269–274

Rahimnejad M, Najafpour GD, Ghoreyshi AA, Shakeri M, Zare H (2011) Methylene blue as electron promoters in microbial fuel cell. Int J Hydrog Energy 36:13335–13341

Rajesh RP, Jadhav DA, Ghangrekar MM (2015) Improving performance of microbial fuel cell while controlling methanogenesis by *Chaetoceros* pretreatment of anodic inoculum. Bioresour Technol 180:66–71

Ramakrishnan U, Bruno B, Swaminathan S (2014) Sequestration of CO_2 by halotolerant algae. J Environ Health Sci Eng 12(81):6

Rismani-Yazdi H, Carver SM, Christy AD, Tuovinen OH (2008) Cathodic limitations in microbial fuel cells: an overview. J Power Sources 180:683–694

Rollefson JB, Stephen CS, Tien M, Bond DR (2011) Identification of an extracellular polysaccharide network essential for cytochrome anchoring and biofilm formation in *Geobacter sulfurreducens*. J Bacteriol 193:1023–1033

Rosenbaum M, Schröder U, Scholz F (2005) In situ electrooxidation of photobiological hydrogen in a photobioelectrochemical fuel cell based on *Rhodobacter sphaeroides*. Environ Sci Technol 39:6328–6333

Rozendal RA, Hamelers HVM, Buisman CJN (2006) Effects of membrane cation transport on pH and microbial fuel cell performance. Environ Sci Technol 40:5206–5211

Sanders R, Caron D, Davidson J, Dennett M, Moran D (2001) Nutrient acquisition and population growth of a mixotrophic alga in axenic and bacterized cultures. Microb Ecol 42:513–523

Santoro C, Li B, Cristiani P, Squadrito G (2013) Power generation of microbial fuel cells (MFCs) with low cathodic platinum loading. Int J Hydrog Energy 38:692–700

Santoro C, Soavib F, Arbizzanib C, Serova A, Kabira S, Carpenterc K, Bretschgerc O, Atanassova P (2016) Co-generation of hydrogen and power/current pulses from supercapacitive MFCs using novel HER iron-based catalysts. Electrochim Acta 220:672–682

Schievano A, Colombo A, Cossetini A, Goglio A, D'Ardes V, Trasatti S, Cristiani P (2017) Single-chamber microbial fuel cells as on-line shock-sensors for volatile fatty acids in anaerobic digesters. Waste Manag (in press)

Schroder U, Nieben J, Scholz F (2003) A generation of microbial fuel cells with current outputs boosted by more than one order of magnitude. Angew Chem Int Ed 42:2880–2883

Sevda S, Dominguez-Benetton X, Weaver HD, Vanbroekhoven K, Sreekrishnan TR, Pant D (2014) Evaluation and enhanced operational performance of microbial fuel cells under alternating anodic open circuit and closed circuit modes with different substrates. Biochem Eng J 90:294–300

Shantaram A, Beyenal H, Veluchamy RRA, Lewandowski Z (2005) Wireless sensors powered by microbial fuel cells. Environ Sci Technol 39:5037–5042

Sharma A, Bhattacharya A, Singh S (2009) Purification and characterization of an extracellular carbonic anhydrase from *Pseudomonas fragi*. Process Biochem 44:1293–1297

Singha S, Jana T, Modestra JA, Kumar AN, Mohan SV (2016) Highly efficient sulfonated polybenzimidazole as a proton exchange membrane for microbial fuel cells. J Power Sources 317:143–152

Smith JA, Tremblay PL, Shrestha PM, Snoeyenbos-West OL, Franks AE, Nevin KP, Lovley DR (2014) Going wireless: Fe(III) oxide reduction without pili by *Geobacter sulfurreducens* strain JS-1. Appl Environ Microbiol 80:4331–4340

Sonawane JM, Adeloju SB, Ghosh PC (2017) Landfill leachate: a promising substrate for microbial fuel cells. Int J Hydrog Energy 42:23794–23798

Srikanth S, Kumar M, Singh D, Singh MP, Das BP (2016) Electro-biocatalytic treatment of petroleum refinery wastewater using microbial fuel cell (MFC) in continuous mode operation. Bioresour Technol 221:70–77

Tee PF, Abdullah MO, Tan IAW, Amin MAM, Nolasco-Hipolito C, Bujang K (2016) Performance evaluation of a hybrid system for efficient palm oil mill effluent treatment via an air-cathode, tubular upflow microbial fuel cell coupled with a granular activated carbon adsorption. Bioresour Technol 216:478–485

Tiwari BR, Noori MT, Ghangrekar MM (2016) A novel low cost polyvinyl alcohol-Nafion borosilicate membrane separator for microbial fuel cell. Mater Chem Phys 182:86–93

Wang L, Zhou X, Zhong S, Zhou N (2010) Novel materials and technologies of microbial fuel cell in environmental engineering. ECS Trans 25:311–333

Wang X, Feng Y, Liu J, Lee H, Li C, Li N, Ren N (2010a) Sequestration of CO_2 discharged from anode by algal cathode in microbial carbon capture cells (MCCs). Biosens Bioelectron 25:2639–2643

Wang H, Lu L, Liu D, Cui F, Wang P (2015) Characteristic changes in algal organic matter derived from *Microcystis aeruginosa* in microbial fuel cells. Bioresour Technol 195:25–30

Watanabe K, Manefiels M, Lee M, Kouzuma A (2009) Electron shuttles in biotechnology. Curr Opin Biotechnol 20:633–641

Wei J, Liang P, Huang X (2011) Recent progress in electrodes for microbial fuel cell. Bioresour Technol 102:9335–9344

Wu XY, Song TS, Zhu XJ, Wei P, Zhou CC (2013) Construction and operation of microbial fuel cell with *Chlorella vulgaris* biocathode for electricity generation. Appl Biochem Biotechnol 171:2082–2092

Xing D, Zuo Y, Cheng S, Regan JM, Logan BE (2008) Electricity generation by *Rhodopseudomonas palustris* DX-1. Environ Sci Technol 42:4146–4151

Xu B, Chen BY, Hsueh CC, Lj Q, Chang CT (2014) Deciphering characteristics of bicyclic aromatics – mediators for reductive decolorization and bioelectricity generation. Bioresour Technol 163:280–286

Yang Y, Xiang Y, Xia C, Wu WM, Sun G, Xu M (2014) Physiological and electrochemical effects of different electron acceptors on bacterial anode respiration in bioelectrochemical systems. Bioresour Technol 164:270–275

Yao S, He YL, Li YS, Xi H (2014) Effect of the membrane electrode assemble design on the performance of single chamber microbial fuel cells. Energy Procedia 61:1947–1951

Yi H, Nevin KP, Kim BC, Franks AE, Klimes A, Tender LM, Lovley DR (2009) Selection of a variant of *Geobacter sulfurreducens* with enhanced capacity for current production in microbial fuel cells. Biosens Bioelectron 24:3498–3503

You J, Walter A, Greenman J, Melhuish C, Ieropoulos I (2015) Stability and reliability of anodic biofilms under different feedstock conditions: towards microbial fuel cell sensors. Sens Biosens Res 6:43–50

Zheng W, Cai T, Huang M, Chen D (2017) Comparison of electrochemical performances and microbial community structures of two photosynthetic microbial fuel. J Biosci Bioeng 124(5):551–558

Zhang X, Cheng S, Liang P, Huang X, Logan BE (2011) Scalable air cathode microbial fuel cells using glass fiber separators, plastic mesh supporters, and graphite fiber brush anodes. Bioresour Technol 102:372–375

Zhang Y, Hu Y, Li S, Sun J, Hou B (2011a) Manganese dioxide-coated carbon nanotubes as an improved cathodic catalyst for oxygen reduction in a microbial fuel cell. J Power Sources 196(22):9284–9289

Zhang J, Zheng P, Zhang M, Chen H, Chen T, Xie Z, Cai J, Abbas G (2013) Kinetics of substrate degradation and electricity generation in anodic denitrification microbial fuel cell (AD-MFC). Bioresour Technol 149:44–50

Zhang J, Li J, Ye D, Zhu X, Liao Q, Zhang B (2014) Tubular bamboo charcoal for anode in microbial fuel cells. J Power Sources 272:277–282

Zhang L, Li J, Zhu X, Ye DD, Fu Q, Liao Q (2017) Response of stacked microbial fuel cells with serpentine flow fields to variable operating conditions. Int J Hydrog Energy 42(45):27641–27648

Zhang Y, Jiang J, Zhao Q, Wang K, Yu H (2018) Analysis of functional genomes from metagenomes: revealing the accelerated electron transfer in microbial fuel cell with rhamnolipid addition. Bioelectrochemistry 119:59–67

Zhao F, Slade RCT, Varcoe JR (2009) Techniques for the study and development of microbial fuel cells : an electrochemical perspective. Chem Soc Rev 38:1926–1939

Zhou M, Chi M, Luo J, He H, Jin T (2011) An overview of electrode materials in microbial fuel cells. J Power Sources 196:4427–4435

Zhou M, Wang H, Hassett JD, Gu T (2013) Recent advances in microbial fuel cells (MFCs) and microbial electrolysis cells (MECs) for wastewater treatment, bioenergy and bioproducts. J Chem Technol Biotechnol 88:508–518

Zhou X, Chen X, Li H, Xiong J, Li X, Li W (2016) Surface oxygen-rich titanium as anode for high performance microbial fuel cell. Electrochim Acta 209:582–590

Zhuang L, Feng C, Zhou S, Li Y, Wang Y (2010) Comparison of membrane- and cloth-cathode assembly for scalable microbial fuel cells: construction, performance and cost. Process Biochem 45(6):929–934

Chapter 3
Plant Microbial Fuel Cell Technology: Developments and Limitations

Santos D. Chicas, Venkataraman Sivasankar, Kiyoshi Omine, Jair Valladarez, and Prabhakaran Mylsamy

3.1 Introduction

For decades societies have depended on fossil fuels for the generation of electricity; however, as fossil fuels are depleted and the demand of energy consumption increases, new renewable energy sources have been developed. Although these new energy sources have addressed environmental concerns and have supported industrialization and economic growth of some countries, energy scarcity and environmental pollution concerns still remain. For instance, the sustainability of biomass energy technologies is questionable due to its competition with food and feed production and high CO_2 emissions due to land-use changes (Helder, et al. 2013a). The need to address environmental pollution, global warming, and energy scarcity has resulted in the search for sustainable energy production and new environmentally friendly methods around the world.

In 1912 Potter proposed the concept of generating electricity by using biofilms as catalysts to convert chemical energy in organic material into electricity (microbial fuel cells); however, due to poor results, further research was discontinued. It was not until the 1990 that microbial fuel cells (MFCs) appeared as alternative energy sources as a response to the demand for no net CO_2-emitting energy sources (Juan R. Trapero et al. 2017). Originally, MFCs were used in wastewater treatment plants to transform chemical energy stored in wastewater into electricity; since then, much progress has been done and the technology has been applied, to what is now known as sediment microbial

S. D. Chicas (✉) · J. Valladarez
Faculty of Science and Technology University of Belize, Belmopan Central Campus, Cayo District, C.A, Belmopan, Belize

V. Sivasankar · K. Omine
Department of Civil Engineering, Nagasaki University, Nagasaki, 8528521, Japan

P. Mylsamy
Post Graduate and Research Department of Botany, Pachaiyappa's College, Chennai 600030, TamilNadu, India

V. Sivasankar et al. (eds.), *Microbial Fuel Cell Technology for Bioelectricity*,
https://doi.org/10.1007/978-3-319-92904-0_3

fuel cells (SMFCs). The SMFCs systems have been used to recover electric power from marine and river beds (Reimers et al. 2001; Tender et al. 2002; Reimers et al. 2006). This system utilizes the natural potential gradient between the sediment and upper oxic water, and electrons released by the microbial oxidation of organic matter flow from the anode to the cathode through an external circuit. In the past decade, MFCs' performance has been enhanced through the selection and modification of electrode material, the advancement of fuel cell configuration, and the adoption of cheap proton exchange membrane with high conductivity efficiency (Liu et al. 2013). Although considerable enhancements have been made, to improve electricity generation of SMFCs, the continuous supply of organic matter is still a challenge to sustain long-term operations. A sustainable alternative to supply organic matter to SMFCs was first introduced by Strik et al. (2008). Strik et al. (2008) used plants that provided organic material to the system through photosynthesis. This new modification of the SMFCs is known as the plant microbial fuel cells (PMFCs). Considering that the PMFCs use photosynthesis to provide organic material to the system and that sunlight is an unlimited source of energy, development of self-sustainable microbial fuel cells that rely on light instead of organics as an energy source has become increasingly popular (Caoet al. 2008; He et al. 2009; Malik et al. 2009). Moreover, PMFCs have been described as sustainable bioelectricity, nondestructive, and carbon-neutral system. This system can be integrated with food production and can be applied at locations like wetlands and green roofs, which are unsuitable for food production (Timmers et al. 2012d). The integration of electricity generation with the local landscape and food crops could eliminate the problem of balancing food and energy production and reduce existing electricity infrastructure requirements as this system could be used in rural areas that have no access to electricity (Strik et al. 2011, 2008).

Since the inception of PMFCs, a lot of progress has been made; thus, this chapter provides a review of the current advances of PMFCs, especially, focusing on anode and cathode materials, the plant types, microbial communities, limitations, and future research.

3.2 General Architecture of a Plant Microbial Fuel Cell

Plant microbial fuel cell (PMFC) is a technology that is solar powered by plant photosynthesis, which is not dependent on direct sunlight (Helder et al. 2013b). PMFC is sustainable because it is renewable, has no competition for arable land, and has a clean conversion without emissions (Wetser et al. 2015). Of the fixed carbon by the plant about 70% of the fixed carbon is transferred to the roots which is released into the rhizosphere (Chiranjeevi et al. 2012). In the rhizosphere, rhizodeposits are realized by plant roots which consist of secretions, excretions, gases, and dead plant material (Helder et al. 2013b). Rhizodeposits are broken down by naturally occurring bacteria around the plant root. The deposition of rhizodeposits by plant roots and the oxidation of these by electrochemical active bacteria to generate electricity are the two fundamental principles of PMFC technology (Timmers et al. 2013a). The main components of a plant microbial fuel cell are vascular plants photosynthesis to fix carbon and produce rhizodeposits from

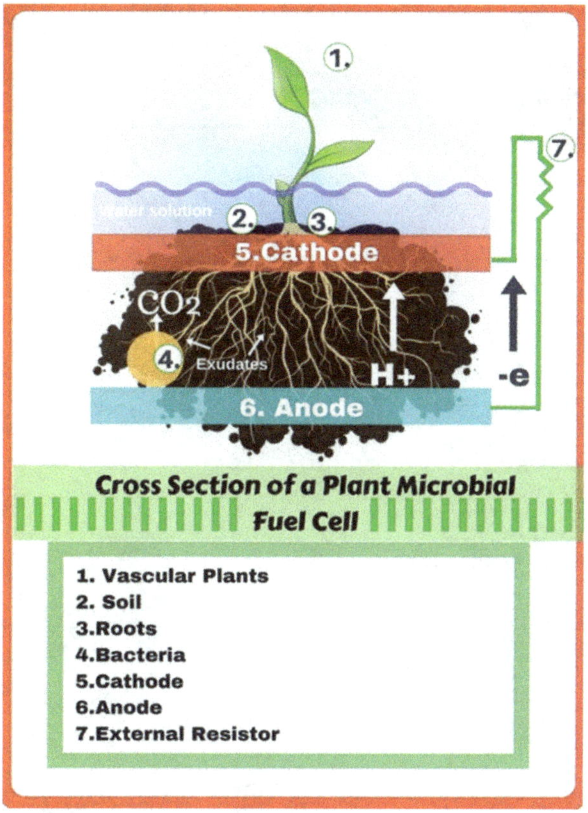

Fig. 3.1 Cross section of a plant microbial fuel cell system

the root systems, which are used in the PMFC as a renewable energy substrate; naturally occurring electrochemical active bacteria living in the plant rhizosphere for oxidation of rhizodeposits; and an anode coupled with a cathode where oxygen, electrons, and protons are reduced to water (Bombelli et al. 2013a; Helder et al. 2013a (Fig. 3.1)).

In the PMFC, the plant roots grow in the anode compartment; the roots provide the electrochemically active bacteria with rhizodeposits. The bacteria oxidize the rhizodeposits into carbon dioxide and protons and donate electrons to the anode (Wetser et al. 2015). Then, the electrons flow through an electrical circuit and power harvester to the cathode compartment where they are consumed, thus, resulting in electrical power generation.

3.3 Anode Materials for Plant Microbial Fuel Cells

There have been great advancements in anode materials designed for improved performance of microbial fuel cell; see Table 3.1 (Hindatu et al. 2017); however, the most frequently used materials in plant microbial fuel cells include carbon (fiber, mesh, mat, felt, brush, granular activated carbon, glassy carbon) and graphite (rod,

Table 3.1 Summary of PMFC configuration and power generation studies 2008–2017

Plant types	Anode fabrication	Cathode fabrication	Growth medium/ substrate	Operating conditions	Electron acceptor	Microbial community	Operation time (days)	Power density ave	Power density max	Current density ave	Current density (low)	Reference
Arundinella anomala	Graphite rod in graphite grains	Graphite felt	Hoagland solution	Climate chamber	O2, Ferricyanide	Bacteria	112	10	22			Helder et al. (2010)
Arundo donax	Graphite rod in graphite grains	Graphite felt	Hoagland solution	Climate chamber	O2, Ferricyanide	Bacteria	112		Break down of system			Helder et al. (2010)
Acorus calamus	Graphite felt	Graphite felt	Pyrene and benzo pyrene-rich water	Climate chamber					na			Yan et al., (2015)
Canna indica	Graphite disk	Carbon cloth	Tap water	Constructed wetland	O2	Geobacteraceae, Anaerolineaceae	90		18		105	Lu et al. (2015)
Eichhornia crassipes	Graphite disks	Graphite disks	Domestic and fermented distillery wastewater	Miniature benthic system					224.93			Venkata Mohan et al. (2011)
Echinochloa glabrescens	Carbon fiber	Stainless steel	Professional medium	Climate chamber					115			Bombelli et al. (2013b)
Glyceria maxima	Graphite granules	Graphite felt	Hoagland solution	Climate chamber	O2	Bacteria	67	4	67	32	153	Strik et al. (2008)
Glyceria maxima	Graphite granules	Graphite felt	Hoagland solution	Climate chamber	O2	*Geobacteraceae*, *Ruminococcaceae*, *Comamonadaceae*			80	164		Timmers et al. (2012a)
Glyceria maxima	Graphite felt	Graphite felt	Ammonium-rich ½ Hoagland solution	Climate chamber					12 (MA)			Timmers et al. (2013a)
Ipomoea aquatica	Granular activated carbon(GAC)	Stainless steel with GAC	Anaerobic sludge from municipal wastewater	Constructed wetland					12.42			Liu et al. (2013)

Lolium perenne	Graphite granules	Carbon felt	Hoagland solution, wastewater with chromium	Green house					55			Habibul et al. (2016)
Oryza sativa	Graphite felt	Graphite felt	NPK fertilizer/ acetate solution	Rice field	O2	*Natronocella, Beijerinckiaceae*	120		6		52	Kaku et al. (2008)
Oryza sativa	Graphite felt	Carbon/ polytetra-fluoroethylene coated	Glucose/ acetate, Bacto yeast/ electrolyte solution	Rice field					19 ± 3.2			Kouzuma et al. (2013)
Oryza sativa	Carbon fiber	Stainless steel	Professional growing medium	Climate chamber					126.3			Bombelli et al. (2013b)
Oryza sativa	Graphite granules	Graphite felt	Vermiculite with Hoagland solution	Climate chamber					72			Arends et al. (2014)
Oryza sativa	Graphite felt	Graphite felt	Soil/fertilizer	Rice field					14.44			Takanezawa et al. (2010)
Oryza sativa	Graphite granules	Graphite granules	Hoagland solution	Greenhouse					33			Chiranjeevi et al. (2012)
Oryza sativa	Graphite felt	Graphite felt	Soil	Rice field					80			Ueoka et al. (2016)
Oryza sativa	Carbon fiber	Carbon fiber	Onada soil	Ambient					23			Moqsud, et al. 2015b)
Pennisetum setaceum	Graphite plate	Graphite plate	Red soil	Ambient	O2	Bacteria	65		163		4.52	Chiranjeevi et al. (2012)
Spartina anglica	Graphite rod in graphite grains	Graphite felt	Hoagland solution	Climate chamber	O2, Ferricyanide	Bacteria	154	21	222			Helder et al. (2010)
Spartina anglica	Graphite granules	Graphite felt	Hoagland solution	Climate chamber	O2	Bacteria	78	22	110	141		Timmers et al. (2010)

(continued)

Table 3.1 (continued)

Plant types	Anode fabrication	Cathode fabrication	Growth medium/ substrate	Operating conditions	Electron acceptor	Microbial community	Operation time (days)	Power density ave	Power density max	Current density ave	Current density (low)	Reference
Spartina anglica	Graphite felt multiple layers	Graphite felt Single layer	Nitrate less ammonium-rich medium	Climate chamber					679 (PGAe)			Wetser et al. (2015)
Spartina anglica	Graphite felt	Graphite felt	Growth medium	Climate chamber					240			Helder et al. (2012b)
Typha latifolia	Carbon felt	Carbon felt, porous air spargers	Sludge from glove manufac-turing company/ gravel for support	Constructed wetland					6.12			Oon et al. (2015)
Sedum album	Graphite fiber felt	Graphite fiber felt and replaced with activated carbon granules (after 120 days startup)	Sand, silt, clay, and organic matter	Plastic pots (green roofs)	O2		360		0.0024			Tapia et al. (2017)
Sedum hybridum	Graphite fiber felt	Graphite fiber felt and replaced with activated carbon granules	Sand, silt, clay, and organic matter	Plastic pots (green roofs)	O2	*Micrococcaceae*	360		0.092	2.71		Tapia et al. (2017)
Sedum kamtschaticum	Graphite fiber felt	Graphite fiber felt and replaced with activated carbon granules	Sand, silt, clay, and organic matter	Plastic pots (green roofs)	O2		360		>0.001			Tapia et al. (2017)

Sedum reflexum	Graphite fiber felt	Graphite fiber felt and replaced with activated carbon granules	Sand, silt, clay, and organic matter	Plastic pots (green roofs)	O2		360		>0.001			Tapia et al. (2017)
Sedum rupestre	Graphite fiber felt	Graphite fiber felt and replaced with activated carbon granules	Sand, silt, clay, and organic matter	Plastic pots (green roofs)	O2		360		0.0155			Tapia et al. (2017)
Sedum sexangulare	Graphite fiber felt	Graphite fiber felt and replaced with activated carbon granules	Sand, silt and clay and organic matter	Plastic pots (green roofs)	O2		360		0.0084			Tapia et al. (2017)
Sedum spurium	Graphite fiber felt	Graphite fiber felt and replaced with activated carbon granules	Sand, silt, clay, and organic matter	Plastic pots (green roofs)	O2		360		>0.001			Tapia et al. (2017)
Oryza sativa	Graphite granule	Graphite granules	Soil or vermiculite with Hoagland solution		Ferricyanide	*Desulfobulbus*, *Geobacteraceae*, *Archaea*	134	21	33	44		De Schamphelaire et al. (2008, 2010)
Phragmites australis	Graphite felt	Graphite felt (ultrafiltration membrane and silicone separator for O2 diffusion)	Peat soil	Wetland	O2		14	22				Wetser et al. (2017)
Spartina anglica	Graphite felt	Graphite felt (ultrafiltration membrane and silicone separator for O2 diffusion)	Salt marsh	Wetland	O2		14	82				Wetser et al. (2017)

(continued)

Table 3.1 (continued)

Plant types	Anode fabrication	Cathode fabrication	Growth medium/ substrate	Operating conditions	Electron acceptor	Microbial community	Operation time (days)	Power density ave	Power density max	Current density ave	Current density (low)	Reference
Brassica juncea	Carbon brush	Carbon cloth (clay mix separator)	Compost soil mix	Pot	O2		30		69.32			Sophia and Sreeja (2017)
Trigonella foenum-graecum	Carbon brush	Carbon cloth (clay mix separator)	Compost soil mix	Pot	O2		30		80.26			Sophia and Sreeja (2017)
Canna stuttgart	Carbon brush	Carbon cloth (clay mix separator)	Compost soil mix	Pot	O2		30		222.54			Sophia and Sreeja (2017)
Oryza sativa	Graphite mat	Graphite mat			O2 (root)				1.3			Chen et al. (2012)
Hydroponic plants (rooted plants grown on the water surface; *Bryophyllum pinnatum*, *Solanum lycopersicum* (tomato), *Oryza sativa* (rice), *Lycopodium* and *Adiantum* (ferns)), submerged plants (*Hydrilla verticillata*, *Myriophyllum*), and self-grown algae	Graphite disks	Graphite disks							110			Chiranjeevi et al. (2013)

Ipomoea aquatica	Granular activated carbon	Granular activated carbon			O2				11.2			Fang et al. (2013)
Myrtillocactus	Glassy C	Glassy C			O2 (BOD)				90			Flexer and Mano (2010)
Spartina anglica	Graphite felt	Graphite felt			Ferricyanide				211			Helder et al. (2012b)
Spartina anglica	Graphite felt	Graphite felt			Ferricyanide				440			Helder et al. (2012b)
Lemna minuta	Carbon felt	Graphite granules			O2				380			Hubenova and Mitov (2012)
Lemna valdiviana	Carbon felt	Graphite granules			O2				140			Hubenova and Mitov (2015)
Glyceria maxima	Graphite granules	Graphite felt			O2				390			Timmers et al. (2012d)
Glyceria maxima	Graphite granules	Graphite felt			O2							Timmers et al. (2012c)
Phragmites australis	Graphite	Graphite			O2				43			Villaseñor et al. (2013)
Canna indica									15.73			Yadav et al. (2012)
Phragmites australis	Graphite plate	Graphite plate			O2				9.4			Zhao et al. (2013)

grain, granules, sheet, disk, felt) (Nitisoravut and Regmi 2017). These materials are selected for being cheap, showing good conductivity, and being stable (Hindatu et al. 2017). Arends et al. (2012) conducted an experiment whereby they compared the suitability of granular carbon as an anode material versus carbon felt and carbon cloth. The results indicated that even though carbon felt obtained the highest current density, carbon granules are more suitable for plants because these accommodated accordingly with the growing plant root, thus, providing new electric connections (Arends et al. 2012). A study conducted by Wang et al. (2016) compared four different anode electrodes, carbon fiber felt (CFF), stainless steel mesh (SSM), graphite rod (GR), and foamed nickel (FN), in a constructed wetland-microbial fuel cell (CW-MFC) and found that different microbial communities populate the different materials used. This is important since the formation of these microbial communities influences the transfer of electrons to the anode (Deng et al. 2012).

3.4 Cathode Materials for Plant Microbial Fuel Cells

The commonly utilized cathode electrodes in PMFC is similar to the ones used in the anode; see Table 3.1; these include carbon (cloth, fiber, felt, brush, granular activated carbon, glassy carbon, carbon/polytetrafluoroethylene coated) and graphite (granules, plate, mat, felt, and stainless steel) (Nitisoravut and Regmi 2017). Oxygen continues to be one of the most commonly used electron acceptors for the cathode reduction reaction. Other electron acceptors have been used, e.g., ferricyanide; however, it is toxic to the environment (Deng et al. 2012).

3.5 Plants Used in MFC Systems

To date, several plants have been used in PMFC studies (See Table 3.1); the selected plants for these studies usually depend on the purpose of the research. These include *Arundinella anomala*, *Acorus calamus*, *Arundo donax*, *Brassica juncea*, *Canna indica*, *Canna stuttgart*, *Echinochloa glabrescens*, *Eichhornia crassipes*, *Glyceria maxima*, hydroponic plants consortium, *Ipomoea aquatica*, *Lemna minuta*, *Lemna valdiviana*, *Lolium perenne*, *Myrtillocactus*, *Oryza sativa ssp. indica*, *Pennisetum setaceum*, *Phragmites australis*, *Sedum album*, *Sedum hybridum*, *Sedum kamtschaticum*, *Sedum reflexum*, *Sedum rupestre*, *Sedum sexangulare*, *Sedum spurium*, *Spartina anglica*, *Trigonella foenum-graecum*, *Typha latifolia*, hydroponic plants (rooted plants grown on the water surface; *Bryophyllum pinnatum*, *Solanum lycopersicum* (tomato), *Oryza sativa* (rice), *Lycopodium* and *Adiantum* (ferns)), submerged plants (*Hydrilla verticillata*, *Myriophyllum*), and self-grown algae. To date, the highest maximum power outputs achieved are from *S. anglica* with 679 mW/m-2. Although the PMFC system keeps improving, the overall conversion of light into electrical energy remains low. However, efforts are underway by a multidisciplinary

European Consortium to optimize electricity production to 3200 mW/m^2, which is the optimal, assuming that the average solar radiation is 150 MW/km^2 in Western Europe. The consortium intends to increase photosynthetic efficiency by 5% and recover 60% of the 70% of the energy (photosynthesis) that is transported to the soil by the plant. This energy recovery will be done by utilizing the MFC system (Strik et al. 2011).

3.6 Microbial Community Found in Plant Microbial Fuel Cells

One aspect of the electrical catalytic activity of the microbes is related to specific microorganisms that are capable of exchanging electrons to the electrodes by catalyzing the oxidation of substrates, either the anode or by using electrons supplied by the cathode or reducing a substrate. Analyses of single-species cultures have indicated that a wide selection of microbial families possess endogenous exoelectrogenic activities, and these include the Alcaligenaceae, Aeromonadaceae, Bacteroidetes, Campylobacteraceae, Clostridiaceae, Desulfuromonadaceae, Enterococcaceae, Geobacteraceae, Pseudomonadaceae, Rhodobacteraceae, Shewanellaceae, and Vibrionaceae (Logan 2009; McCormick et al. 2015; Zhang et al. 2012). Importantly, the anodes of PMFC are ideal for the growth of bacteria that don't produce electricity; thus, it is necessary to develop growth inhabitation strategies of these bacteria in order to increase power production in PMFC.

3.7 Improvements, Limitations, and Future Research for Plant Microbial Fuel Cells

World energy demand is increasing; thus, new technologies are needed in order to meet the increasing energy demand in a sustainable way. Much focus has been placed in the plant microbial fuel cell (PMFC) technology which could produce sustainable bioelectricity; however, there are many limiting factors that need to be addressed in order to make this technology attractive as a renewable and sustainable energy source (Helder et al. 2012b). Since 2008 a lot of research has been made to improve the power output of PMFC. Takanezawa et al. (2010) found that external load, cathode modification with platinum catalysts, and anode position affected the power output (Takanezawa et al. 2010). Chen et al. (2012) reduced internal resistance and improved power output by decreasing the distance between the cathode and the anode by using a biocathode, which used plant roots excreted oxygen (Chen et al. 2012). Helder et al. (2012b) also tried to minimize internal resistance and increase power output by designing a flat-plate PMFC; however, anodic resistance was high due to mass transfer limitation or substrate limitation. To overcome the problem of substrate limitation, they proposed that exudate should be converted into electricity more efficiently, more exudates should be produced by plants, or other rhizodeposits should be used. They also suggested that future research should focus in reducing the anode height because their results indicated that the middle- and bottom-level anodes generated less electricity than the top one (Helder et al. 2012b). Likewise, Timmer et al. (2013b) minimized internal membrane resistance by changing the transport direction of cations. This was achieved by using a bicathode setup. However, further studies are needed to understand the effects of change from cathode 1 and 2 on the anode resistance and integrate the bicathode setup into the flat-plate PMFC (Timmers et al. 2013b). Moreover, other researchers have suggested that in order to optimize energy recovery of a PMFC, the plant selection should focus on high root biomass production combined with low suggested that electricity production (Timmers et al. 2012c). Helder et al. (2012b) suggested that plant-growth medium of the PMFC can be engineering to increase power output by influencing the coulombic efficiency, mixed potential in the anode, and rhizodeposition from plants (Helder et al. 2012a).

Natural weather parameters, additional organic matter, pH, and electrical conductivity have a significant influence on bioelectricity generation in plant microbial fuel cell (Moqsud et al. 2015a). In order to make a reliable, sustainable, and weather-independent electricity production system of the PMFC, cathode stability and cold insulation of anode and cathode should be improved (Helder et al. 2013b). Bacterial communities also influence PMFC power generation; thus, adequate and healthy bacterial population should considerably enhance the power generation of a PMFC (Salinas-Juarez et al. 2016). In an effort to understand the effects of bacterial communities on power generation, bacterial characterizations have been made. Ahn et al. (2014) found that when current was produced, the abundances of 16S rRNAs sequences showed low similarities to the previously characterized bacteria; thus they proposed the isolation of these bacteria and characterization of their electrochemical properties, which can help to optimize the performance of PMFC (Ahn et al. 2014). Cabezas et al. (2015) also found that when current was produced, changes in the bacterial and archaeal community compositions occurred, and factors such as plant presence and inoculum have a role in determining the composition of active microorganisms on anodes. They proposed further experiments, using a stable-isotope-probing approach in order to get more insight into the interaction between root exudates and anode-reducing bacteria (Cabezas et al. 2015). Lu et al. (2015) found that oligotrophic growth condition could benefit PMFC current production because this condition limited the competition from denitrifying bacteria or different methanogens. They suggested that growth inhabitation strategies aimed at hydrogenotrophic methanogens can be developed in order to increase power production in PMFC (Lu et al. 2015).

At this stage it is too early to compare the PMFC with other renewable energy technologies since it is a waterlogged system and still under development and it has limited power output. Some researchers are currently expanding the applicability of PMFC to arid and semiarid conditions. Tapia et al. (2017) concluded that power generation using a PMFC was possible in semiarid conditions; nevertheless, the power generated was lower than that of flooded PMFC. Tapia et al. also proposed that PMFC could be used as a potential indicator of soil water content in semiarid areas where water-use efficiency is needed (Tapia et al. 2017). In order to assess the feasibility of the large-scale application of PMFC, Wetser et al. (2017) implemented the tubular design of plant microbial fuel cells in wetlands. The results indicated that electricity generation was not optimal due to oxygen crossover from cathode to anode. According to Wetser et al. (2017), power density can be increased by improving the PMFC design limiting crossover of oxygen and substrate (Wetser et al. 2017). Helder et al. (2013a) suggested that electricity production is not the unique selling point of the PMFC but the opportunity of combining its electricity production with other applications.

Although much improvement has been made on PMFC as it relates to both the biological components and device architecture, future work is needed to improve and understand the incorporation of electrodes into the rhizosphere landscape, complex nature of the exudates, times of maximal and minimal exudation, exudation rates, anodic and cathodic reaction rates, presence of suitable microorganisms and microbe-exudate interactions, and large-scale application (Bombelli et al. 2013a).

References

Ahn JH, Jeong WS, Choi MY, Kim BY, Song J, Weon HY (2014) Phylogenetic diversity of dominant bacterial and archaeal communities in plant-microbial fuel cells using rice plants. [Research Support, Non-U.S. Gov't]. J Microbiol Biotechnol 24(12):1707–1718

Arends JBA, Blondeel E, Tennison SR, Boon N, Verstraete W (2012) Suitability of granular carbon as an anode material for sediment microbial fuel cells. J Soils Sediments 12(7):1197–1206. https://doi.org/10.1007/s11368-012-0537-6

Arends JBA, Speeckaert J, Blondeel E, De Vrieze J, Boeckx P, Verstraete W, Rabaey K, Boon N (2014) Greenhouse gas emissions from rice microcosms amended with a plant microbial fuel cell. Appl Microbiol Biotechnol 98(7):3205–3217. https://doi.org/10.1007/s00253-013-5328-5

Bombelli P, Iyer DM, Covshoff S, McCormick AJ, Yunus K, Hibberd JM, Fisher AC, Howe CJ (2013a) Comparison of power output by rice (Oryza sativa) and an associated weed (Echinochloa glabrescens) in vascular plant bio-photovoltaic (VP-BPV) systems. [Research Support, Non-U.S. Gov't]. Appl Microbiol Biotechnol 97(1):429–438. https://doi.org/10.1007/s00253-012-4473-6

Bombelli P, Iyer DMR, Covshoff S, McCormick AJ, Yunus K, Hibberd JM, Fisher AC, Howe CJ (2013b) Comparison of power output by rice (Oryza sativa) and an associated weed (Echinochloa glabrescens) in vascular plant bio-photovoltaic (VP-BPV) systems. Appl Microbiol Biotechnol 97(1):429–438. https://doi.org/10.1007/s00253-012-4473-6

Cabezas A, Pommerenke B, Boon N, Friedrich MW (2015) Geobacter, Anaeromyxobacter and Anaerolineae populations are enriched on anodes of root exudate driven microbial fuel cells in rice field soil. Environ Microbiol 7(10):1758–2229. https://doi.org/10.1111/17582229.12277

Chen Z, Huang YC, Liang JH, Zhao F, Zhu YG (2012) A novel sediment microbial fuel cell with a biocathode in the rice rhizosphere. [Evaluation Studies Research Support, Non-U.S. Gov't]. Bioresour Technol 108:55–59. https://doi.org/10.1016/j.biortech.2011.10.040

Chiranjeevi P, Mohanakrishna G, Mohan SV (2012) Rhizosphere mediated electrogenesis with the function of anode placement for harnessing bioenergy through CO2 sequestration. Bioresour Technol 124:364–370. https://doi.org/10.1016/j.biortech.2012.08.020

Chiranjeevi P, Chandra R, Mohan SV (2013) Ecologically engineered submerged and emergent macrophyte based system: an integrated eco-electrogenic design for harnessing power with simultaneous wastewater treatment. Ecol Eng 51(Supplement C):181–190. https://doi.org/10.1016/j.ecoleng.2012.12.014

De Schamphelaire L, Van Den Bossche L, Hai SD, Höfte M, Boon N, Rabaey K, Verstraete W (2008) Microbial fuel cells generating electricity from rhizodeposits of rice plants. Environ Sci Technol 42(8):3053–3058. https://doi.org/10.1021/es071938w

De Schamphelaire L, Cabezas A, Marzorati M, Friedrich MW, Boon N, Verstraete W (2010) Microbial community analysis of anodes from sediment microbial fuel cells powered by rhizodeposits of living rice plants. Appl Environ Microbiol 76(6):2002–2008. https://doi.org/10.1128/AEM.02432-09

Deng H, Chen Z, Zhao F (2012) Energy from plants and microorganisms: progress in plant-microbial fuel cells. [Research Support, Non-U.S. Gov't Review]. ChemSusChem 5(6):1006–1011. https://doi.org/10.1002/cssc.201100257

Fang Z, Song HL, Cang N, Li XN (2013) Performance of microbial fuel cell coupled constructed wetland system for decolorization of azo dye and bioelectricity generation. Bioresour Technol 144:165–171. https://doi.org/10.1016/j.biortech.2013.06.073

Flexer V, Mano N (2010) From dynamic measurements of photosynthesis in a living plant to sunlight transformation into electricity. Anal Chem 82(4):1444–1449. https://doi.org/10.1021/ac902537h

Habibul N, Hu Y, Wang YK, Chen W, Yu HQ, Sheng GP (2016) Bioelectrochemical chromium(VI) removal in plant-microbial fuel cells. Environ Sci Technol 50(7):3882–3889. https://doi.org/10.1021/acs.est.5b06376

Helder M, Strik DPBTB, Hamelers HVM, Kuhn AJ, Blok C, Buisman CJN (2010) Concurrent bio-electricity and biomass production in three plant-microbial fuel cells using Spartina anglica, Arundinella anomala and Arundo donax. Bioresour Technol 101(10):3541–3547. https://doi.org/10.1016/j.biortech.2009.12.124

Helder M, Strik DP, Hamelers HV, Kuijken RC, Buisman CJ (2012a) New plant-growth medium for increased power output of the plant-microbial fuel cell. [Research Support, Non-U.S. Gov't]. Bioresour Technol 104:417–423. https://doi.org/10.1016/j.biortech.2011.11.005

Helder M, Strik D, Hamelers B, Buisman C (2012b) The flat-plate plant-microbial fuel cell: the effect of a new design on internal resistance. *Biotechnol Biofuels 5:70.* https://doi.org/10.1186/1754-6834-5-70

Helder M, Chen W-S, van der Harst EJM, Strik DPBTB, Hamelers HBVM, Buisman CJN, Potting J (2013a) Electricity production with living plants on a green roof: environmental performance of the plant-microbial fuel cell. Biofuels Bioprod Biorefin 7(1):52–64. https://doi.org/10.1002/bbb.1373

Helder M, Strik DPBTB, Timmers RA, Raes SMT, Hamelers HVM, Buisman CJN (2013b) Resilience of roof-top plant-microbial fuel cells during Dutch winter. Biomass Bioenergy 51:1–7. https://doi.org/10.1016/j.biombioe.2012.10.011

Hindatu Y, Annuar MSM, Gumel AM (2017) Mini-review: anode modification for improved performance of microbial fuel cell. Renew Sust Energ Rev 73:236–248. https://doi.org/10.1016/j.rser.2017.01.138

Hubenova Y, Mitov M (2012) Conversion of solar energy into electricity by using duckweed in direct photosynthetic plant fuel cell. Bioelectrochemistry 87(Supplement C):185–191. https://doi.org/10.1016/j.bioelechem.2012.02.008

Hubenova Y, Mitov M (2015) Enhanced metabolic and redox activity of vascular aquatic plant Lemna valdiviana under polarization in direct photosynthetic plant fuel cell. Bioelectrochemistry 106(Part A):226–231. https://doi.org/10.1016/j.bioelechem.2014.07.007

He Z, Kan J, Mansfeld F, Angenent LT, Nealson KH (2009) Self-sustained phototrophic microbial fuel cells based on the synergistic cooperation between photosynthetic microorganisms and heterotrophic bacteria. Environ Sci Technol 43:1648–1654

Kaku N, Yonezawa N, Kodama Y, Watanabe K (2008) Plant/microbe cooperation for electricity generation in a rice paddy field. [article]. Appl Microbiol Biotechnol 79(1):43–49. https://doi.org/10.1007/s00253-008-1410-9

Kouzuma A, Kasai T, Nakagawa G, Yamamuro A, Abe T, Watanabe K (2013) Comparative metagenomics of anode-associated microbiomes developed in rice paddy-field microbial fuel cells. PLoS One 8(11):e77443. https://doi.org/10.1371/journal.pone.0077443

Liu S, Song H, Li X, Yang F (2013) Power generation enhancement by utilizing plant Photosynthate in microbial fuel cell coupled constructed wetland system. Int J Photoenergy 2013:1–10. https://doi.org/10.1155/2013/172010

Logan BE (2009) Exoelectrogenic bacteria that power microbial fuel cells. Nat Rev Microbiol 7:375. https://doi.org/10.1038/nrmicro2113

Lu L, Xing D, Ren ZJ (2015) Microbial community structure accompanied with electricity production in a constructed wetland plant microbial fuel cell. Bioresour Technol 195:115–121. https://doi.org/10.1016/j.biortech.2015.05.098

McCormick AJ, Bombelli P, Bradley RW, Thorne R, Wenzel T, Howe CJ (2015) Biophotovoltaics: oxygenic photosynthetic organisms in the world of bioelectrochemical systems. Energy Environ Sci 8(4):1092–1109. https://doi.org/10.1039/C4EE03875D. https://doi.org/10.1039/C4EE03875D

Moqsud MA, Hannan MA, Omine K (2015a) Assessment of factors influencing bioelectricity generation in paddy plant microbial cells. Global Adv Res J Agric Sci 4(12):840–850

Moqsud MA, Yoshitake J, Bushra QS, Hyodo M, Omine K, Strik D (2015b) Compost in plant microbial fuel cell for bioelectricity generation. Waste Manag 36:63–69. https://doi.org/10.1016/j.wasman.2014.11.004

Malik S, Drott E, Grisdela P, Lee J, Lee C, Lowy DA, Gray S, Tender LM (2009) A self-assembling self-repairing microbial photoelectrochemical solar cell. Energy Environ Sci 2:292–298

Nitisoravut R, Regmi R (2017) Plant microbial fuel cells: a promising biosystems engineering. [review]. Renew Sust Energ Rev 76:81–89. https://doi.org/10.1016/j.rser.2017.03.064

Oon YL, Ong SA, Ho LN, Wong YS, Oon YS, Lehl HK, Thung WE (2015) Hybrid system up-flow constructed wetland integrated with microbial fuel cell for simultaneous wastewater treatment and electricity generation. Bioresour Technol 186:270–275. https://doi.org/10.1016/j.biortech.2015.03.014

Reimers CE, Tender LM, Fertig S, Wang W (2001) Harvesting energy from the marine sediment-water interface. Environ Sci Technol 35(1):192–195.

Reimers CE, Girguis P, Stecher HA III, Tender LM, Ryckelynck N, Whaling P (2006). Microbial fuel cell energy from an ocean cold seep. Geobiology 4:123–136. https://doi.org/10.1111/j.1472-4669.2006.00071.x

Salinas-Juarez MG, Roquero P, Duran-Dominguez-de-Bazua Mdel C (2016) Plant and microorganisms support media for electricity generation in biological fuel cells with living hydrophytes. Bioelectrochemistry 112:145–152. https://doi.org/10.1016/j.bioelechem.2016.02.007

Sophia AC, Sreeja S (2017) Green energy generation from plant microbial fuel cells (PMFC) using compost and a novel clay separator. Sustain Energy Technol Assess 21:59–66. https://doi.org/10.1016/j.seta.2017.05.001

Strik DPBTB, Hamelers HVM, Snel JFH, Buisman CJN (2008) Green electricity production with living plants and bacteria in a fuel cell. Int J Energy Res 32(9):870–876. https://doi.org/10.1002/er.1397

Strik DPBTB, Timmers RA, Helder M, Steinbusch KJJ, Hamelers HVM, Buisman CJN (2011) Microbial solar cells: applying photosynthetic and electrochemically active organisms. [review]. Trends Biotechnol 29(1):41–49. https://doi.org/10.1016/j.tibtech.2010.10.001

Takanezawa K, Nishio K, Kato S, Hashimoto K, Watanabe K (2010) Factors affecting electric output from rice-paddy microbial fuel cells. [Research Support, Non-U.S. Gov't]. Biosci Biotechnol Biochem 74(6):1271–1273. https://doi.org/10.1271/bbb.90852

Tapia NF, Rojas C, Bonilla CA, Vargas IT (2017) Evaluation of sedum as driver for plant microbial fuel cells in a semi-arid green roof ecosystem. Ecol Eng 108:203–210. https://doi.org/10.1016/j.ecoleng.2017.08.017

Tender LM, Reimers CE, Stecher III HA, Holmes DE, Bond DR, Lowy DA, Pilobello K, Fertig SJ, Lovley DR (2002) Harnessing microbially generated power on the seafloor. Nat Biotechnol 20(8):821–825

Timmers RA, Strik DPBTB, Hamelers HVM, Buisman CJN (2010) Long-term performance of a plant microbial fuel cell with Spartina anglica. Appl Microbiol Biotechnol 86(3):973–981. https://doi.org/10.1007/s00253-010-2440-7

Timmers RA, Rothballer M, Strik DPBTB, Engel M, Schulz S, Schloter M, Hartmann A, Hamelers B, Buisman C (2012a) Microbial community structure elucidates performance of glyceria maxima plant microbial fuel cell. Appl Microbiol Biotechnol 94(2):537–548. https://doi.org/10.1007/s00253-012-3894-6

Timmers RA, Strik DP, Arampatzoglou C, Buisman CJ, Hamelers HV (2012b) Rhizosphere anode model explains high oxygen levels during operation of a Glyceria maxima PMFC. [Research Support, Non-U.S. Gov't]. Bioresour Technol 108:60–67. https://doi.org/10.1016/j.biortech.2011.10.088

Timmers RA, Strik DPBTB, Arampatzoglou C, Buisman CJN, Hamelers HVM (2012c) Rhizosphere anode model explains high oxygen levels during operation of a Glyceria maxima PMFC. Bioresour Technol 108:60–67. https://doi.org/10.1016/j.biortech.2011.10.088

Timmers RA, Strik DPBTB, Hamelers HVM, Buisman CJN (2012d) Characterization of the internal resistance of a plant microbial fuel cell. Electrochim Acta 72:165–171. https://doi.org/10.1016/j.electacta.2012.04.023

Timmers RA, Strik DPBTB, Hamelers HVM, Buisman CJN (2013a) Electricity generation by a novel design tubular plant microbial fuel cell. Biomass Bioenergy 51:60–67. https://doi.org/10.1016/j.biombioe.2013.01.002

Timmers RA, Strik DPBTB, Hamelers HVM, Buisman CJN (2013b) Increase of power output by change of ion transport direction in a plant microbial fuel cell. Int J Energy Res 37(9):1103–1111. https://doi.org/10.1002/er.2957

Trapero JR, Horcajada L, Linares JJ, Lobato J (2017) Is microbial fuel celltechnology ready? An economic answer towards industrialcommercialization, Appl. Energy 185:698–707.

Ueoka N, Sese N, Sue M, Kouzuma A, Watanabe K (2016) Sizes of anode and cathode affect electricity generation in Rice Paddy-field microbial fuel cells. J Sustain Bioenergy Syst *06*(01):6. https://doi.org/10.4236/jsbs.2016.61002

Venkata Mohan S, Mohanakrishna G, Chiranjeevi P (2011) Sustainable power generation from floating macrophytes based ecological microenvironment through embedded fuel cells along with simultaneous wastewater treatment. Bioresour Technol 102(14):7036–7042. https://doi.org/10.1016/j.biortech.2011.04.033

Villaseñor J, Capilla P, Rodrigo MA, Cañizares P, Fernández FJ (2013) Operation of a horizontal subsurface flow constructed wetland - microbial fuel cell treating wastewater under different organic loading rates. Water Res 47(17):6731–6738. https://doi.org/10.1016/j.watres.2013.09.005

Wang J, Song X, Wang Y, Abayneh B, Ding Y, Yan D, Bai J (2016) Microbial community structure of different electrode materials in constructed wetland incorporating microbial fuel cell. Bioresour Technol 221:697–702. https://doi.org/10.1016/j.biortech.2016.09.116

Wetser K, Sudirjo E, Buisman CJN, Strik DPBTB (2015) Electricity generation by a plant microbial fuel cell with an integrated oxygen reducing biocathode. Appl Energy 137:151–157. https://doi.org/10.1016/j.apenergy.2014.10.006

Wetser K, Dieleman K, Buisman C, Strik D (2017) Electricity from wetlands: tubular plant microbial fuels with silicone gas-diffusion biocathodes. Appl Energy 185:642–649. https://doi.org/10.1016/j.apenergy.2016.10.122

Yadav AK, Dash P, Mohanty A, Abbassi R, Mishra BK (2012) Performance assessment of innovative constructed wetland-microbial fuel cell for electricity production and dye removal. Ecol Eng 47(Supplement C):126–131. https://doi.org/10.1016/j.ecoleng.2012.06.029

Yan Z, Jiang H, Cai H, Zhou Y, Krumholz LR (2015) Complex interactions between the Macrophyte Acorus Calamus and microbial fuel cells during pyrene and benzo[a]pyrene degradation in sediments. Sci Rep 5:10709. https://doi.org/10.1038/srep10709 https://www.nature.com/articles/srep10709#supplementary-information

Zhang J, Zhang E, Scott K, Burgess JG (2012) Enhanced electricity production by use of reconstituted artificial consortia of estuarine Bacteria grown as biofilms. Environ Sci Technol 46(5):2984–2992. https://doi.org/10.1021/es2020007

Zhao Y, Collum S, Phelan M, Goodbody T, Doherty L, Hu Y (2013) Preliminary investigation of constructed wetland incorporating microbial fuel cell: batch and continuous flow trials. Chem Eng J 229:364–370. https://doi.org/10.1016/j.cej.2013.06.023

Chapter 4
Current Advances in Paddy Plant Microbial Fuel Cells

Kiyoshi Omine, Santos D. Chicas, and Venkataraman Sivasankar

4.1 Introduction

Microbial fuel cells (MFCs) are bioelectrochemical cells that convert microbial reducing power into electrical energy that is green (Logan and Regan 2006). Attempts have been made to apply MFC systems to recover electric power from marine and riverbeds termed as sediment MFCs (SMFCs) (Schamphelaire et al. 2008). MFC as a hybrid composting method, which reuses kitchen waste as raw material, has also been proposed (Moqsud et al. 2014).

Plant microbial fuel cells (PMFCs) are a recently developed technology that uses organic rhizodeposits, comprising of root exudates and dead root cells, as the electron donor for heterotrophic microorganisms in the plant rhizosphere (Strik et al. 2008). Plants excrete photosynthesized organic compounds from roots. Those organics are used by microbes for electricity generation in PMFCs. PMFCs are remarkably sustainable because they have a clean conversion without emissions and have no competition for arable land or nature. PMFCs can also be implemented in rice paddy fields combining food and electricity production and so circumventing the competition with food production (Kaku et al. 2008).

Graphite is often used as cathode material in a PMFC. However, the reduction of oxygen on graphite is slow and limits the power output of the PMFC. Electrocatalysts like platinum are able to catalyze the reduction of oxygen. The high costs and the potential poisoning compounds in the solution make platinum undesired to be applied in the PMFC.

K. Omine (✉) · V. Sivasankar
Geo-Tech Laboratory, Department of Civil Engineering, Graduate School of Engineering, Nagasaki University, Nagasaki, 8528521, Japan
e-mail: omine@nagasaki-u.ac.jp

S. D. Chicas
Faculty of Sciences and Technology, University of Belize, Belmopan Central Campus, Cayo District, C.A, Belize

V. Sivasankar et al. (eds.), *Microbial Fuel Cell Technology for Bioelectricity*,
https://doi.org/10.1007/978-3-319-92904-0_4

In this study, a performance of paddy plant microbial fuel cell (PMFC) is evaluated by experiments using container of bucket and PET bottle. Two types of electrodes, namely, carbon fiber and activated bamboo charcoal, are used on paddy PMFC. Influences of electrode material and existence of iron wire attached to anode on voltage generation are investigated. The influence of connections in series or parallel, using small-sized PET bottle (500 mL), is also investigated.

4.2 Test Materials and Methods

Two types of electrodes were used on the PMFC. Carbon fibers do not have any negative effect on the growth of paddy roots (Moqsud et al. 2015). However, it is considered that carbon fiber is not suitable for the anode, because the roots of paddy are closely attached with the carbon fiber, thus making it difficult to be removed. In this study, carbon fiber (Toho Rayon Co., Ltd., Tokyo) was used only as a cathode. Activated bamboo charcoal (KPC Co., Ltd., Shiga Prefecture, Japan) was used as anode or cathode. These electrodes are good at conducting electricity with an electrical resistance of 5 ohms. Figure 4.1 shows the electrode materials of cathode and anode. The activated bamboo charcoal sizes were around 120 × 50 mm and 50 × 40 mm, and the carbon fiber mass was 10 g; these were connected to a stainless wire. Power generation of soil microbial fuel cell (SMFC) using organic waste increases by wrapping iron wire on the activated bamboo charcoal anode (Moqsud et al. 2013). The activated bamboo charcoal with iron wire was also used in this experiment to investigate its effect on power generation of PMFC.

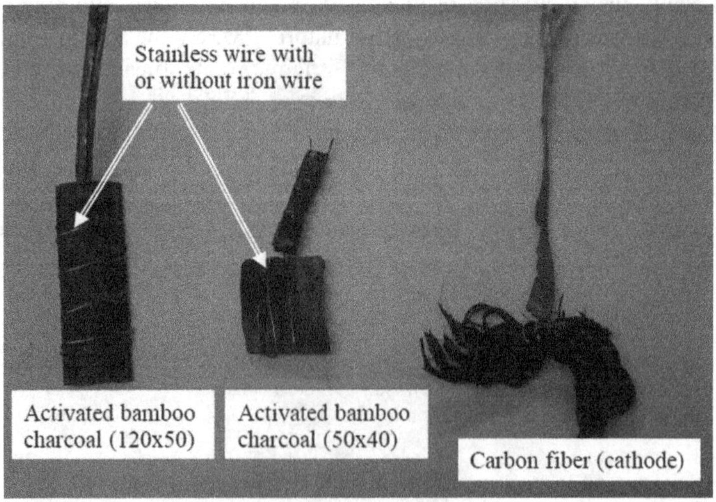

Fig. 4.1 Electrode materials of cathode and anode

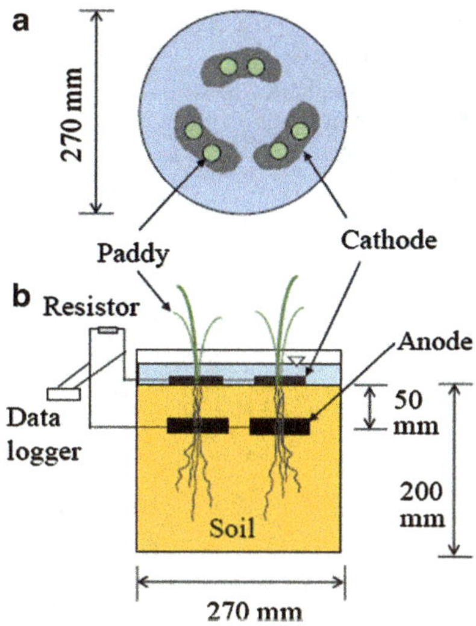

Fig. 4.2 Plan (a) and cross section (b) of the experimental device using bucket of 13 L

Table 4.1 Test conditions on paddy plant MFC using bucket

No.	Plant	Fertilizer	Cathode	Anode
1	Without	Without	Carbon fiber	Activated bamboo charcoal
2	Paddy	Chemical	Carbon fiber	Activated bamboo charcoal
3	Paddy	Organic	Carbon fiber	Activated bamboo charcoal
4	Paddy	Organic	Carbon fiber	Activated bamboo charcoal
5	Paddy	Organic	Carbon fiber	Activated bamboo charcoal with iron wire
6	Paddy	Organic	Carbon fiber	Activated bamboo charcoal with iron wire
7	Paddy	Organic	Carbon fiber	Activated bamboo charcoal with iron wire
8	Paddy	Organic	Carbon fiber	Activated bamboo charcoal with iron wire
9	Paddy	Organic	None	None

No.5–No.8: same conditions

Schematic diagram on experimental device of the PMFC using bucket of 13 L is illustrated in Fig. 4.2. Test conditions of the PMFC are shown in Table 4.1. The soil was prepared by mixing clayey soil, sandy soil, culture soil, and leaf mold. Eight buckets were prepared for the PMFCs. Bucket No. 1 was not planted for comparing the electricity generation with or without plant. There was no fertilizer mixed in this bucket. Bucket No. 2 was prepared by mixing chemical fertilizer of 5 g into the soil. Bucket Nos. 3–8 were prepared by mixing organic fertilizer of 30 g. The carbon fiber was used for all buckets as cathode. The activated bamboo charcoal was used for bucket Nos. 1–4. The activated bamboo charcoal with iron wire was used for bucket Nos. 5–8 for the purpose of increasing the performance of the PMFC. Bucket Nos. 3–4 and bucket Nos. 5–8 were prepared as replications of the same conditions, respectively.

Fig. 4.3 Plan (a) and cross section (b) of the experimental device with PET bottle of 0.5 L

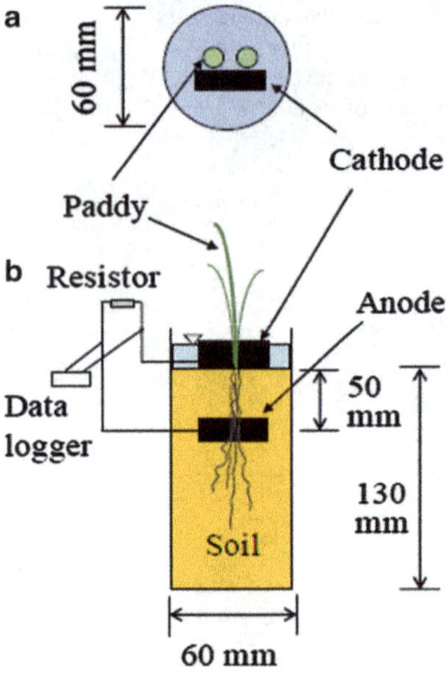

Table 4.2 Test conditions on paddy plant MFCs of five PET bottles' connection in series or parallel

	Plant	Fertilizer	Cathode	Anode
Case 1	Paddy	Organic	Activated bamboo charcoal	Activated bamboo charcoal
Case 2	Paddy	Organic	Activated bamboo charcoal	Activated bamboo charcoal with iron wire

Three anodes made of the activated bamboo charcoal in size of 120 × 50 mm were inserted into the soil, and three cathodes made of the carbon fiber in mass of 10 g were placed on a surface of the soil. The anode area covers around 0.006 m^2 inside the soil of the PMFC. The anode was set approximately 50 mm below the surface of the soil, while the cathode was placed immediately above the soil surface, but under the water. These electrodes were connected via lead wires. Both the anode and cathode were connected to a data logger. The data logger is set to measure the voltage in every 5-min interval.

Additionally bucket No. 9 with organic fertilizer and without electrode was prepared. The rice plants were planted in the soil in each bucket except for bucket No. 1. Black rice (ancient rice) was selected, because the black rice is resistant to disease and easy to grow.

In order to investigate the influence of electrode material and voltage generation by difference of connection, small-sized paddy plant MFC was prepared. Schematic diagram on experimental device of the PMFC using PET bottle of 500 mL is illustrated in Fig. 4.3. Test conditions of the PMFC are shown in Table 4.2. The same soils described above mixed with 6 g of organic fertilizer were used in each PMFC.

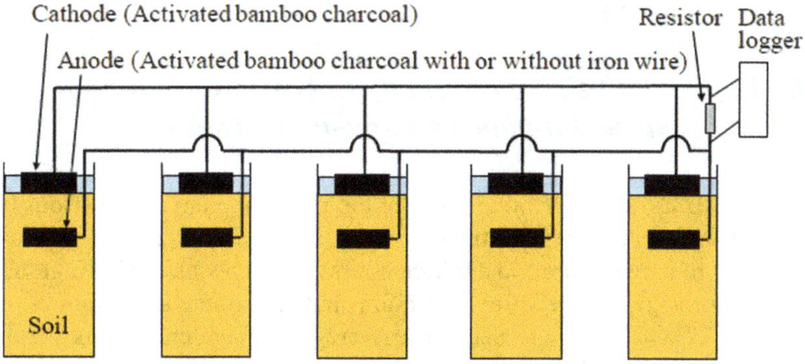

Fig. 4.4 Plant MFCs of connection in parallel using five PET bottles of 0.5 L

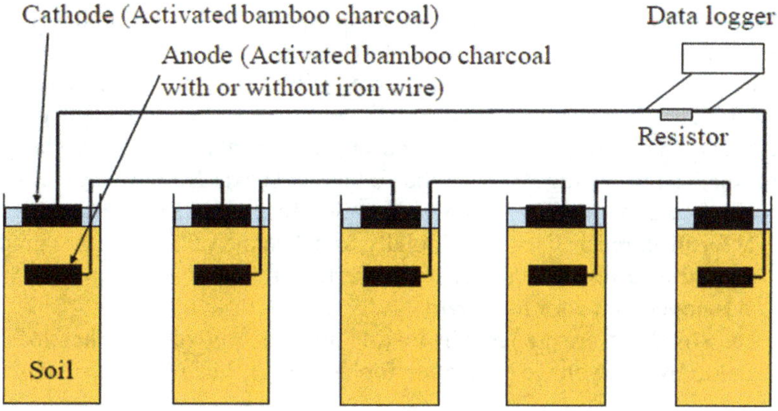

Fig. 4.5 Plant MFCs of connection in series using five PET bottles of 0.5 L

The activated bamboo charcoal was used for both electrode materials, anode and cathode. Two types of PMFCs using anodes with and without iron wire, Case 1 and Case 2, were prepared, respectively.

The activated bamboo charcoal in size of 50 × 40 mm was inserted into the soil, and the same size of the electrode was placed on the surface of the soil. The anode area covers around 0.002 m^2 inside the soil of the PMFC. The anode was set approximately 50 mm below the surface of the soil, while the cathode was placed immediately above the soil surface, but under the water. Electrodes of five PMFCs were connected in parallel or series via lead wires to a data logger. Figures 4.4 and 4.5 show PMFCs of the connection in parallel or series using five PET bottles of 500 mL, respectively. The data logger is set to measure the voltage in every 5-min interval. The black rice plant was planted in the soil.

4.3 Results and Discussion

4.3.1 Experiment Using Bucket of 13 L with Carbon Fiber and Activated Bamboo Charcoal as Electrodes

Paddy plant MFCs were performed during the rice cropping season (from June to August) in Nagasaki University Bunkyo-machi campus, Japan. Figure 4.6 illustrates the variations of temperature and daylight hours during this period in Nagasaki City (Japan Meteorological Agency). Maximum and minimum temperatures in this period were 37 and 16 °C, respectively. Average temperature in this period was 26.5 °C. Average daylight hour was 6.4 h. On the whole, there were many sunny days. The weather condition during the study period was good for growing paddy plants.

Figure 4.7 illustrates the variation of voltage generation with time in rice PMFCs in different buckets. Rice seedlings were planted to the buckets on June 9, 2017, and data were collected from June 17, 2017. The voltage generation on the case of without paddy plant (No. 1) increased gradually from June 15 and reached to 0.2 V. After that, the value increased and decreased. It was observed that algae grew in the bucket. This might have occurred due to the presence of nutrients in the culture soil. The result suggests that the voltage of bucket No. 1 was generated due to the presence of algae. The voltage generation for the case of bucket No. 2 with chemical fertilizer reached to 0.5 V initially, and then the value decreases gradually. This occurred as a result of the chemical fertilizer which works quickly, but the effect does not continue for long term.

Bucket Nos. 3–4 were prepared by mixing organic fertilizer as replications of the same conditions. The voltage generation for bucket No. 3 increased up to 0.37 V, but

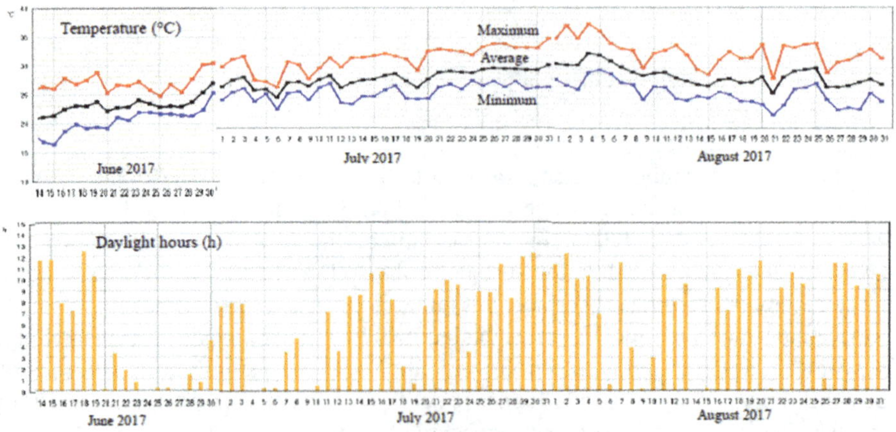

Fig. 4.6 Variations of temperature and day light hours in Nagasaki City (Japan Meteorological Agency)

after that, it dropped down. This might have occurred due to a defective connection. On the other hand, the voltage for bucket No. 4 increased gradually, and the value changed considerably. Maximum voltage reached at 0.83 V on July 31, 2017. It was the highest so far in PMFC research (Moqsud et al. 2015; Liu et al. 2013; Strik et al. 2008).

Bucket Nos. 5–8 with the activated bamboo charcoal with iron wire were prepared by mixing organic fertilizer as replications of the same conditions. The voltage for bucket Nos. 5–8 increased gradually, and the results were also similar. Maximum voltage reached at 0.68 V on August 24, 2017. It was observed that the voltage for bucket Nos. 5–8 is more stable when compared with that of bucket Nos. 3–4. However, it is not easy to get stable voltage on PMFCs.

Figure 4.8 shows the growth of paddy plant with time in different buckets. The growth of the plant for all buckets increased gradually, and the length became more than 800 mm. Growing speed of bucket Nos. 3–8 with organic fertilizer is relatively high comparing that of bucket No. 2 with chemical fertilizer. It was also observed that the growth of bucket Nos. 5–8 with iron wire is enhanced. It may be considered that iron was supplied to the plant as nutrition.

Figure 4.9 shows growth of paddy plant in different buckets. Additionally, paddy seedling was planted to bucket No. 9 with organic fertilizer and without electrodes. It was found that the electrodes do not influence the growth of paddy plant.

Figure 4.10 illustrates variation of voltage with duration and influence of solar radiation during July 14–24, 2017. Bucket Nos. 1, 2, 4, and 5 were selected. The solar radiation shows clear peak value in daytime, and the value becomes zero during nighttime. The voltage of all buckets also showed clear peak value in daytime. Bucket No. 5 with organic fertilizer and iron wire showed high voltage when compared with that of other buckets in this period. It is considered that iron wire contributed to the increase in voltage. However, it was found that bucket Nos. 1, 2, and 4 kept the voltage between 0.1 and 0.2 V even at nighttime. It is not certain why these differences occurred.

4.3.2 Experiment Using PET Bottle of 500 mL with Activated Bamboo Charcoal for Anode and Cathode

PMFCs using PET bottle of 500 mL were performed for investigating influences of electrode material and voltage generation by difference of connection. Activated bamboo charcoal was used for both electrode materials, anode and cathode. Two types of anodes with and without iron wire were prepared. Five PMFCs were connected in series at first, and these PMFCs were reconnected in parallel later. Figure 4.11 illustrates variation of voltage on PMFCs in a series connection during July 15–22, 2017, and the influence of solar radiation. Average voltage values of PMFCs in series connection with or without iron wire were 3.12 V and 1.19 V, respectively. It is obvious that the voltage generation of plant MFCs increased by

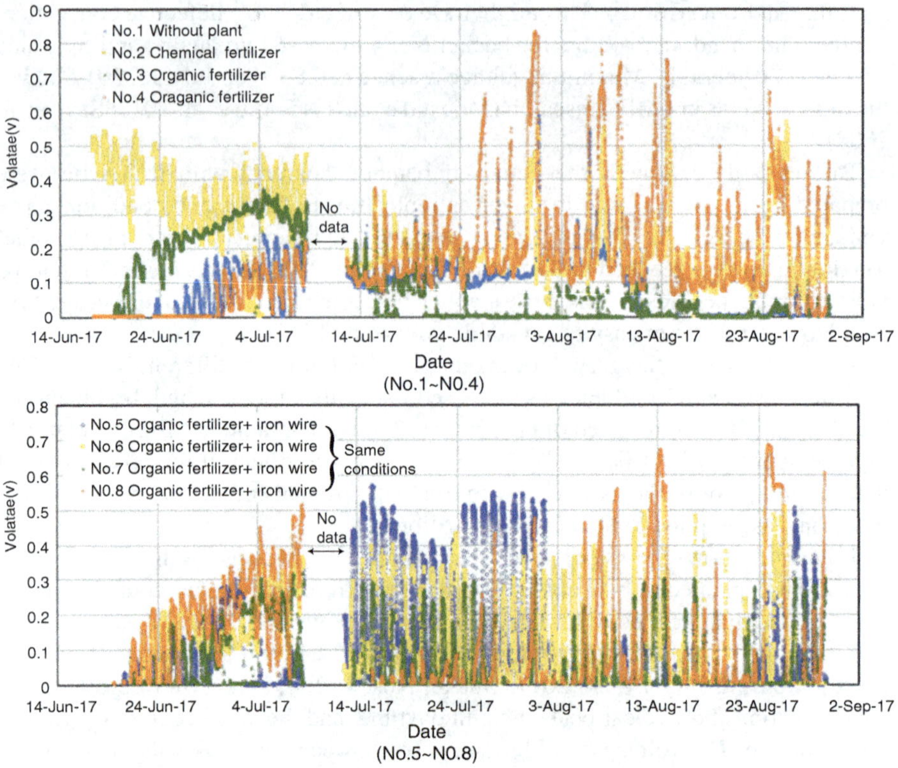

Fig. 4.7 Variation of voltage generation with time in different buckets

using anode with iron wire. This resulted in a voltage increase. A similar effect of iron wire was also found in soil MFC using organic wastes (Moqsud et al. 2013). Furthermore, it is remarkable that PMFCs generated voltage continuously even at nighttime. Sometime the voltage changed, but no effect of solar radiation was observed.

Figure 4.12 illustrates variation of voltage on PMFCs in parallel connection during July 25 to August 6, 2017. Similar trends in influences of iron wire and material of cathode were also found on PMFCs in parallel connection.

Average voltage values of PMFCs in parallel connection with or without iron wire were 0.517 V and 0.332 V, respectively. Average voltage of PMFCs in parallel connection was lower than that in series connection. Ideally, a voltage of five batteries in series connection becomes five times of that of a battery, and a voltage of batteries in parallel connection is the same. The voltage ratios of series and parallel connections on PMFCs using cathode with or without iron wire were 6.0 and 3.6, respectively. The results indicate that the voltage generation of PMFCs increased effectively by using iron wire.

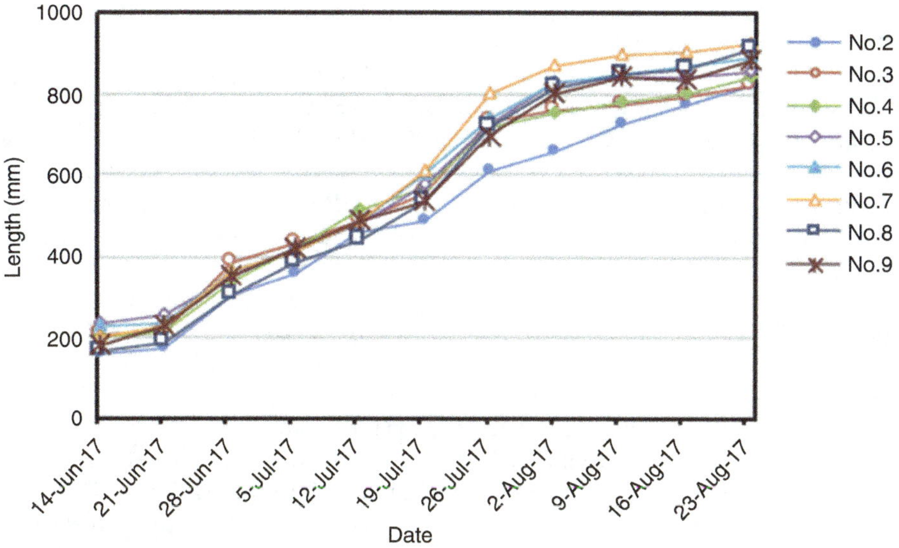

Fig. 4.8 Length of paddy plant with time in different buckets

Fig. 4.9 Growth of paddy plants in different buckets

Figure 4.13 shows the growth of paddy plant using PET bottle of 500 mL in the case of anode with iron wire. Figure 4.14 illustrates the length of paddy plant with time and influence of anode with or without iron wire. The values were measured as an average of five PET bottles of 500 mL for each case. Both paddy plants grew gradually to approximately 600 mm by the end of July 2017. After that, the growth of the plants slowed down. The rice plants using bucket grew more than 800 mm in length as shown in Fig. 4.8. It is therefore considered that the lower growth of the stem is due to the small size of the PET bottle, which limited root growth. Finally, the length of the paddy plant in the PMFCs with iron wire was longer than without iron wire. Despite the small-sized condition, it was determined that PMFCs using PET bottle can generate relatively large voltage.

Electrode output is measured in volts (*V*) against time. The current *I* in amperes (A) is calculated using Ohm's law, $I = V/R$, where *V* is the measured voltage in volts (V) and *R* is the known value of the external load resistor in ohms. From this it is possible to calculate the electric power output *P* in watts (W) of PMFCs by taking the product of the voltage and current, i.e., $P = I \times V$. For obtaining a maximum power of PMFCs, the values of voltage are measured using different resistances.

Figure 4.15 shows the relationship between voltage and current in the PMFCs of five PET bottles in a series connection on July 20, 2017. It was found that the relationship was almost linear. The intercept and inclination of the line represent electromotive force and internal resistance for the MFCs, respectively. It represents that the PMFCs with a good performance indicate high electromotive force. The test results obtained from Fig. 4.15 are given in Table 4.3. The electromotive force of five PMFCs in series connection with iron wire was 3157 mV. The internal resistance of PMFCs was relatively high, because PMFCs were connected in series. Maximum electric power is calculated from the linear relationship between voltage and current. The maximum power per anode area is 40.3 mW/m^2 for the PMFC with iron wire.

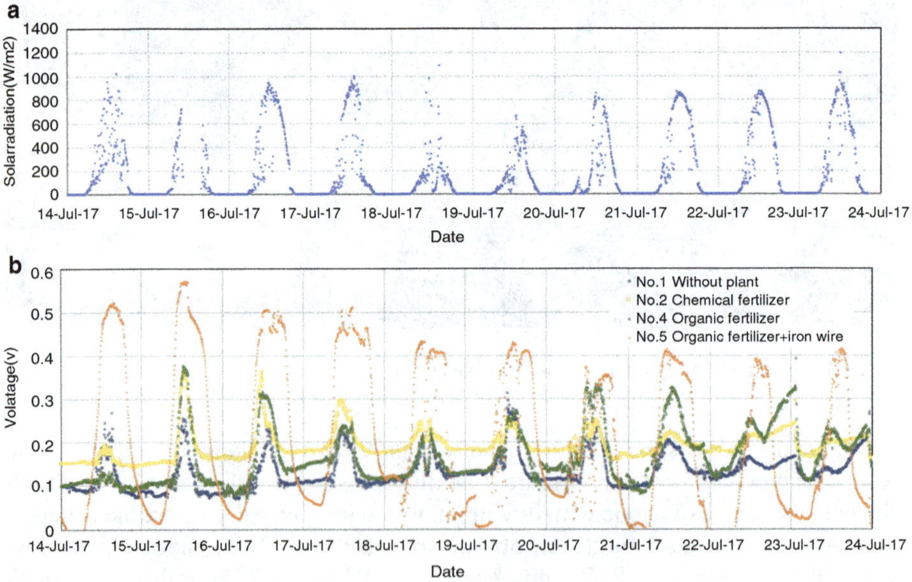

Fig. 4.10 Variation of voltage and duration and influence of solar radiation. (a) Solar radiation; (b) voltage radiation

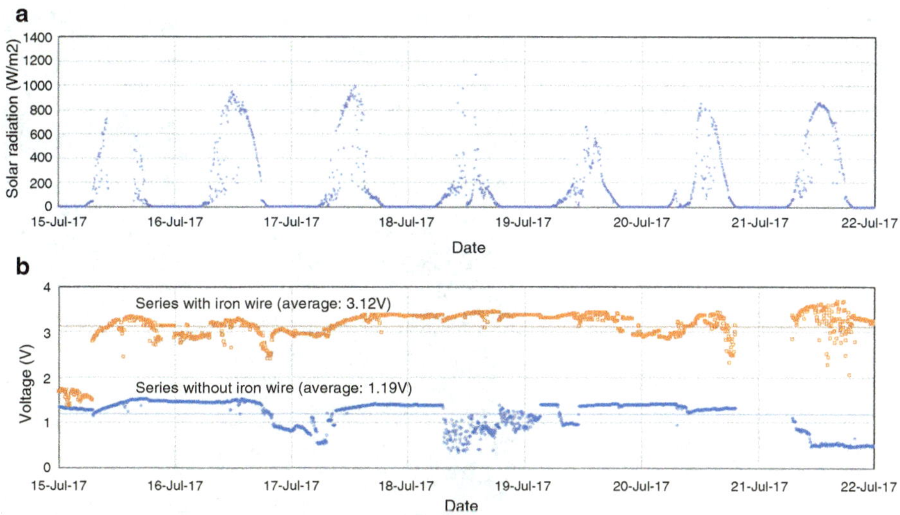

Fig. 4.11 Variation of voltage with duration on PMFCs in series connection using anode with or without iron wire and influence of solar radiation. (a) Solar radiation; (b) voltage radiation

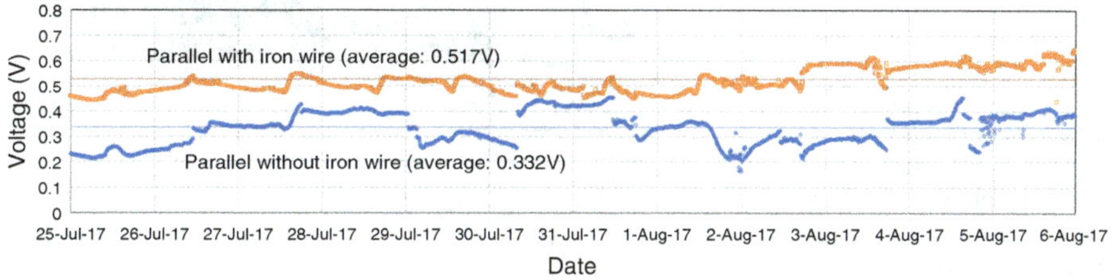

Fig. 4.12 Variation of voltage with duration on PMFCs in parallel connection and influence of anode with or without iron wire

4.4 Conclusions

The following conclusions were obtained from this study.

In the experiments of the paddy plant MFCs with anode of activated bamboo charcoal and cathode of carbon fiber using bucket of 13 L:

1. The result suggests that the voltage of the case without plant was generated due to the presence of algae. The voltage generation of the case with chemical fertilizer increased fast and reached to 0.5 V. It is considered that chemical fertilizer works quickly but the effect does not continue for long.
2. The voltage of the case with organic fertilizer increased gradually, and maximum voltage reached was 0.83 V. It was the highest so far in PMFC research. The voltage of the case with organic fertilizer and using anode with iron wire increased gradually, and maximum voltage reached was 0.68 V. It was observed that the

Fig. 4.13 Growth of paddy plant using PET bottle of 0.5 L

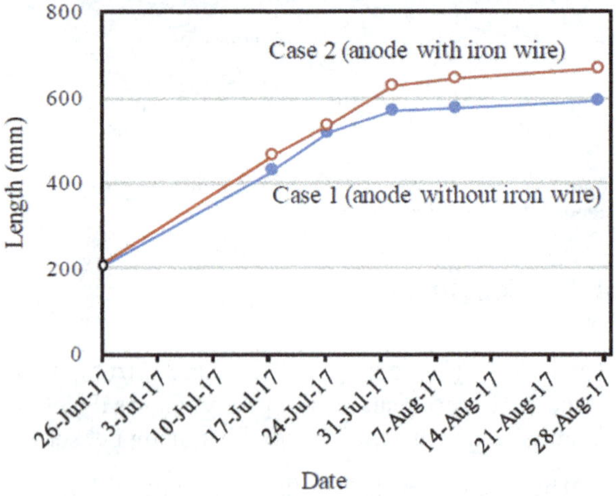

Fig. 4.14 Length of paddy plant with time and influence of anode with or without iron wire (average of five PET bottles of 500 mL for each case)

voltage of this case is more stable compared with that of the case without iron wire. However, the voltage on the PMFCs depends on sunlight.

3. It was found that the electrodes in PMFCs do not influence the growth of paddy plant. It was also observed that the growth of paddy plant is promoted when iron wire is used. This suggests that iron was supplied to the plant as nutrition.

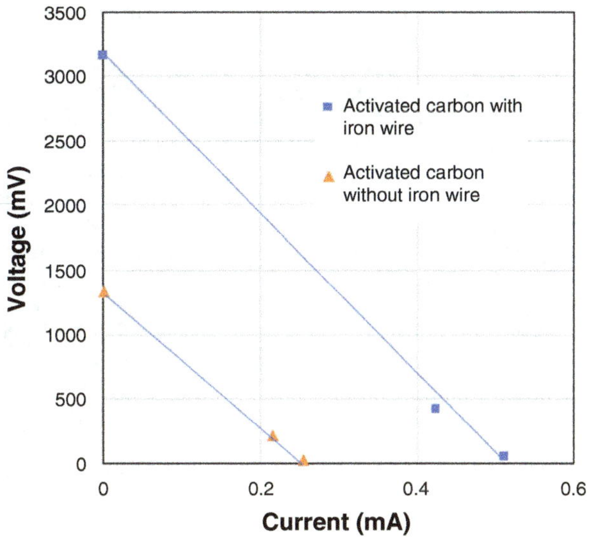

Fig. 4.15 Relationship between current and the voltage of paddy MFCs

Table 4.3 Test results on PMFCs in five PET bottles' series connection

	Electromotive force	Internal resistance (Ω)	Maximum power per area of anode (mW/m^2)
Case 1	1334	5252	8.5
Case 2	3157	6178	40.3

In the experiments of the paddy PMFCs with activated bamboo charcoal in both anode and cathode using PET bottle of 500 mL:

4. Five PMFCs were connected in series or parallel. It was observed that the voltage generation of plant MFCs increases by using anode with iron wire. It became more than 2.5 times in series connection.
5. It was remarkable that voltage of PMFCs generates continuously even at nighttime. Sometime the voltage changed, but no effect of solar radiation was observed.
6. The maximum power per anode area of 40.3 mW/m^2 was obtained on the PMFCs with iron wire in series connection.

Acknowledgement The authors thank Nagasaki University, Nagasaki, Japan.

References

Kaku N, Yonezawa N, Kodama Y, Watanabe K (2008) Plant/microbe cooperation for electricity generation in a rice paddy field. Appl Microbiol Biotechnol 79:43–49

Liu S, Song H, Li X, Yang F (2013) Power generation enhancement by utilizing plant photosynthate in microbial fuel cell coupled constructed wetland system. Int J Photoenergy 2013:1–11

Logan BE, Regan JM (2006) Electricity producing bacterial communities in microbial fuel cells. Trends Microbiol 14:512–518

Moqsud MA, Omine K, Yasufuku N, Hyodo M, Nakata Y (2013) Microbial fuel cell (MFC) for bioelectricity generation from organic wastes. Waste Manag 33:2465–2469

Moqsud MA, Omine K, Yasufuku N, Bushra QS, Hyodo M, Nakata Y (2014) Bioelectricity from kitchen garbage and bamboo wastes. Waste Manag Res 32:124–130

Moqsud MA, Yoshitake J, Bushra QS, Hyodo M, Omine K, Strik D (2015) Compost in plant microbial fuel cell for bioelectricity generation. Waste Manag 36:63–69

Schamphelaire DL, Van de Bossche L, Dang HS, Höfte M, Boon N, Rabaey K, Verstraete W (2008) Microbial fuel cells generating electricity from rhizodeposits of rice plants. Environ Sci Technol 42:3053–3058

Strik D, Hamelers HVM, Snel J, Buisman CJ (2008) Green electricity production with living plants and bacteria in a fuel cell. Int J Energy Res 32:870–876

Chapter 5
Algal Microbial Fuel Cells—Nature's Perpetual Energy Resource

Lavanyasri Rathinavel, Deepika Jothinathan, Venkataraman Sivasankar, Paul Agastian, and Prabhakaran Mylsamy

5.1 Current Scenario

The world's rapidly growing population is leading to increased energy demands worldwide. The population explosion and the rapid consumption of limited oil reserves is increasing carbon dioxide levels in the atmosphere, thus leading to global warming. Climate change is another, greater, threat to humans and the environment. Therefore, the demand for energy and its social consequences are leading researchers to look for substitutes for existing energy sources (Satyanarayana et al. 2011). Much wide-ranging research is being carried out to find possible energy solutions. The technology called microbial fuel cells (MFCs), where bacteria and other microbes generate electricity from waste and biomass, has gained the attention of researchers for its attractive features.

L. Rathinavel · P. Mylsamy (✉)
Post Graduate and Research Department of Botany, Pachaiyappa's College, Chennai, 600030, Tamil Nadu, India

D. Jothinathan
Department of Life Sciences, Central University of Tamil Nadu, Thiruvarur, 610005, Tamil Nadu, India

V. Sivasankar
Department of Civil Engineering, Nagasaki University, Nagasaki, 8528521, Japan

P. Agastian
Department of Plant Biology and Biotechnology, Loyola College, Nungambakkam, Chennai, 600034, Tamil Nadu, India

V. Sivasankar et al. (eds.), *Microbial Fuel Cell Technology for Bioelectricity*,
https://doi.org/10.1007/978-3-319-92904-0_5

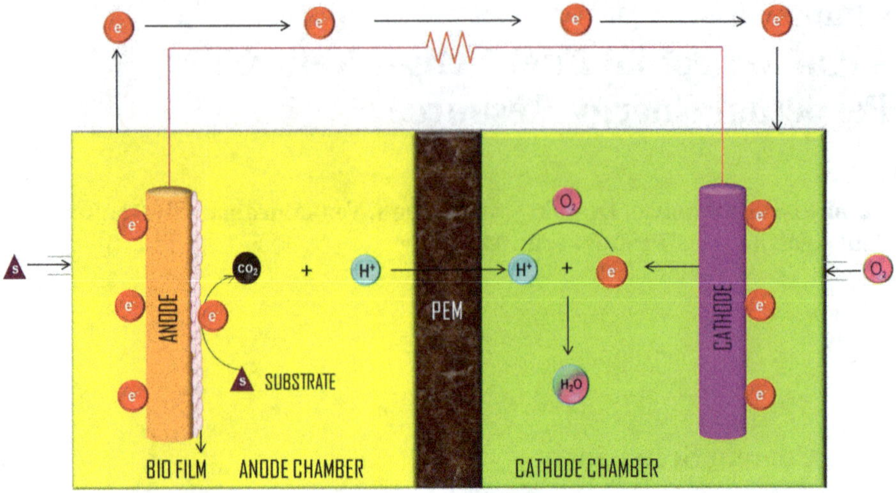

Fig. 5.1 Schematic diagram of basic microbial fuel cell (MFC)

5.1.1 Microbial Fuel Cells (MFCs)

Microbial fuel cells (MFCs) are a rapidly emerging technology, where electricity is generated from the microbial metabolization of substances; during this process oxidation-reduction occurs, releasing electrons through which the electricity is generated. MFCs contain two chambers, an anode and a cathode (Fig. 5.1), which are separated by a proton exchange membrane (PEM) (You et al. 2006). With this technology, microorganisms metabolize organic substances in the anode chamber, producing protons and electrons. The electrons migrate to the anode and reach the cathode via a circuit that is connected externally, while protons from the anode chamber are transferred to the cathode chamber via the PEM that is present between the anode and cathode (Oh et al. 2004).

The electrons and protons combine, with the reduction of oxygen to water taking place in the cathode chamber. MFCs have multiple gas inflows and outflows (Sevda et al. 2013). The cathode chamber has oxygen inlets that greatly affect the electricity output produced by the MFC. The oxygen source provided to the cathode chamber differs depending on the type of MFC used. For single-cell MFCs, atmospheric air is used, while mechanical aeration is used for dual-cell MFCs. Carbon dioxide is the main gaseous end product, and glucose and acetate or wastewater are used as a substrate (Freguia et al. 2007). The cathode chamber has an alkaline condition, which increases the absorption of carbon dioxide from the anode. This condition develops because of the accumulation of hydroxide ions, resulting from oxygen reduction at the cathode (Rozendal et al. 2006). In practice, there are limiting factors in the applications of MFC for oxygen gas delivery and carbon dioxide gas accumulation; these limitations can be overcome by the use of efficient and sustainable catalysts for the cathode reaction (El Mekawy et al. 2013).

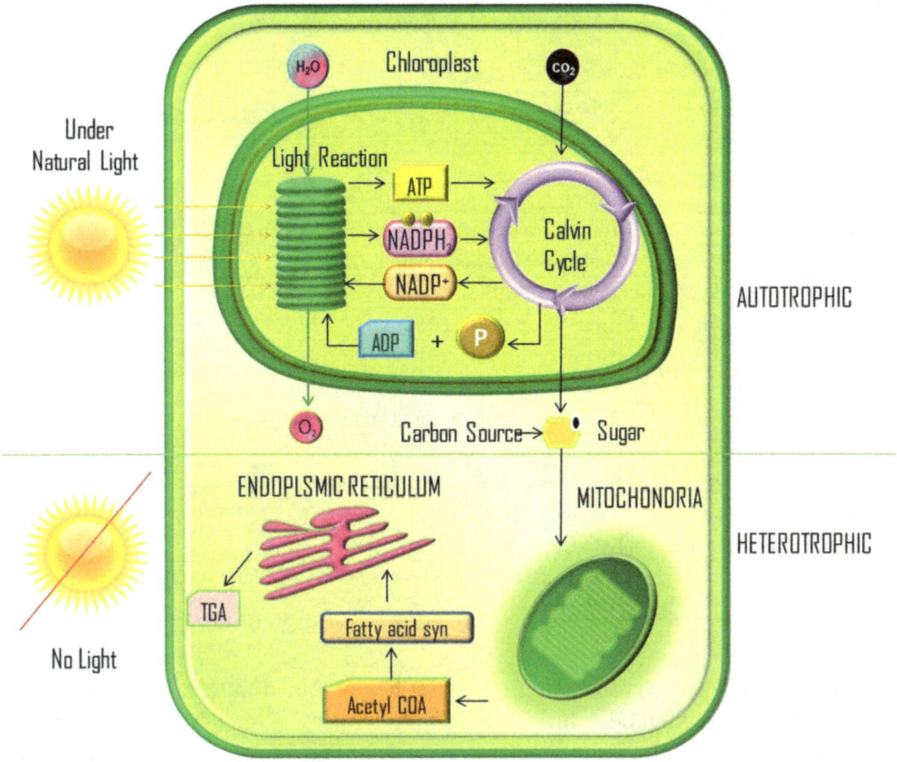

Fig. 5.2 Schematic representation of autotrophic and heterotrophic growth systems in algae

5.1.2 Algae

Algae are chlorophyll-containing organisms that range in size from microscopic and unicellular to very large and multicellular. Some algae are autotrophic in nature, deriving their own food from their surroundings in the form of sunlight. Algae have a distinctive role in maintaining the food chain and oxygen supply on Earth. Moreover, they have a high growth rate and high carbon dioxide fixation rate (Cheng et al. 2006a, b). Algae play a vital role in transforming solar energy into different forms of biochemical energy by their photosynthetic throughput (Mohan et al. 2011). Photosynthesis is the complex biological redox reaction that occurs in algae, by which they utilize solar energy to produce oxygen, carbohydrates, and other compounds. There are two different algal growth types, autotrophic and heterotrophic (Karube 1992). The growth system of algae that use carbon dioxide as a carbon source in the presence of light energy or in an illuminated environment is termed autotrophic (Fig. 5.2), while algae that grow in the absence of light, in photobioreactors (PBRs), by utilizing a carbon dioxide source from substrates provided in the culture medium, are heterotrophic.

Table 5.1 Different types of algae used as substrates in photosynthetic microbial fuel cells (PMFCs)

Algal species used in single-chambered PMFCs	
Species	Reference
Chlamydomonas reinhardtii	Nishio et al. (2013)
Chlorella vulgaris	Sharon B Velasquez et al. (2009)
Cyanobacteria	Yong Yuan et al. (2011) and Zhao. et al. (2012)
Ulva lactuca	Sharon B Velasquez et al. (2009)
Algal species used in dual-chambered PMFCs	
Microcystis aeruginosa	Huan Wang et al. (2012)
Chlorella vulgaris	Huan Wang et al. (2012)
Arthrospira maxima	Inglesby et al. (2012)
Scenedesmus obtusus	Rashid et al. (2013) and Cui et al. (2014)
Laminaria saccharina	Gadhamshetty et al. (2013)
Scenedesmus obliquus	Kondaveeti et al. (2014) and Hur et al. (2014)
Chlorella vulgaris	Lakaniemi et al. (2012)
Dunaliella tertiolecta	Lakaniemi et al. (2012)
Mixed algae	Strik et al. (2008), De Schamphelaire et al. (2009), and Huan Wang et al. (2012)

Autotrophic and heterotrophic modes can be combined to form a mixed culture (mixotrophic) growth mode, through which photosynthetic metabolism and respiratory metabolism function simultaneously to assimilate organic carbon and carbon dioxide (Lee 2004). Different types of algal species (Xiao et al 2014) used as substrates in photosynthetic MFCs (PFMCs) are listed in Table 5.1. The heterotrophic growth mode has an added advantage, since it allows the use of any type of bioreactor, with no specific design being necessary. In heterotrophic mode, the growth rate of the algal biomass is very high, along with the production of ATP. Also, the nitrogen yield and lipid content are very much higher than in the autotrophic mode. However, heterotophic algal cultures have several drawbacks in that the microalgal species used are limited. The energy expense is high when organic substrates are supplemented in a heterotrophic system are also subject to contamination with other microorganisms (Yang et al. 2000).

5.1.3 Experimental Setup of MFCs

For more than a decade it is has been believed that microorganisms could generate electricity, but only in recent years has the technique been instigated in the laboratory (Barua et al. 2010). MFCs are capable of utilizing microorganisms as a catalyst for converting the chemical energy of feed stocks into electricity (Aelterman et al. 2006). MFCs are complex microbial ecosystems where the redox reaction is part of

Table 5.2 Different types of donors and acceptors used in PMFCs

Donor at anode chamber	Acceptor at cathode chamber	Products obtained	Reference
Process: Oxidation	Process: Reduction		
Algal species	Potassium ferricyanide	Electricity	Strik et al. (2008)
Water	Potassium ferricyanide	Electricity	Thorne et al. (2011)
Water	Oxygen	Electricity	Zou et al. (2009)
Water and glucose	Potassium ferricyanide	Electricity	Yagishita et al. (1997)
Sediment material	Oxygen	Electricity	He et al. (2009)
Trypticase soy broth	Proton	Electricity	Qian et al. (2010)
Wastewater	Oxygen	Algal biomass + electricity	Xiao Z et al. (2012)
Marine sediment material	Oxygen	Glucose and oxygen+ electricity	Malik et al. (2009)
Organic acids and alcohols	Potassium ferricyanide	Hydrogen+ electricity	Rosenbaum et al. (2005b)
Succinate and propionate	Oxygen	Hydrogen+ electricity	Cho et al. (2008), Strik et al. (2010)

the microbial metabolism rather than being mediated by an inorganic catalyst (Gruning et al. 2014). Generally MFCs contain two chambers: an anode chamber and a cathode chamber, which are separated by a PEM. An anaerobic biofilm is formed on the electrode in the anode chamber, where oxidation of the substrate results in the release of protons and electrons. The protons are transferred from the anode to the cathode via the PEM. The electrons produced on the anode move to the cathode via an external circuit. The electrons reduce electron acceptor in the cathode chamber (Rabaey et al. 2005a, b). MFCs are constructed with different kinds of materials and with different configurations. Temperature and pH conditions vary depending upon the algal species used in the reactors. Other parameters, such as reactor size, electrode surface area, electron acceptors, and operating times, differ in each model. Different kinds of anodes and cathodes that act as donors and acceptors are listed in Table 5.2.

5.2 Electrode Materials

5.2.1 Properties of Electrode Materials

The performance of the MFC depends mainly on the choice of electrode material, as the adhesion of the microbes, transfer of electrons, and efficiency of the electrochemical substance depend on this material. To measure power production, carbon-based materials (carbon fiber, carbon felt, carbon cloth) are used. Logan (2010) reported that cathode materials should have the catalytic properties that are essential for oxygen reduction. Criteria for the selection of materials are different for anodes and cathodes, but there are certain properties that both should possess in general, as listed below.

Porosity and Surface Area The power output of the MFC is controlled by the surface area of the electrode. The loss in ohms is directly proportional to the electrode resistance. By decreasing the resistance the surface area can be increased, although the volume remains the same. This increase in surface area increases the efficiency of the MFC. Wang et al. (2011) and Rismani et al. (2008) reported that large numbers of reaction sites were provided by a large surface area; both these groups have also reported that electrical conductivity is greatly affected by the pore size of the electrode material.

Electrical Conductivity Biofilm present on the anode contains microbes that release electrons, and later these electrons travel through an external circuit. Electrode materials with higher electrical conductivity have lower resistance. To facilitate the transfer of electrons, the interfacial impedance has to be low. Natarajan et al. (2004) reported that a triple phase boundary reaction was facilitated by ionic conductivity at the cathode.

Durability and Stability Reduction and oxidation conditions in MFC increase the volume of material and results in decomposition. The electrode material's durability is increased when it has high surface roughness, but this might result in contamination. Hence, with an electrode that has high surface roughness, the MFC's long-term performance would be reduced. Mustakeem et al. (2015), reported that electrode materials should be durable in both acidic and basic media.

Accessibility and Cost The setup cost of an MFC depends on the cost of the electrode material used. When an MFC is about to be commercialized the cost of the material should be low and the material should be easily available. Platinum is an expensive metal that is non-durable and non-sustainable. Accordingly, in future, metal materials such as composites will be alternatives for expensive electrode material. The anode material should be biocompatible. Mustakeem et al. (2015) suggested that material with higher biocompatibility would adhere to the microbes, and consequently the life of the MFC would be increased.

5.3 Materials Used for the Anode

Anode materials should be very conductive, biocompatible, and chemically stable. The most versatile electrode material is carbon, which is available in different forms, such as graphite plates/rods/granules (Fig. 5.3) and fibrous materials in the

Fig. 5.3 Different types of graphite anode materials

form of carbon paper, carbon cloth, carbon foam, carbon felt and carbon fiber (Fig. 5.4). Graphite plates and rods are considered to be the simplest materials and the best anode electrode material because they are inexpensive, their handling is very easy, and their surface area is very defined. Park et al. (1999) and Gil et al. (2003) used graphite felt as electrodes because of its large surface area. He et al. (2005a, b) reported that even reticulate vitreous carbon material, which is very compact, can be used to achieve a greater surface area.

5.4 Materials Used for the Cathode

Park et al. (2003) reported that ferricyanide (K3 [Fe (CN) 6]) was the most popular electron acceptor used in MFCs owing to its good performance, and Rabaey et al. (2005a, b) reported that ferricyanide had lower potential than plain carbon when used for the cathode. However, the major disadvantage of ferricyanide is that oxygen cannot be sufficiently reoxidized, requiring regular replacement of the catholyte. In MFCs, the most suitable electron acceptor is oxygen, because of its oxidation potential and because it is easily available and free of cost and water is formed as an end product. The performance of an MFC may depend on the choice of cathode material, the selection of which is based entirely on the application required.

Fig. 5.4 Different types of carbon anode materials

5.5 Membranes

A membrane is essential for the transfer of protons and hydrogen ions from the anode to the cathode, inhibiting the electrons from hydrogen atoms; the PEM is such a membrane. There are different kinds of PEMs, such as bipolar membranes, cation exchange membranes (CEMs), and anion exchange membranes (AEMs). Reimers et al. (2001) reported that fluorinated polymer was the best base material for CEMs. With respect to optimum proton conductivity, sulfonic acid groups are used in membranes in proton-exchange membrane fuel cells. Owing to their lower thermal durability and low conductivity of hydroxyl ions, hydrocarbon polymer backbones and quaternary ammonium groups are the best base for AEMs.

5.6 Integration of Algae in MFCs

During flow chain reactions, photosynthetic organisms undergo charge separation and discharge electrons and protons, with a synergic effect taking place between heterotrophic microorganisms and algae. The heterotrophic microorganisms metabolize the organic matter substrate, degrading it, and produce oxygen and bicarbonates, which are metabolized by the algae, using solar energy. Kruzic et al. (2009)

integrated an aeration system to replace a sustainable photosynthetic one. When algae are growing in the cathode chamber of an MFC, electricity is produced by a photosynthetic process (Juang et al. (2012)). McGowan et al. (2000) reported that the substrate is oxidized at the anode when the algae in the cathode are the electron source, and the carbon dioxide is reduced to biomass. For electron shuttling, a mediator is used in the cathode chamber through which the electrons flow from anode to cathode. The electrons from anode enters the catholyte to reduce oxidized state of mediator and enter the algae to release the electron and later gets oxidized again. The shuttled electrons are consumed by the algal cells that grow during the metabolic pathways by which carbon dioxide is transformed to biomass and oxygen. The oxidized mediator is released by the algal cell into the media and this cycle is repeated again were the mediator gets reduced again by the electrons within the catholyte Powell et al. (2009). When illumination was applied, a biochemical reaction took place in both the anode and cathode chambers, as explained by Zhou et al. (2012), and shown below:

$$C_6H_{12}O_6 + 6H2O \rightarrow 6CO_2 + 24H^+ + 24e \tag{5.1}$$

$$6CO_2 + 12H^+ + 12e \rightarrow C_6H_{12}O_6 \ (\text{biomass}) + 3O_2 \tag{5.2}$$

$$C_6H_{12}O_6 + 6O_2 \rightarrow 6CO_2 + 6H_2O \tag{5.3}$$

Under illumination, algal species undergo a photosynthetic process to produce biomass and organic matter. Oxygen consumption by algae takes place in the dark to oxidize the organic matter, through which energy is obtained (Del Campo AG et al. 2013a, b, c) (Eq. 5.3).

Certain photosynthetic bacteria, such as *Spirulina platensis*, are used as catalysts at the anode. Without the help of any mediator, electrochemical potential is maintained by the biofilm that is formed around the electrode; this biofilm can accept the generated electrons.

5.7 Different Types of PMFC Configurations

Technology using solar energy is now the focus of great attention with the ecological management of energy resources. In the past 10 years many innovative technologies have been developed to convert solar energy into bioelectricity with bio-electrochemical systems. In the absence of artificial mediators, photovoltaic devices can be used to separate photosynthetic and heterotrophic energy production. PMFCs consist of an anode and cathode; the cathode contains a biofilm surrounded by photosynthetic microorganisms in which photosynthesis takes place. At the end of photosynthesis these microorganisms act as electron donors and produce different kinds of metabolites, while carbon dioxide is removed. Increasing the power density is the most challenging task for improving the configuration of PMFCs.

5.8 Coupled PMFCs

A coupled PMFC is an integrated system consisting of a bi-anode-chamber MFC where carbon dioxide is pumped directly into the photo bioreactor that is coupled with the MFC (Fig. 5.5). This type of MFC functions in the absence of an ion exchange membrane; hence, it is very cost effective and simple in structure. Strik et al. (2008) constructed an MFC using two electrodes separated by a CEM. The MFC was connected to an illuminated PBR to grow algae by supplying air through a sparger. Algae grown under light illumination undergo photosynthesis, and energy conversion takes place to form a biomass of electrochemically active microorganisms in the anode compartment, through which electricity is produced. A photosynthetic algal MFC works on a principle based on the selected type of algae and microorganisms employed in an open system, without any toxic intermediaries. This model has generated electricity obtained as a result of catalysis for about 100 days.

Similarly to the results reported above, Powell et al. (2009) demonstrated a photosynthetic cathode as one part of an MFC employing *Chlorella vulgaris*; the other part of the MFC was an anode that employed yeast with a fermentative quality. This

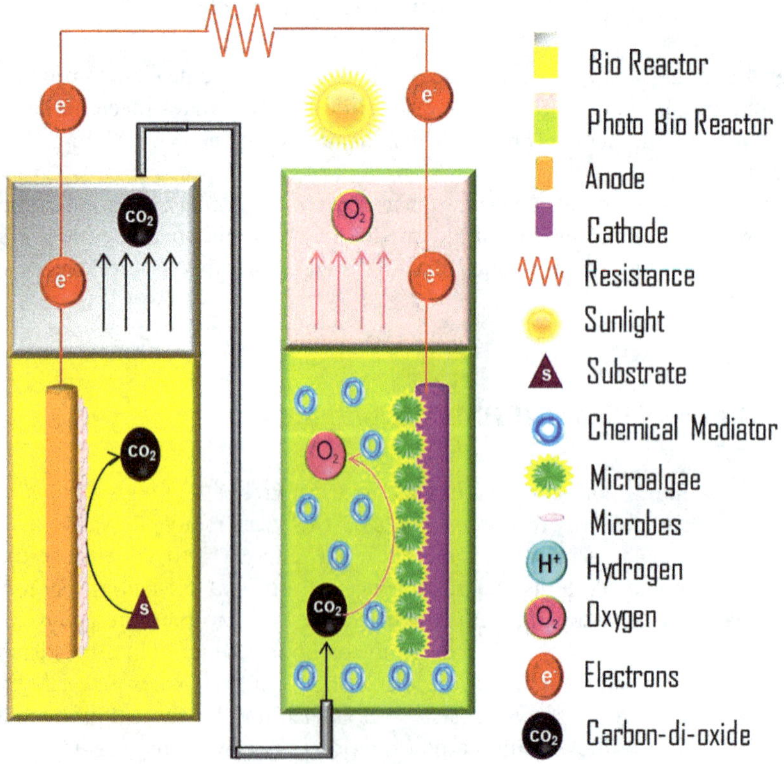

Fig. 5.5 Schematic diagram of coupled photosynthetic MFC (PMFC)

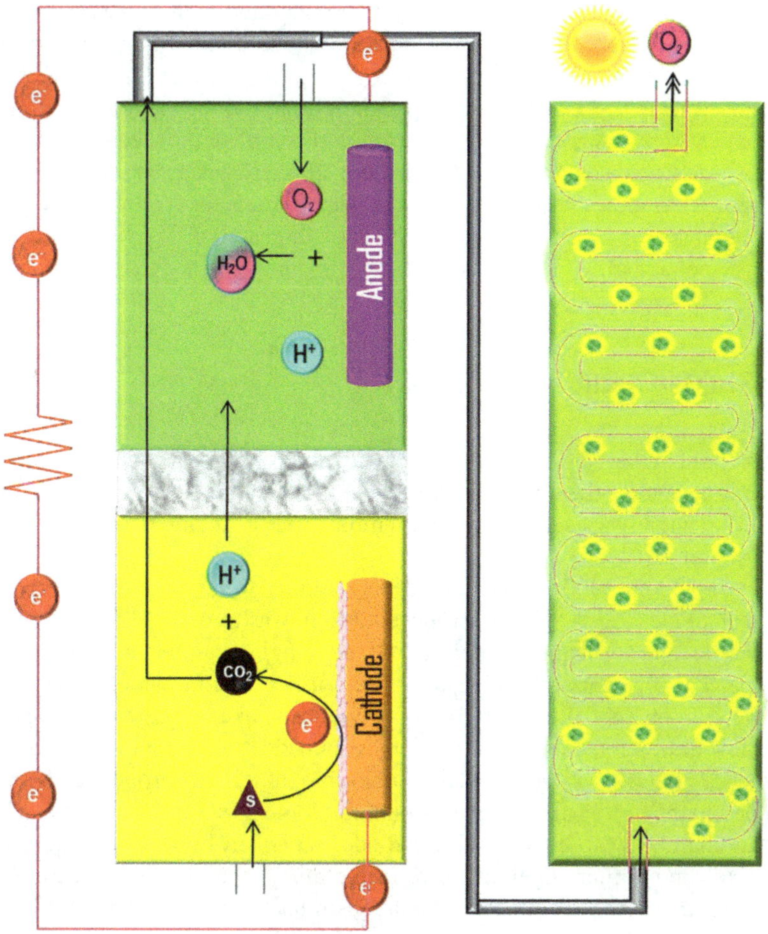

Fig. 5.6 Schematic diagram of coupled PMFC – upflow MFC-based design

model was said to be a coupled MFC. The cathode cell was designed to generate power and metabolize the carbon dioxide emission from bioethanol plants, whereas the anode cell was designed in such a way that it was illuminated by sunlight and aerated with feed and air consisting of 10% carbon dioxide passed to the cell cultures, with electron shuffle between the electrode and the yeast. Jiang et al. (2013) proposed a similar design for a coupled MFC using a PBR and an upflow MFC where the effluent was pumped continuously into the PBR (Fig. 5.6). Microalgae under continuous illumination were employed for this experiment. The coupled MFC was made using plastic cylinders, and the electrode was a carbon fiber brush. The anode and cathode chambers were separated using glass and wool beads. This model of an integrated PBR and MFC was designed for wastewater treatment and power generation.

Silvaggi (2016) proposed an MFC system integrated with an algal bioreactor in which synthetic wastewater was fed into the anode compartment, where organic compounds were biologically degraded to generate electrons. The generated electrons moved from the anode electrode (carbon brush) to the cathode electrode (carbon cloth), where oxygen reduction occurred to complete the electrical circuit. The treated wastewater was discharged into a transitional beaker, and this solution was then supplied to the cathode compartment (algal bioreactor), where algae grew and produced dissolved oxygen to support the cathode reaction. The final effluent (containing suspended algal cells) was discharged from the cathode compartment.

5.9 Single-Chambered PMFCs

In a single-chambered MFC, photosynthetic microorganisms were employed, as these microbes have the ability to shuttle electrons to the electrode with no mediators; this design was said to be a membrane-less single-chambered MFC (El Mekawy A et al. (2014).

Fu et al. (2009, 2010) proposed a design similar to the one noted above, using blue green algae (Fig. 5.8). This proposed design was to be used for power generation. The design consists of a non-membrane single chamber with an anode and electrode. The algae act as a biocatalyst and form a biofilm, which creates electropotential. Under light illumination, photosynthesis takes place and in dark conditions a respiration reaction takes place, by which electric current is generated.

Chandra et al. (2012) and Venkata Subhash et al. (2013) proposed another type of single-chambered PMFC, termed a photobiological fuel cell. This type of PMFC has dual chambers, an anode and a cathode, separated by a PEM. These authors used mixotrophic microalgal cultures. In the mixotrophic culture medium, where both autotrophic and heterotrophic metabolism take place, these algae form a biofilm, utilizing carbon dioxide, and they act as a carbon source.

Similarly, Nishio et al. (2013) used a synergetic approach, by introducing a mixed culture of bacterial and microalgal cells, which improves the performance of single-chambered PMFCs. Their design used a general MFC with a portable biobattery. This system has the ability to produce certain organic byproducts, such as acetate, as a result of assimilation carried out by the bacteria, with electricity produced at the end. The highlight of their work was to recharge the MFC and extend the operation time. Photosynthetic reaction was achieved by illuminating light and dark conditions, which seems to be a reversible process that recharges the MFC to prolong the operation time.

Lin et al. (2013) designed an MFC with no membrane or mediator. Different materials were used for the anode and cathode. Gold mesh was used as the anode and carbon cloth as the cathode. They used *Spirulina platensis*, which aggregated in the anode and formed a biofilm. The biofilm was tested for chlorophyll content, which seemed to be very high; this high chlorophyll content was an added advantage for generating high voltage and high power density.

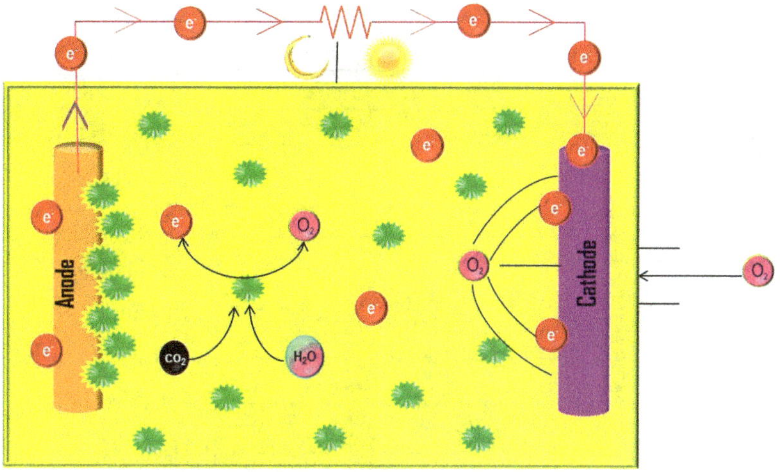

Fig. 5.7 Schematic diagram of a single-chambered MFC

Hai-ming and Jiang (2016) proposed a combination of MFCs with microalgal cultivation for bioelectricity generation and domestic wastewater treatment, using a device (Fig. 5.7) in which bacteria were employed as catalysts to oxidize organic matter as well as to generate electrical current. A sediment MFC (SMFC) was constructed with an anaerobic tube glued to the top of the chamber. The tube was sealed with a butyl rubber stopper and a perforated plastic screw cap. A platinum-coated carbon cloth and carbon fiber brush were used as the cathode and anode electrodes, respectively, for the SMFC, and the electrodes were connected to a copper wire through an external resistance. A stainless steel sheet was used in the cathode as a current collector. The brush anode was placed on the other side of the chamber with its end located 1 cm from the cathode (Fig. 5.8) Hai-ming and Jiang (2016).

Cylindrical-chambered MFCs are very effective for chemical oxygen demand (COD) removal from wastewater, but are not effective for nitrogen and phosphorus removal. Alternatively, microalgae can effectively remove nitrogen and phosphorus from wastewater. To improve the efficiency of wastewater treatment, a combined system consisting of an MFC and microalgal cultivation was developed, and the effectiveness of the system for wastewater treatment and electricity generation was evaluated Hai-ming and Jiang (2016).

5.10 Dual-Chambered PMFCs

A dual-chambered PMFC that uses algae for the synthesis of oxygen in the cathode chamber is the most preferred design. This design contains an ion exchange membrane to separate the two chambers. Different kinds of experiments have been carried out using these dual-chambered PMFCs, as described below.

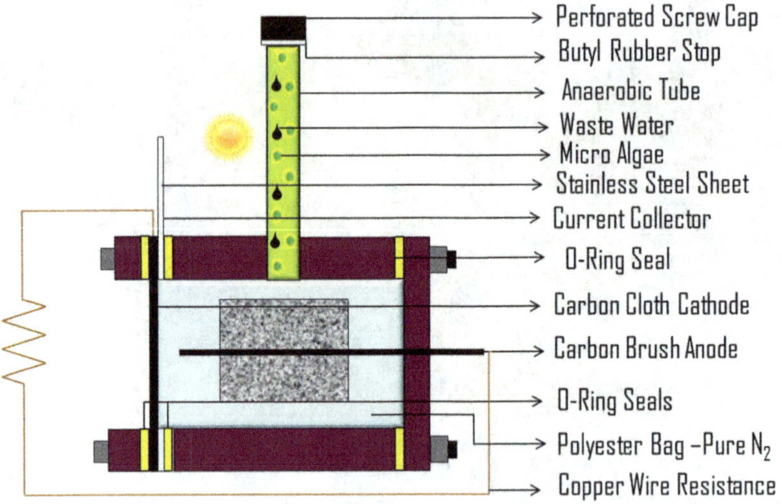

Fig. 5.8 Schematic diagram of single-chambered photosynthetic microbial fuel cell (SC-PMFC)

Rodrigo et al. (2009) designed a model dual-chambered PMFC (Fig. 5.9) in which microalgae, under illumination for 12 h per day, were used in the cathode chamber. The anode chamber, which contained bacteria, emitted carbon dioxide. In this design a vent is constructed at the top of the two chambers and connected with a pipe. The emitted carbon dioxide travels through this vent from the anode to the cathode and the microalgae utilize this carbon dioxide for growth during photosynthesis. As a result, biomass production of microalgae is also achieved.

Powell et al. (2009) used *C. vulgaris* for a comparative experiment. The algae were employed in the cathode chamber as an electron acceptor. They were also responsible for carbon dioxide removal. To determine biomass production, a sealed glass bulb was filled with a known volume of nutrient medium and carbon dioxide, along with the *C. vulgaris*. Evaluation of the cell yield was calculated by using the concentration of *C. vulgaris* cells and carbon dioxide. These authors' experiment resulted in very high cell growth, at the rate of 3.6 mg/L-h, and a reasonable power density was achieved.

Yadav (2009) constructed a dual-chambered MFC using a cylindrical plastic jar of 500 ml capacity, with 450 ml of synthetic wastewater being fed into the anode chamber; the same volume was fed into the cathode chamber. The two chambers were connected with a tube and separated by a PEM (Nafion 117; Manufacturer: Sigma Aldrich, USA). The whole experimental setup was placed under continuous illumination with fluorescent light (Philips spiral fluorescent light lamps, 15 W) to provide light for photosynthesis by algal beads. The preliminary investigation showed that the entrapped algal beads underwent constant photosynthesis and maintained the dissolved oxygen concentration in the cathode chamber solution at

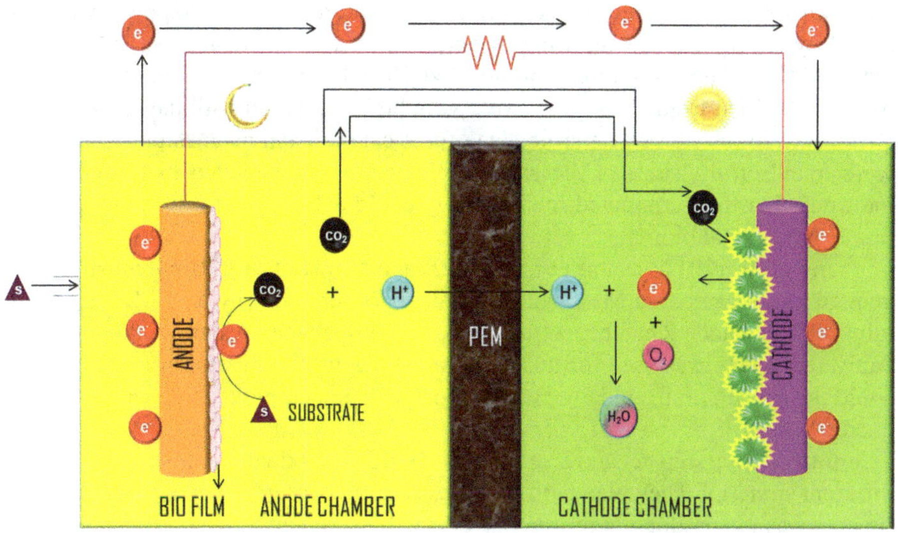

Fig. 5.9 Schematic diagram of dual-chambered photosynthetic microbial fuel cell (DC-PMFC)

around 4.0 mg/l, which is reasonably good for a successful MFC. This MFC produced power in the range of 3.97.53E × 10^{-6} W, power density in the range of 0.238 mW/m^2, and current density in the range of 1.05 mA/m. 48% reduction in COD was also observed after 5 days of experimentation.

Ramanathan et al. (2011) proposed a dual-chambered PMFC for studying nine marine microalgae: *Isochrysis sp.*, *Nannochloropsis sp.*, *Dicrateria sp.*, *Chaetoceros calcitrans*, *Pavlova sp.*, *Synechocystis sp.*, *Dunaliella sp.*, *Chlorella salina*, and *Tetraselmis gracilis*. These algae were used for generating electricity directly from biodegradable compounds.

Mitra and Hill (2011) proposed an MFC design consisting of an autotrophic cathode with *C. vulgaris* and an anode consisting of fermentative *Saccharomyces cerevisiae*, and they evaluated this system for electricity production. The system was connected with various levels of resistance to characterize and evaluate the power generation capacity and study the voltage dynamics. To study the effect of algal cell density and energy production, a recycle system was introduced into the cathode. The experimental output with respect to the cell density was 437 to 2140 mg/L. Higher the cell density resulted in higher power production of about 0.6 mW/m^2 with 5000 Ω as loading resistance.

Lakaniemi et al. (2012) carried out a similar experiment with other algae in a dual-chambered PMFC; they used freshwater microalgae (*C. vulgaris*) and marine microalgae (*Dunaliella tertiolecta*). This experiment evaluated the production of algal biomass to be used as a feed stock for the production of the electricity within a dual chamber at a temperature of 37 °C. The inoculum for the anode chamber was obtained from the sewage waste of a municipal sludge digester. Inoculums were nutritionally maintained for two different algal cultures. Maximum power was generated by continuous subculturing of enriched anaerobic organisms. Butanol was obtained from the algal biomass of the anode. The level of power generated and the butanol obtained from *C. vulgaris* were very high compared with the results for *D. tertiolecta.* In the slurry of marine algae some calcium and magnesium precipitates were found on the sides of the cathode. The authors concluded that their results indicated that their combined methodology could achieve high bioenergy production from an algal biomass.

Juang et al. (2012), Zhou et al. (2012), and Gajda et al. (2015) constructed a general dual-chambered MFC with the chambers separated by a PEM. The inoculum for the anode chamber was obtained from a wastewater treatment plant. Activated sludge was used for this experiment. Light illumination was excluded to avoid the growth of algae. However, microalgae were employed as a catalyst in the anode.

Raman K et al. (2012) and Lan JC et al. (2013) used dual-chambered MFCs in a different stratergy. They planned a three level of process to be carried out. The first process was the production of microalgae and bacterial cultures. In the second process, mechanical aeration was applied to the microalgal culture. Finally, MFCs illumination was increased mildly. All these three stratergies were experimented to evaluate the power generation obtained through each condition.

Singhvi et al. (2013) proposed a dual-chambered salt bridge MFC for a detoxification process. They studied the effects of algae in detoxifying water contaminated with chromium VI. The device they used for the experiments showed great efficiency for chromium removal, with 98% removed within 96 h at pH 2. The acidic pH condition helped in removing the chromium and in COD removal, as well as aiding open circuit potential and power density. This system proved to have high efficiency for bioremediation as well as power production.

Wu et al. (2013a, b) developed a tubular PBR, using *C. vulgaris* in the cathode compartment to produce oxygen. Two different types of cathode materials were used in this experiment. To evaluate the efficacy of the MFC with algae in the cathode, the MFC was tested with both light and dark cycles. Their results indicated that the algae they used could be effective oxygenators. The lifespan of the algae seemed to be reduced when they were continuously illuminated.

Luimstra et al. (2013) proposed a PMFC design that could be used for algal screening and electricity generation. Disposable polystyrene bottles were used to prepare the anode chamber, where simple carbon coating was applied. This chamber was utilized for algal growth. This design has unique features, such as screening the algae and analyzing and isolating the microorganisms that have electrogenic activity. Several types of bacteria that were isolated were shown to possess electrogenic activity.

Using a photosynthetic algal MFC, He et al. (2013a, b) employed *C. vulgaris* as an immobilized culture in the cathode compartment to treat wastewater and aid in the generation of electricity and biomass production. The conditions with respect to the immobilization of the algae, as well as the matrix concentration and the inoculum concentration, were studied in detail.

Campo et al. (2013a, b, c) proposed a design of MFC assisted at the cathode. Mechanical aeration was not provided to the cathode chamber. Hence, there was a requirement for oxygen, which was achieved by using *C. vulgaris*. The cathode was illuminated for 12 h every day. It took about 25 days to reach the standard conditions required for the evaluation. The rate of dissolved oxygen and the cell voltage were evaluated daily. The results indicated that the dissolved oxygen rate was not constant throughout the day, with the maximum being reached when the process was carried out in the dark. The cell voltage and the oxygen profile remained the same throughout the experimental period. Half an hour was required for the supply of carbon dioxide to be stabilized and for the system to begin working. In the acclimation stage, the power density seemed to be increased by about 13.5 mW m^2.

Gajda et al. (2013) showed that oxygen was produced in an illumination-dependent manner in photosynthetic organisms which helped to raise the generation of power by 42%. Further studies revealed that the use of a biotic cathode showed a response to light and raised the generation of power by 48% compared with that for an abiotic cathode.

Gadhamshetty et al. (2013) used a dual-chambered MFC for a batch-fed method, employing *Laminaria saccharina* as an electron donor, with mixed cultures acting as a biocatalyst in the anode chamber. The cultures were studied with three pre treatment conditions such as, 1. autoclave treatment 2. microwave irradiation and 3. No-Treatment. To control the performance of the dual-chambered MFC, a control set up was used to fix the baseline of the MFC.

Rashid et al. (2013) generated electricity using activated sludge and an algal biomass. The MFC anode was inoculated with the activated sludge. Different concentrations of the algal biomass were dried and tested. The concentration of algal biomass required to produce a voltage higher than 0.89 V was 5 g/L, and the power density was found to be 1.78 W/m^2. The output was found to be comparatively low without pretreatment. The algal biomass was tested as a substrate after oil extraction, but power output was very low. Hence, this work shows that using the whole algal biomass enhances energy production.

Kondaveeti et al. (2014) used a renewable algal biomass, *Scenedesmus obliquus*, as a substrate for generating electricity in dual-chambered MFCs. From a polarization test, it was found that the maximum power density with the pretreated algal biomass was 102 mW/m^{-2} (951 mW/m^{-3}) at a current generation of 276 mA/m^{-2}. The main organic compounds in the algal oriented biomass were lactate and acetate, and these were mainly used for electricity generation. Other byproducts, such as propionate and butyrate, were formed in negligible amounts.

Hur et al. (2014) utilized the spectroscopic changes observed in algal-derived organic matter to evaluate MFC function. Technically, variations were found in less dense component and proteins comprised in large-size. During the period of electricity generation fluorescent compounds decomposed. These authors have also reported that extracellular organic matter shows a very low ultraviolet (UV) absorption rate. Smaller-sized compounds that absorb UV seemed to decompose by themselves in the initial stages, as found by the performance of size exclusion chromatography. The protein and polysaccharide substrates were examined by Fourier transform infrared spectroscopy, which showed two structures that are very dominant in algal-derived organic matter in the microbial fuel system.

Kakarla et al. (2014) proposed a dual-chambered MFC that used algae as an oxygenator. Plain carbon paper was used as anode electrode. A carbon fiber brush and plain carbon paper were used as cathode electrodes for a comparitive study. The carbon fiber brush in the MFC cathode exhibited a voltage of 0.21 ± 0.01 V, whereas the plain carbon paper cathode had an output voltage of 0.06 ± 0.005 V. The carbon fiber brush showed a higher power output than that of the plain carbon paper.

Gouveia et al. (2014) were determined to extract pigments from microalgae. They used *C. vulgaris* in the cathode compartment and a bacterium in the anode. This study was done under different light intensities, and maximum power was attained when the light intensity was 96 lE/ (m^2 s), for which the power generated was about 62.7 mW/m^2. The authors reported that increasing the light intensity tended to increase the power production. The impact of light intensity also showed positive potential for carotenogenesis with respect to the pigments produced by the microalgae.

Cui et al. (2014) attempted to grow microalgae simultaneously in the two chambers of a dual-chambered MFC. The substrate used at the anode was a dead microalgal biomass. The carbon dioxide generated at the anode was utilized for the growth of microalgae at the cathode. This was a comparative study between an algal-fed MFC and an acetate-fed MFC. For 0.5 g/L microalgal powder, the maximum power density was 1926 ± 21.4 mW/m^2 and a coulombic efficiency of 6.3 ± 0.2% was achieved. Microalgal growth could not be sustained in the acetate-fed MFC, which lacked a carbon dioxide supply.

Gajda et al. (2015) described the potential of algal biomass production along with the treatment of wastewater and power generation, using a complete biotic MFC. Current was generated by an anaerobic biofilm that was present in the anode half-cell. Biomass was formed by the oxygen reduction reaction that took place with the help of phototrophic biofilm. Algal growth in the cathode chamber was monitored and parameters for power production were assessed and comparatively analyzed. The generation of electricity activated the crossover of cations and helped in the formation of an algal biomass. Later the harvested algal biomass was reused in a closed system.

Chang Xu et al. (2015) demonstrated two different MFC models using algae. Their system was constructed with graphite or carbon electrodes and had no mediators. The first model had an anode chamber inoculated with microalgae and the cathode chamber was filled with potassium ferricyanide. In the second model, microalgae were inoculated in both anode and cathode at various conditions. *Chlorella pyrenoidosa*, which acts as an electron donor, was used in both chambers. The results indicated that higher electricity production was achieved using the first model, under low light intensity. The high algal density in ? limited the production of electricity. 4-Nitroaniline was used to increase the permeability of the algal cells, thus increasing the open circuit voltage in return. Proton leak-promoting agents such as resveratrol and 2,4-dinitrophenol acting on the mitochondria of the algal cells increased the bioelectricity production of the algal MFCs.

5.11 Sediment MFCs (SMFCs)

In SMFCs, power generation can be produced naturally by employing an anode in the sediment, while the cathode is immersed in the water and lies above the sediment. This kind of experimental setup is defined as an SMFC (Fig. 5.10). Reimers et al. (2006) and Schamphelaire et al. (2008) have called this type of system a benthic MFC. Two kinds of reactions take place in SMFCs—redox reactions and cathodic reactions. Organic molecules are oxidized by microorganisms in the sediment in what is called redox reactions, whereas the reduction reaction of electron acceptors is similar to that of oxygen dissolved in water.

Another SMFC model was proposed by Jeon et al. (2012) (Fig. 5.11). Their design has an anode and a cathode placed on opposite sides of a cylindrical plastic chamber made of poly acrylic plastic. Graphite felt is placed in both the anode and the cathode. The electrodes of both chambers are externally connected using a copper wire. The setup of the anode chamber was fixed, as follows. Initially the sediment was placed in the chamber where the anode was fixed to the middle of the sediment. Later, the anode was covered using sterilized sand. To collect the gas that is generated from the anode placed in the sediment, a funnel-shaped glass collector is fixed on the sediment surface and connected to a fixed sample bag for gas collection.

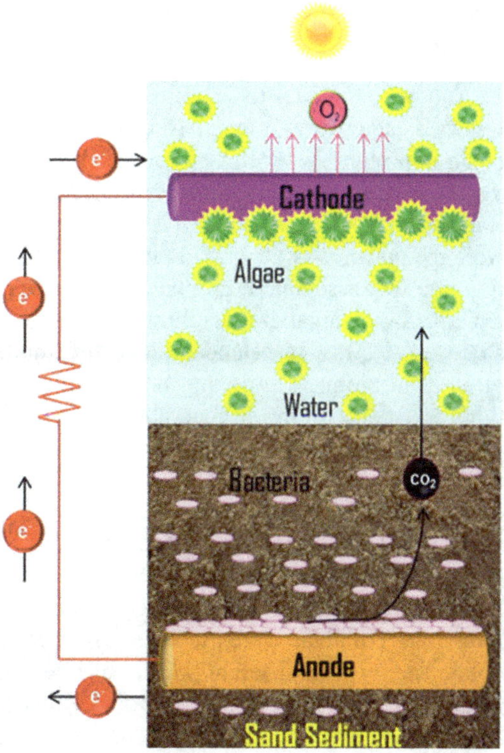

Fig. 5.10 Schematic diagram of photosynthetic sediment microbial fuel cells (PSMFCs)

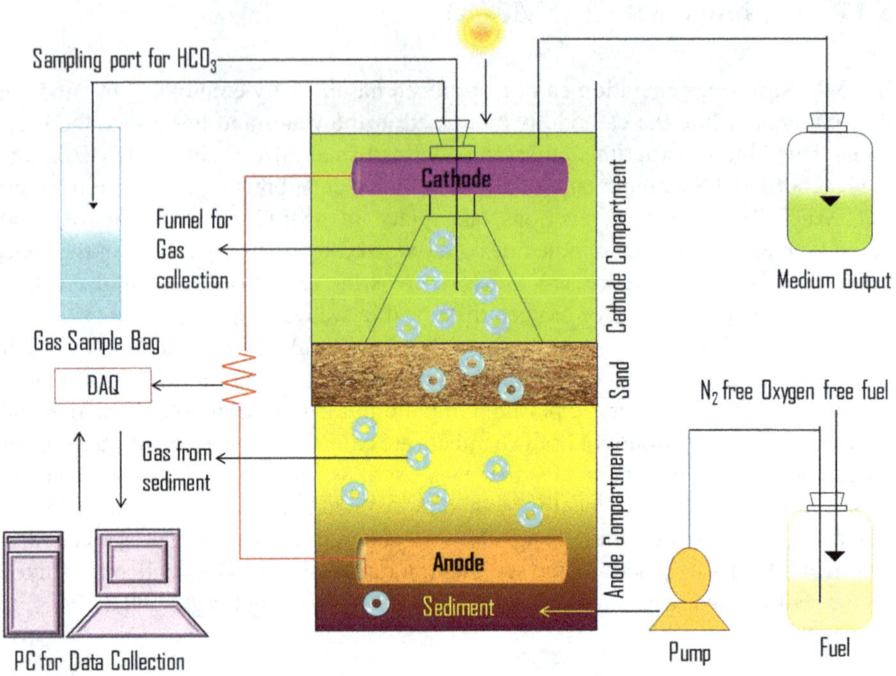

Fig. 5.11 Schematic diagram of algal culture system using PSMFC

Another parameter to customize the production of an algal biomass in an SMFC was evaluated using *C. vulgaris*. Generally, the current generated was considered to be an important factor for the rate of increase of carbon dioxide generation. Also, the production of methane was inhibited as a result of power generation. Hence to evaluate the similar efficacy *C. vulgaris* was employed in the cathode chamber were the power was generated under10 Ω resistance. Of note, the biomass production rate was associated only with the power generated through the SMFC. In this experiment the algal dry weight was reported to be 420 mg/L and the current generated was 48.5 mA/m^2. Hence, this SMFC model was considered for the production of an algal biomass, utilizing the carbon dioxide produced by the oxidation reactions as a result of power generation.

5.12 Twelve-Reactor Algal Fuel Cells

Electricity production was investigated using single-chambered MFCs in which different types of algae were used. The algae were used in powder form in the MFC to obtain energy with different power densities. Sharon B Velasquez et al. (2009) used a 12-reactor MFC, with each reactor having a volume of 25 ml. The 12 MFCs were operated according to different strategies. Four of the MFCs functioned as closed circuit systems, four functioned as open circuit systems, and the other four were constructed as anaerobic reactors with an end-plate sealing. Logan and Regan (2006) used a graphite fiber brush in the anode in both MFCs and anaerobic reactors. Cheng and Logan (2007) used ammonia gas at high temperature to treat a graphite fiber brush, and constructed the cathode following the methodology of Cheng et al. (2006a, b), in which method the cathode was prepared using platinum as a catalyst, with four layers for diffusion. Although the materials in a mixed culture are in a non-sterile condition, the materials that were used by Cheng et al. (2006a, b) were sterilized in an autoclave at 121 °C for 15 min. Comparatively, *Ulva lactuca* was completely degraded, whereas *C. vulgaris* generated more power with respect to the mass substrate. The power density obtained by *C. vulgaris* was 277 W/m^3 and *U. lactuca* produced a power density of about 215 W/m^3. A linear sweep voltammetry method was used to obtain the polarization curves to interpret the power densities obtained through the different cycles. At the end of the process, the microbes grown in the reactors were evaluated for fingerprint analysis, which reported that only 11% of these microbes were similar to the cultures that had been inoculated. Finally Cheng et al. (2006a, b) suggested that these types of multiple MFC reactors help in producing a renewable source of energy.

5.13 Nine-Cascade Algal Fuel Cells

X.A. Walter et al. (2015) used a design comprising nine MFCs. A sequential mode of operation was carried out using the nine cascades. A downstream mode was used to feed the output to the consecutive cascades. The results of this setup mode were also studied by Ioannis Ieropoulos et al. (2008) and Winfield et al. (2012), who reported that this downstream feeding setup provided excellent utilization of the organic substrate and generated a higher current density, because of the shorter diffuser distance. The construction design is explained in detail as follows. Black acrylic material was used to construct the anode compartment. This specific material was selected to avoid the growth of phototrophic microorganisms. The connecting tubes were constructed using the same material, for the same reason. The anode and cathode were both made of carbon fibers; the anode compartment had a volume of 4.5 ml and the fiber material measured 64 cm^2, whereas the cathode fiber material measured about 160 cm^2. Continuous flow of tap water at 5 ml/min acted as a catholyte. The anode and cathode electrodes were submitted to a three-dimensional transformation, exposing a surface area of 3.3 cm^2.The terracotta membrane used in this design has a hard surface area of about 6.8 cm^2 and thickness of about 2 mm. The amount of water absorbed (% of weight) by terracotta membranes was 9.1% ± 0.3% Winfield et al. (2013). Each MFC was connected with light - tight - gas-gap drippers. This method was used to avoid current conduction via the fluids from each unit and to keep the whole unit free from electricity for manual monitoring. The anode compartment consisted of continuously grown *Synechococcus leopoliensis* culture. Phototrophs are digested using a pre-digester, which produces oxygen in return. This nine-cascade MFC system, with the help of a fresh culture, could produce a power voltage of 42 W/m^3. Certain parameters of this system, such as its long-term stability, will have to be optimized in future.

5.14 Anode Assistance with Phototrophic Microorganisms

Zou et al. (2009) and Pisciotta et al. (2011) reported that PMFCs which employ photosynthetic microorganisms in the anode chamber undergo a photocatalytic water reaction by which electrons are generated. Generally, PMFCs differ from the normal type of MFCs, which produce electricity as a result of the oxidation of organic compounds. Algae-assisted anodes have an electrochemical catalytic capacity that is used to generate electricity. A simple schematic representation of an algae-assisted PMFC is shown in Fig. 5.12). Different algal species employed in anodes are listed in Table 5.3.

5.15 Anode-Assisted Electrochemical Catalysis

Anode-assisting phototrophic microorganisms such as heterotrophic bacteria use organic carbon as a carbon source. Different types of bacteria employed in anodes had different outcomes (Xing et al. 2008). *Rhodopseudomonas palustris* is a

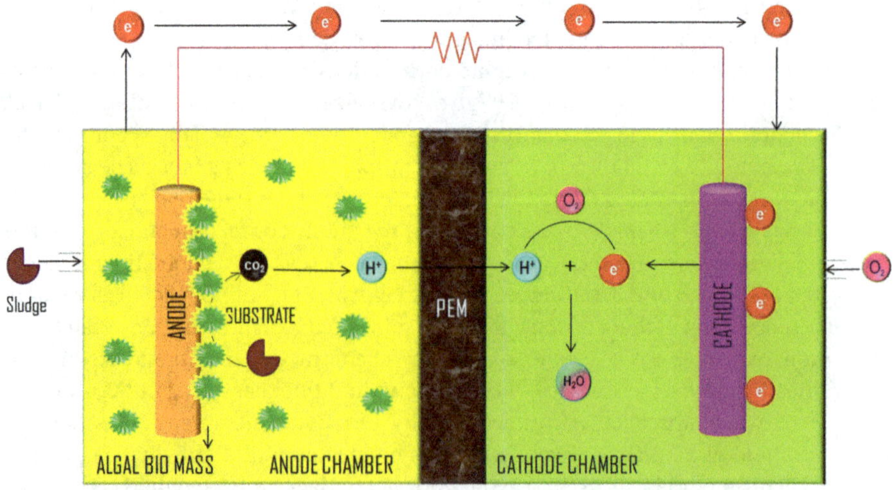

Fig. 5.12 Schematic diagram of phototrophic microorganism assisting the anode process. From Rashid et al. (2013), with permission from?

Table 5.3 Algal species assisting in the anode chamber

Algal species used in single-chambered PMFCs	
Species	Reference
Rhodobacter sphaeroides	Cho et al. (2008)
Rhodopseudomonas palustris	Xing et al. (2008)
Chlamydomonas reinhardtii	Nishio et al. (2013)
Mixed algae	Chandra et al. (2012), Subhash et al. (2013), and Malik et al. (2009)
Algal species used in dual-chambered PMFCs	
Chlorobium limicola	Badalamenti et al. (2013)
Rhodopseudomonas palustris	Inglesby et al. (2012)
Mixed culture	Cao et al. (2008) and Badalamenti et al. (2013)

phototrophic non-sulfur bacterium. When this bacterium assists in the anode it shows high activity, and electron transfer to the anode electrode is executed indirectly. This strain is used widely because of its utilization of various organic compounds in wastewaters and domestic wastes. Compared with *Rhodobacter, Rhodopseudomonas* is considered to be the dominant type of bacteria for producing electricity when soluble electron mediators are used to assist in the anode chamber. Cao et al. (2008) observed that illuminating the chambers had a positive effect on the production of electricity. Inglesby and Fisher (2012) revealed that *R. palustris* consumed the whole cell of the cyanobacterium *Arthrospira maxima* to generate electricity in two types of MFCs. Morishima et al. (2007) reported that hydrogen was obtained as a product of organic oxidation in anode assisted MFC, which affected the electricity production. So they carried out gene manipulation in such a way that a gene-manipulated *R. palustris* suppressed the production of hydrogen, resulting in a high-performance MFC with higher electricity production. Chandra et al. (2012) evaluated the efficacy of the application of mixed phototrophic bacteria by using these bacteria to assist in an anode chamber. They revealed that, in mixotrophic PMFCs, electricity production was higher in illuminated than in dark conditions, since the oxygenic phototrophs were dominant. Similarly, Subhash et al. (2013) generated electricity using mixotrophic microalgae in the anode as a biocatalyst. They reported that this kind of mixotrophic system generated electricity at a low output. Badalamenti et al. (2013) reported *Chlorobia* as very dominant phototrophs that assisted in the anode chambers. With reference to these various results, it is clear that different kinds of phototrophic bacteria can assist in the anode, where they act as key factors responsible for electricity generation.

5.16 Substrates as End Products

As a result of the phototrophic activities in MFCs, energy-rich compounds are produced by these phototrophic microorganisms and are later converted to electricity. He et al. (2005a, b) used *Rhodobacter capsulatus* as a substrate in a dual-chamber MFC linked with a PBR. This approach was further investigated and simplified by Rosenbaum et al.(2005a, b) and Cho et al. (2008), who used phototrophic microorganisms to assist in an anode-based fuel cell. Further, Rosenbaum et al. (2005a, b) used *Escherichia coli* in dark fermentation and *R. sphaeroides* in photo fermentation to utilize the organic compounds by which these microbes produce hydrogen, which later generates electricity. However, the hydrogen produced seemed to be very low compared with the amount of oxidized hydrogen. Hydrogen pressure in the chamber will decrease the production of hydrogen. The oxidation of hydrogen is carried out by using platinum as a metal catalyst. Without using platinum as a catalyst, electricity can be generated using *Rhodopseudomonas spp*., rather than *R. sphaeroides* (Cho et al. 2008). Malik et al. (2009) generated electricity using cyanobacteria that produce glucose, which is utilized by the microbes present in the anode. Nisho et al. (2013) used *Geobacter sulfurreducens* in a phototrophic MFC where the microbe utilized the formate produced by *C. reinhardtii*, which aids in generating electricity. Badalamenti et al. (2014) used two different bacteria—*Chlorobium* and *Geobacter*—both bacteria were used as monocultures and co-cultures for electricity generation.

5.17 Cathode Assistance with Phototrophic Microorganisms

Using photosynthetic microorganisms in the cathode chamber of an MFC has many benefits, such as biomass production, carbon dioxide reduction, and the supply of oxygen. Algal species assisting in the cathode chamber are listed in Table 5.4.

5.18 Oxygen Production

An attractive feature of the cathode process is the oxygen production that occurs owing to mechanical aeration, which utilizes a large amount of energy. Xiao Z et al. (2012) and Wu et al. (2013a, b) undertook research on MFCs with light illumination that generated electricity using photosynthetic microorganisms. Compared with results for mechanical aeration, the dissolved oxygen concentration remained high were as the dissolved oxygen concentration was affected by the illumination condition (Campo et al. 2013a, b, c). Kokabian et al. (2013) reported that a method of desalination at the cathode, using *C. vulgaris* microalgae, had better results than the use of an abiotic cathode. He et al. (2013a, b), for their studies, utilized a pure

Table 5.4 Algal species assisting in the cathode chamber

Algal species used in single-chambered MFCs	
Species	Reference
Chlorella vulgaris	Fei Zhang et al. (2011)
Algal species used in dual-chambered PMFCs	
Chlorella vulgaris	Gouveia et al. (2014), Campo et al. (2013a, b, c), Huan Wang et al. (2012), and Powell et al. (2009)
Desmodesmus sp. A8	Wu et al. (2014)
Microcystis aeruginosa IPP	Cai et al. (2013)
Mixed culture	Xiao Z et al. (2012), Lobato et al. (2013), Juang et al. (2012), and Cao et al. (2008)
Algal species used in three-chambered PMFCs	
Chlorella vulgaris	Kokabian et al. (2013)

culture of *C. vulgaris* which produced oxygen in the cathode electrode; this oxygen was accepted by the electrons from the cathode electrode. Powell et al. (2009) employed *C. vulgaris* in the cathode and concluded that, in the presence of an electron mediator, *C. vulgaris* exhibited the property of an electron acceptor. Cao et al. (2009) and Lyautey et al. (2011) used a mixed culture to investigate electrochemical activities. They were unable to differentiate the electron transfer roles shown by different phototrophic microorganisms.

5.19 Carbon Dioxide Utilization

Photosynthetic microorganisms use carbon dioxide as a carbon source. Reduction of carbon dioxide takes place via photosynthesis. Wang et al. (2010) designed an MFC called a microbial carbon capture cell. This microbial carbon capture cell employs photosynthetic microorganisms at the cathode to utilize the carbon dioxide produced from the anode region as a result of the oxidation of organic compounds. The carbon dioxide generated at the anode is absorbed by the cathode for *C. vulgaris* growth, with no evidence of carbon dioxide shown in the headspace of the cathode compartment. Cui et al. (2014) used *C. vulgaris* for their study in the cathode compartment. Their studies revealed that the carbon dioxide supply was affected by the concentration of organic compounds in the anode. They suggested that developing microbial carbon capture cells would help to propel new MFC technology that could neutralize carbon. Zhou et al. (2012) also used *C. vulgaris*, in the form of sodium alginate and calcium chloride beads . They reported that 88% maximum power density was achieved by the algae immobilized on the beads than the suspended algae. Similarly, He et al. (2014) used an immobilization technique, using a matrix and optimized conditions such as cross linking time and initial innoculum concentration resulted in a 258% increase of power density compared to the previous optimization by which maximum power density obtained was 88% only.

5.20 Production of Biomass

MFCs are also used for biomass production, employing a photosynthetic cathode with an electrode and biomass suspended in the cathode solution, but quantification of the biomass produced is very challenging. Cao et al. (2009), in their studies, focused on the biomass that is suspended in the cathode compartment. Their study revealed that the biomass on the cathode electrode contained a high level of lipid. Gouveia et al. (2014) used a dual-chambered MFC and achieved a greater biomass concentration than the previous method with 2800 mgL^{-1}, but it was affected by the hydraulic retention time (HRT), with a long HRT helping to accumulate more biomass. Hyeon Jin Jeon et al. (2012), using a multiple-feed batch-operated MFC, achieved a high algal biomass with an HRT of 410 days. Xiao Z et al. (2012) used a continuous-feed batch system, but they integrated the photo-bioelectrochemical system with an HRT of 3 days, which resulted in a low biomass concentration. Gouveia et al. (2014) extracted pigments from the algal biomass, which is very rich in carotenoids. The pigment composition is affected by the light intensity. Christi (2007) produced photosynthetic microorganisms that were used for energy production. Fei Zhang et al. (2011) achieved a high algal biomass concentration using a single-chambered SMFC. Energy production from an algal biomass, using an MFC, is very attractive. Many new strategies have to be developed to evaluate the factors that consume energy during the process and during algal biomass production (Fig. 5.13).

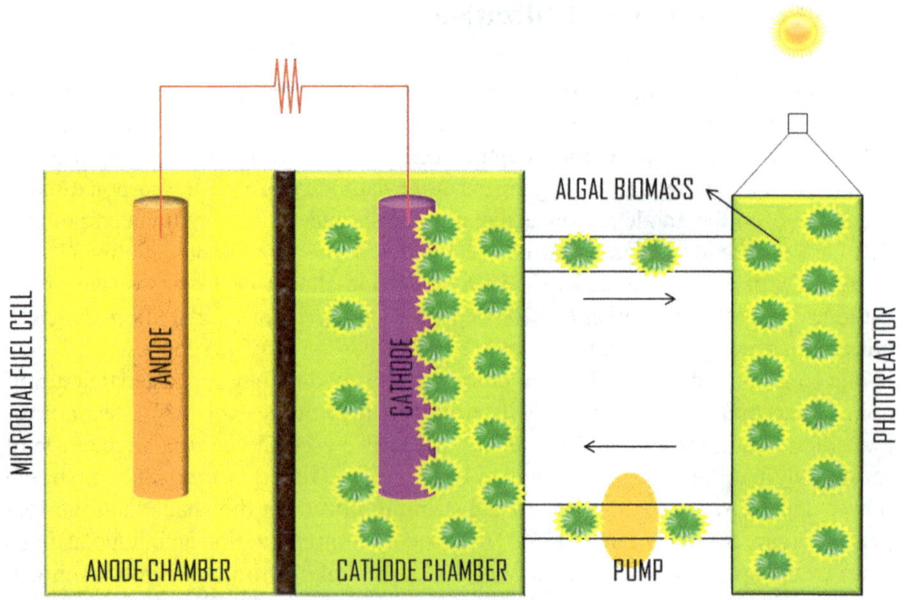

Fig. 5.13 Schematic diagram of phototrophic microorganism assisting the cathode process

5.21 Treatment of Wastewater

One important factor in choosing an MFC is selection for wastewater treatment, especially to remove contaminants, as well as to evaluate the performance of the MFC. Some MFCs have a cathode containing an organic solution that mimics wastewater fed into the anode, and the removal of organic compounds has been successfully achieved with such MFCs. Detailed studies are not reported with regard to the removal of organic compounds using algae at the cathode. Hyeon Jin Jeon et al. (2012) used an upflow-type MFC for growing algae in the cathode, but sufficient information about the algae employed for contaminant removal was not reported. Another study, by Li Xiao et al. (2012), reported the reduction of phosphorus and nitrogen concentrations with a cathode-assisted MFC. Subhadra et al. (2011) identified a large water foot print which was a key challenge for commercialized algal bioreactors. Olguín (2012) reported that growing algae would have dual benefits, such as biomass production and contaminant removal. Generally, treating wastewater in the cathode will certainly stimulate heterotrophic bacterial growth, although the organic compounds will be electron donors, reacting with the cathode electrode, a factor that would impair the generation of electricity. Generally, wastewater treated at the anode is fed into the cathode, where nutrients for the growth of algae are provided by the treated wastewater; these nutrients are also associated with the removal of organic residues.

5.22 Illumination Effects

Photosynthetic microorganisms grow with the help of illumination, depending on factors such as the intensity and duration of the illumination. Wu et al. (2014) and He et al. (2013a, b) have reported that electricity generation increases when there is an increase in illumination, with an associated increase in dissolved oxygen production. Xiao Z et al. (2012) performed a comparative study, of light and dark conditions, showing that the dark period significantly decreased electricity and biomass production; these authors also explained the important factors of the dark period for PMFCs. Juang et al. (2012), in their research, reported high electricity production with low light intensity. Their studies also suggest the importance of light intensities, photosynthetic microorganisms, and the protocol for operating conditions using algae assisted at cathodes in MFCs.

5.23 Challenges and Prospects

Production of electricity with MFCs has been achieved by using phototrophic microorganisms. In-depth knowledge about the challenges with respect to the application of MFCs will help to define the research and focus on the reported issues. The most challenging aspect of algal MFCs is to solve the technical problems of microalgal processing in the MFC. In the cathode chamber of the MFC algae require a large surface area for illumination.

It is challenging to use photosynthetic products such as hydrogen and organic compounds for the generation of electricity by MFCs, where biomass production is very much higher than the energy produced. As a result of photosynthesis, oxygen is produced in algal-assisted MFC anodes, with illumination and the design and operation of the reactor presenting great challenges. Photo hydrogen production utilizes organic compounds that act as substrates. Mixed cultures may affect phototrophic activity in MFCs.

As yet, there are no scientifically proven procedures for the large-scale production of energy by MFCs. Sediment MFCs can be employed for remote sensor powering, where the intensity of light inhibits the cell growth. When the density of the algae is too high, light penetration is too low, and this can disturb cell growth. Bombelli et al. (2011) reported that algal MFCs can produce electricity through biological pathways by converting light energy into electrical energy. The biomass thus obtained is organic and has zero carbon. Carbon dioxide, oxygen, and biomass production, as well as consumption, has to be balanced in the system.

Researchers are making efforts to enhance the power generation output of MFCs, and very high output could also increase the efficiency of algal cultivation. Using photosynthetic microorganisms directly for electricity production is not the best option, as the cell walls are resistant to hydrolysis. With anaerobic digestion, more energy is recovered was as MFCs which has less advantage compared to the aerobic digestion. The carbon dioxide supply inside the MFC has reduced the cost of aeration for algae. Further studies need to investigate the illumination effects in anode-assisted electrochemical catalysis.

Xing et al. (2008) reported that illumination in MFCs was not required for current generation, although research by Cao et al. (2008) reported that illumination improved the generation of electricity. Microalgae grow under various conditions where carbon dioxide recovery requires the self-growth for algae cultivation which is limited with the supply of light resources. Lin et al. (2015) experimented with a large eight-chamber photocathode with varied light intensity, and they have also employed an open cultivation method where light variation is not required.

For algal MFCs, both biomass cultivation and electricity generation are strategies employed. Compared with suspended algae, algal bead cultures increase the supply of oxygen, but the growth of algal bead cultures is slow and these cultures produce a very low biomass compared with that produced by suspended algae. These limitations enhances a high production reduces the costs for algal MFC. At the anode, substrate is utilized, and consistency of the performance of algal MFCs is low. The biofilm at the anode has great efficiency with low density in surface. The biofilm is resistant to the transfer of electrons; many research studies of biofilms have been reported, but only those with pure cultures of photosynthetic microorganisms have been noted. However, mixed cultures are generally used for the treatment of wastewater and they are applied practically. The power output of algal MFCs cannot be higher than the value afforded by the biofilm on the anode, which is only a few hundred milliwatts per square meter. The anode biofilm produces carbon dioxide, which is consumed by algal cells in the cathode chamber in the presence of light. The oxygen thus yielded acts as an electron acceptor for the cathode chamber.

Algal MFCs are considered to be a platform for biochemical and biofuel production utilizing wastewater organic substances. The design of some MFCs with PBRs is advantageous. The PBR can be in any configuration, such as flat or tubular. Much research is being carried out to implement algal MFCs as technical devices for algal biomass production and for electricity generation. Future perspectives of algal-MFC systems are described in the following section.

5.24 Future Perspectives of PMFCs

To address some of the challenges described above, Li and Zhen (2014) have proposed models of MFC technology that employ phototrophic microorganisms as a substrate in the anode and oxygen supply in the cathode. They expect that these models will lead to further investigations of photosynthetic microorganisms and MFCs.

The first proposed model involves a system where the algal biomass is degraded by using light energy and converted to electric energy. The model consists of three units: an MFC, an anaerobic digester, and a PBR (Fig. 5.14). The PBR is used for algal biomass production by photosynthesis. The produced biomass is placed in the anaerobic digester, which produces biogas. The algal cells are digested and then they are imported into the anode in the MFC. Bioenergy is produced from this system, where anaerobic digesters produce biogas that is further used by the MFC and directly produces electricity.

The second proposed model focuses on the cathode-assisted photosynthetic microorganisms that are used for wastewater treatment. These organisms are placed in either a closed or an open tubular bioreactor (Fig. 5.15). MFCs are integrated in the algal bioreactors (Xiao Z et al. 2012). Wastewater is fed into the MFC for degradation and the degraded effluent is discharged through an outlet to the algal bioreactors, which support the growth of algae. Closed tubular reactors seem to be more efficient, but are costlier than open channel reactors, which provide very low algal production but are easy to maintain. Closed tubular reactor systems are used for small-scale applications, whereas open-channel reactor systems are used for large-scale production.

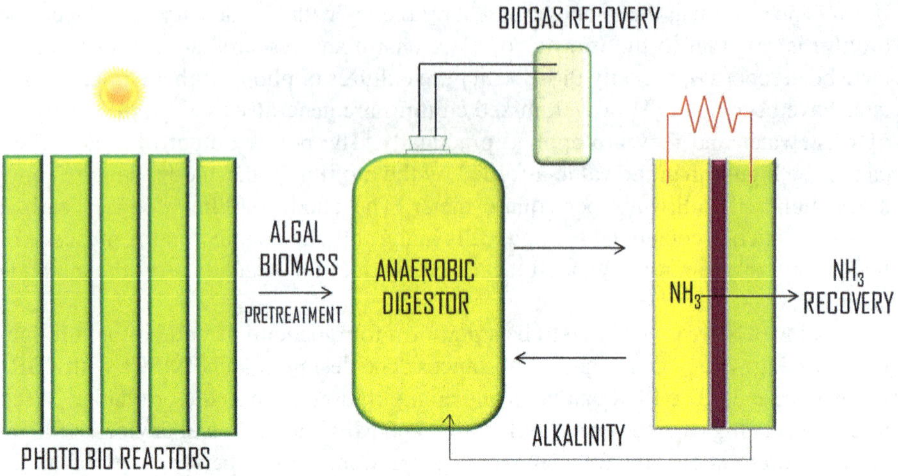

Fig. 5.14 Photo-MFC paradigms

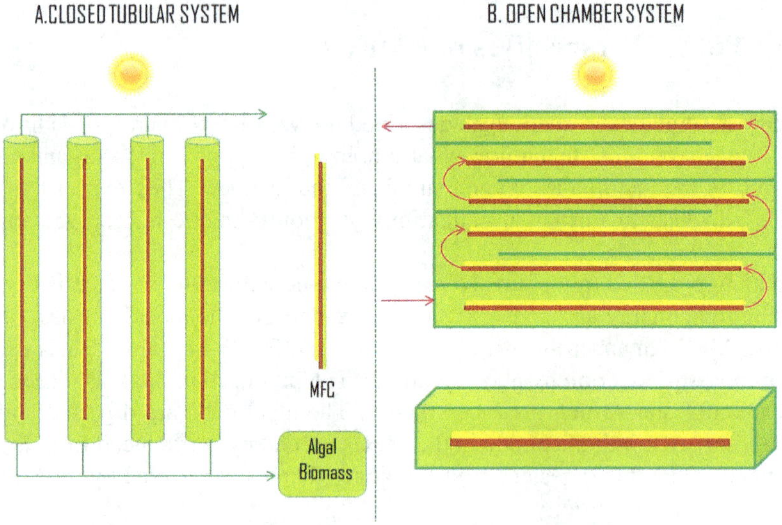

Fig. 5.15 Algal bioreactor paradigm—open and closed MFC systems

5.25 Conclusion

There have been significant developments and technological advances in MFC processes using microalgae. The advantage of incorporating PMFCs is to generate electricity. Owing to the low conversion efficiency, in certain systems algae are used as a substrate. Advances in algal MFC applications will lead to the development of a device that links microalgal cultivation, using a cathodic chamber and a conventional anodic chamber, and employs electron donors as fuel, providing a new pathway for converting light energy into electrical energy, with the production of less carbon dioxide. Oxygen production is an added advantage in the cathode reaction where biomass accumulation takes place. Wastewater treatment is accomplished by using MFCs, and, with further upgrades to MFC systems, photosynthetic microorganisms should be developed that show synergistic cooperation similar to that occurring with anaerobic digestion, and MFCs will be integrated with algal bioreactors, similar to algal ponds. In future, the development of algal fuel cells will have a substantial effect on the production of algal biomass, which can be utilized for commercial benefit in various fields.

References

Aelterman P, Rabaey K, Clauwaert P, Verstraete W (2006) Microbial fuel cells for wastewater treatment. Water Sci Technol 54(8):9–15

Badalamenti JP, Torres CI, Krajmalnik-Brown R (2013) Light-responsive current generation by phototrophically enriched anode biofilms dominated b green sulfur bacteria. Biotechnol Bioeng 110:1020–1027

Badalamenti JP, Torres CI, Krajmalnik-Brown R (2014) Coupling dark metabolism to electricity generation using photosynthetic co cultures. Biotechnol Bioeng 111:223–231

Barua PK, Deka D (2010) Electricity generation from biowaste based microbial fuel cells. Int J Energy Inform Commun 1:1

Bombelli P et al (2011) Quantitative analysis of the factors limiting solar power transduction by Synechocystis sp. PCC 6803 in biological photovoltaic devices. Energy Environ Sci 2011(4):4690–4698

Cai P-J, Xiao X, He Y-R, Li W-W, Zang G-L, Sheng G-P et al (2013) Reactive oxygen species (ROS) generated by cyanobacteria act as an electron acceptor in the biocathode of a bioelectrochemical system. Biosens Bioelectron 39:306–310

Cao X, Huang X, Boon N, Liang P, Fan M (2008) Electricity generation by an enriched phototrophic consortium in a microbial fuel cell. Electrochem Commun 10:1392–1395

Cao X, Huang X, Liang P, Boon N, Fan M, Zhanga L (2009) A completely anoxic microbial fuel cell using a photo-biocathode for cathodic carbon dioxide reduction. Energy Environ Sci 2:498–501

Chandra R, Venkata Subhash G, Venkata Mohan S (2012) Mixotrophic operation of photo-bio electro catalytic fuel cell under an oxygenic microenvironment enhances the light dependent bioelectrogenic activity. Bioresour Technol 109:46–56

Cheng S, Logan BE (2007) Ammonia treatment of carbon cloth anodes to enhance power generation of microbial fuel cells. Electrochem Commun 9(3):492–496

Cheng L, Zhang L, Chen H, Gao C (2006a) Carbon dioxide removal from air by microalgae cultured in a membrane-photo bioreactor. Sep Purif Technol 50:324–329

Cheng S, Liu H, Logan BE (2006b) Increased performance of single-chamber microbial fuel cells using an improved cathode structure. Electrochem Commun 8(3):489–494

Cho YK, Donohue TJ, Tejedor I, Anderson MA, McMahon KD, Noguera DR (2008) Development of a solar-powered microbial fuel cell. J Appl Microbiol 104:640–650

Christi Y (2007) Biodiesel from microalgae. Biotechnol Adv 25:294–306

Correa-Duarte MA, Wagner N, Rojas-Chapana J, Morsczeck C, Thie M, Giersig M (2004) Fabrication and biocompatibility of carbon nanotube-based 3D networks as scaffolds for cell seeding and growth. Nano Lett 4(11):2233

Cui Y, Rashid N, Hu N, Rehman MSU, Han J-I (2014) Electricity generation and microalgae cultivation in microbial fuel cell using microalgae-enriched anode and bio-cathode. Energy Convers Manag 79:674–680

De Schamphelaire L, Rabaey K, Boeckx P, Boon N, Verstraete W (2008) Outlook for benefits of sediment microbial fuel cells with two bio-electrodes. Microb Biotechnol 1:446–462
Del Campo AG, Cañizares P, Rodrigo MA, Fernández FJ, Lobato J (2013) Microbial fuel cell with an algae-assisted cathode: a preliminary assessment. J Power Sources 242:638–645
El Mekawy A, Hegab HM, Dominguez-Benetton X, Pant D (2013) Internal resistance of microfluidic microbial fuel cell: challenges and potential opportunities. Bioresour Technol 142:672–682
El Mekawy A, Hegab HM, Vanbroekhoven K, Pant D (2014) Techno productive potential of photosynthetic microbial fuel cells through different configurations. Renew Sust Energ Rev 39 (2014) 617–627
Freguia S, Rabaey K, Yuan Z, Keller J (2007) Electron and carbon balances in microbial fuel cells reveal temporary bacterial storage behavior during electricity generation. Environ Sci Technol 41:2915–2921
Fu C, Su C, Hung T, Hsieh C, Suryani D, Wu W (2009) Effects of biomass weight and light intensity on the performance of photosynthetic microbial fuel cells with *Spirulina platensis*. Bioresour Technol 100:4183–4186
Fu C, Hung T, Wu W, Wen T, Su C (2010) Current and voltage responses in instant photosynthetic microbial cells with *Spirulina platensis*. Biochem Eng J 52:175–180
Gadhamshetty V, Belanger D, Gardiner C-J, Cummings A, Hynes A (2013) Evaluation of Laminaria-based microbial fuel cells (LbMs) for electricity production. Bioresour Technol 127:378–385
Gajda I, Greenman J, Melhuish C, Ieropoulos I (2013) Photosynthetic cathodes for microbial fuel cells. Int J Hydrog Energy 38:11559–11564
Gajda I, Greenman J, Melhuish C, Ieropoulos I (2015) Self-sustainable electricity production from algae grown in a microbial fuel cell system Biomass Bioenergy 82:87–93
Ge Z, Zhang F, Grimaud J, Hurst J, He Z (2013) Long-term investigation of microbial fuel cells treating primary sludge or digested sludge. Bioresour Technol 136:509–514
Gil GC, Chang IS, Kim BH, Kim M, Jang JK, Park HS, Kim HJ (2003) Operational parameters affecting the performance of a mediator-less microbial fuel cell. Biosens Bioelectron 18:327–334
Gouveia L, Neves C, Sebastião D, Nobre BP, Matos CT (2014) Effect of light on the production of bioelectricity and added-value micro algae biomass in a photo- synthetic alga microbial fuel cell. Bioresour Technol 154:171–177
Gruning A, Beecroft NJ, Avignone-Rossa C (2014) Metabolic composition of anode community predicts electrical power in microbial fuel cells. Retrieved 1 January 2015 from: https://archive.org/details/biorxiv-10.1101-002337
Hai-ming, Jiang (2016) Combination of microbial fuel cells with microalgae cultivation for bioelectricity generation and domestic wastewater treatment. Environ Eng Sci 34:489–495
He D, Bultel Y, Magnin J-P, Roux C, Willison JC (2005a) Hydrogen photosynthesis by Rhodobacter capsulatus and its coupling to a PEM fuel cell. J Power Sources 141:19–23
He Z, Minteer SD, Angenent LT (2005b) Electricity generation from artificial wastewater using an up flow microbial fuel cell. Environ Sci Technol 39:5262–5267
He Z, Kan J, Mansfeld F, Angenent LT, Nealson KH (2009) Self-sustained phototrophic microbial fuel cells based on the synergistic cooperation between photosynthetic microorganismsand heterotrophic bacteria. Environ Sci Technol 43(5):1648–1654
He H, Zhou M, Yang J, Hu Y, Zhao Y (2013a) Simultaneous waste water treatment, electricity generation and biomass production by an immobilized photosynthetic algal microbial fuel cell. Bioprocess Biosyst Eng 37:873–880
He H, Zhou M, Yang J, Youshuang H, Zhao Y (2013b) Simultaneous wastewater treatment, electricity generation and biomass production by an immobilized photosynthetic algal microbial fuel cell. Bioprocess Biosyst Eng 37:873–880 00449-013-1058-4
He Z et al (2014) Applications and perspectives of phototrophic microorganisms for electricity generation from organic compounds in microbial fuel cells. Renew Sust Energ Rev 37:550–559
Heister E, Brunner EW, Dieckmann GR, Jurewicz I, Dalton AB (2013) Are carbon nanotubes a natural solution? Applications in biology and medicine. ACS Appl Mater Interfaces 5(6):1870

Hur J, Lee B-M, Choi K-S, Min B (2014) Tracking the spectroscopic and chromatographic changes of algal derived organic matter in a microbial fuel cell. Environ Sci Pollut Res 21:2230–2239

Ieropoulos I, Greenman J, Melhuish C (2008) Microbial fuel cells based on carbon veil electrodes: stack configuration and scalability. Int J Energy Res 32(13):1228–1240

Iijima S (1991) Helical microtubules of graphitic carbon. Nature 354:6348–6356

Inglesby AE, Fisher AC (2012) Enhanced methane yields from an aerobic digestion of Arthrospira maxima biomass in an advanced flow-through reactor with an integrated re circulation loop microbial fuel cell. Energy Environ Sci 5:7996–8006

Inglesby AE, Beatty DA, Fisher AC (2012) Rhodopseudomonas palustris purple bacteria fed Arthrospira maxima cyanobacteria: demonstration of application in microbial fuel cells. RSC Adv 2:4829–4838

Jeon HJ, Seo K-w, Lee SH, Yang Y-H, Kumaran RS, Kim S, Hong SW, Choi YS, Kim HJ (2012) Production of algal biomass (Chlorella vulgaris) using sediment microbial fuel cells. Bioresour Technol 109:308–311

Jiang H, Luo S, Shi X, Dai M, Guo R (2013) A system combining microbial fuel cell with photo bioreactor for continuous domestic waste water treatment and bioelectricity generation. J Cent South Univ 20:488–494

Juang DF, Lee CH, Hsueh SC (2012) Comparison of electrogenic capabilities of microbial fuel cell with different light power on algae grown cathode. Bioresour Technol 123:23–29

Kakarla R, Min B (2014) Evaluation of microbial fuel cell operation using algae as an oxygen supplier: carbon paper cathode vs. carbon brush cathode. Bioprocess Biosyst Eng. https://doi.org/10.1007/s00449-014-1223-4

Karube I, Takeuchi T, Barnes DJ (1992) Modern biochemical engineering, vol 46. Springer, Berlin/Heidelberg

Kokabian B, Gude VG (2013) Photosynthetic microbial desalination cells (PMDCs) for clean energy, water and biomass production. Environ Sci Processes Impact 15:2178–2185

Kondaveeti S, Choi KS, Kakarla R, Min B (2014) Microalgae Scenedesmus obliquus as renewable biomass feedstock for electricity generation in microbial fuel cells (MFCs). Front Environ Sci Eng. 8:784–791. https://doi.org/10.1007/s11783-013-0590-4

Kruzic AP, Kreissl JF (2009) Natural treatment and on site systems. Water Environ Res 81:1346–1360

Kymakis E, Amaratunga GAJ (2002) Single wall carbon nanotube conjugated polymer photovoltaic devices. Appl Phys Lett 80(1):112

Lakaniemi A-M, Tuovinen OH, Puhakka JA (2012) Production of electricity and butanol from microalgal biomass in microbial fuel cells. BioEnerg Res 5:481–491

Lan JC-W, Raman K, Huang C-M, Chang C-M (2013) The impact of mono chromatic blue and red LED light upon performance of photo microbial fuel cells (PMFCs) using Chlamydomonas reinhardtii transformation F5 as biocatalyst. Biochem Eng J 78:39–43

Lee Y-K (2004) Algal nutrition heterotrophic carbon nutrition. In: Richmond A (ed) Handb microalgal cult biotechnol appl phycol. Blackwell Publishing, Oxford, UK, p 116

Li W-W, Han-Qing Y, He Z (2014) Towards sustainable wastewater treatment by using microbial fuel cells-centered technologies. Energy Environ Sci 7:911

Li Xiao EB, Young JA, Berges ZH (2012) Integrated photo-bioelectrochemical system for contaminants removal and bioenergy production. Environ Sci Technol 46(20):11459–11466

Lin C-C, Wei C-H, Chen C-I, Shieh C-J, Liu Y-C (2013) Characteristics of the photosynthesis microbial fuel cell with a *Spirulina platensis* biofilm. Bioresour Technol 135:640–643

Liu T, Rao L, Yuan Y, Zhuang L (2015) Bioelectricity generation in a microbial fuel cell with a self-sustainable photo cathode. Sci World J 2015:1–8

Lobato J, del Campo AG, Fernández FJ, Cañizares P, Rodrigo MA (2013) Lagooning microbial fuel cells: a first approach by coupling electricity-producing micro-organisms and algae. Appl Energy 110:220–226

Logan BE (2010) Scaling up microbial fuel cells and other bioelectrochemical systems. Appl Microbiol Biotechnol 85(6):1665

Logan BE, Regan JM (2006) Electricity-producing bacterial communities in microbial fuel cells. Trends Microbiol 14:512–518

Logan BE, Hamelers B, Rozendal R, Schroder U, Keller J, Freguia S, Aelter man P, Verstraete W, Rabaey K (2006) Microbial fuel cells: methodology and technology. Environ Sci Technol 40(7):5181–5192
Luimstra VM et al (2013) A cost-effective microbial fuel cell to detect and select for photosynthetic electrogenic activity in algae and cyanobacteria. J Appl Phycol. https://doi.org/10.1007/s10811-013-0051-2
Lyautey E, Cournet A, Morin S, Bouletreau S, Etcheverry L, Charcosset JY (2011) Electro activity of phototrophic river biofilms and constitutive cultivable bacteria. Appl Environ Microbiol 77:5394–5401
Malik S, Drott E, Grisdela P, Lee J, Lee C, Lowy DA et al (2009) A self-assembling self-repairing microbial photo electro chemical solar cell. Energy Environ Sci 2:292–298
McGowan JG, Connors S (2000) WINDPOWER: a turn of the century review. Annu Rev Energy Environ 25:147–197
Mitra P, Hill GA (2011) Continuous microbial fuel cell using a photoautotrophic cathode and a fermentative anode. Can J Chem Eng 90:1006–1010
Mohan SV, Devi MP, Mohanakrishna G, Amarnath N, Babu ML, Sarma PN (2011) Potential of mixed microalgae to harness biodiesel from ecological water-bodies with simultaneous treatment. Bioresour Technol 102:1109–1117
Morishima K, Yoshida M, Furuya A, Moriuchi T, Ota M, Furukawa Y (2007) Improving the performance of a direct photosynthetic/metabolic bio-fuel cell (DPBFC) using gene manipulated bacteria. J Micromech Microeng 17:S274–S279
Mustakeem M et al (2015) Electrode materials for microbial fuel cells: nanomaterial approach. Mater Renew Sustain Energy 4:22
Natarajan D, Van Nguyen T (2004) Effect of electrode configuration and electronic conductivity on current density distribution measurements in PEM fuel cells. J Power Sources 135(1):95
Nishio K, Hashimoto K, Watanabe K (2013) Light/electricity conversion by defined co cultures of Chlamydomonas and Geobacter. J Biosci Bioeng 115:412–417
Oh S, Min B, Logan BE (2004) Cathode performances a factor in electricity generation in microbial fuel cells. Environ SciTechnol 38:4900–4904
Olguín EJ (2012) Dual purpose microalgae bacteria based systems that treat waste-water and produce biodiesel and chemical products within a bio refinery. Biotechnol Adv 30:1031–1046
Park DH, Zeikus JG (2003) Improved fuel cell and electrode designs for producing electricity from microbial degradation. Biotechnol Bioeng 81:348–355
Park DH, Zeikus JG (2009) Utilization of electrically reduced neutral red by Actinobacillus succinogenes: physiological function of neutral red in membrane-driven fumarate reduction and energy conservation. J Bacteriol 181:2403–2410
Pisciotta JM, Zou Y, Baskakov IV (2011) Role of the photosynthetic electron transfer chain in electrogenic activity of cyanobacteria. Appl Microbiol Biotechnol 91:377–385
Powell EE, Mapiour ML, Evitts RW, Hill GA (2009) Growth kinetics of Chlorella vulgaris and its use as a cathodic half-cell. Bioresour Technol 100:269–274
Powell EE, Mapiour ML, Evitts RW, Hill GA (2013) For multiple functionalities in microbial fuel cells. Bioprocess Biosyst Eng 36:1913–1921
Qian F, Wang G, Li Y (2010) Solar-Driven microbial photoelectrochemical cells with a nanowire photocathode. Nano Lett 10(11):4686-4691
Rabaey K, Boon N, Hofte M, Verstraete W (2005a) Microbial phenazine production enhances electron transfer in biofuel cells. Environ Sci Technol 39:3401–3408
Rabaey K, Clauwaert P, Aelterman P, Verstraete W (2005b) Tubular microbial fuel cells for efficient electricity generation. Environ Sci Technol 39:8077–8082
Raman K, Lan JC-W (2012) Performance and kinetic study of photo microbial fuel cells (PMFCs) with different electrode distances. Appl Energy 100:100–105
Ramanathan G, Birthous RS, Abirami D, Highcourt D (2011) Efficacy of marine microalgae as exoelectrogen in microbial fuel cell system for bioelectricity generation. World J Fish Marine Sci 3(1):79–87
Rashid N et al (2013) Enhanced electricity generation by using algae biomass and activated sludge in microbial fuel cell. Sci Total Environ 456–457:91–94

Reimers CE, Tender LM, Fertig S, Wang W (2001) Harvesting energy from the marine sediment-water interface. Environ Sci Technol 35:192–195
Reimers CE, Girguis P, Stecher HA, Tender LM, Ryckelynck N, Whaling P (2006) Microbial fuel cell energy from an ocean cold seep. Geobiology 4:123–136
Rismani-Yazdi H, Carver SM, Christy AD, Tuovinen OH (2008) Cathodic limitations in microbial fuel cells: an overview. J Power Sources 180(2):683
Rodrigo MA, Cañizares P, García H, Linares JJ, Lobato J (2009) Study of the acclimation stage and of the effect of the biodegradability on the performance of a microbial fuel cell. Bioresour Technol 100:4704–4710
Rosenbaum M, Schroder U, Scholz F (2005a) In situ electro oxidation of photobiological hydrogen in a photo bioelectro chemical fuel cell based on Rhodobacter sphaeroides. Environ Sci Technol 39:6328–6333
Rosenbaum M, Schroder U, Scholz F (2005b) Utilizing the green alga Chlamydomonas reinhardtii for microbial electricity generation: a living solar cell. Appl Microbiol Biotechnol 68:753–756
Rozendal RA, Hamelers HVM, Buisman CJN (2006) Effects of membrane cation transport on pH and microbial fuel cell performance. Environ Sci Technol 40:5206–5211
Satyanarayana KG, Mariano AB, Vargas JVC (2011) A reviews on microalgae, a versatile source for sustainable energy and materials. Int J Energy Res 35:291–311
Schamphelaire D et al (2009) Revival of the biological sunlight to biogas energy conversion system. Biotechnol Bioeng 103:296–304
Sevda S, Dominguez-Benetton X, Vanbroekhoven K, Sreekrishnan TR, Pant D (2013) Characterization and comparison of the performance of two different separator types in air–cathode microbial fuel cell treating synthetic waste-water. Chem Eng J 228:1–11
Sharma T, Mohana Reddy AL, Chandra TS, Ramaprabhu S (2008) Development of carbon nanotubes and nanofluids based microbial fuel cell. Int J Hydrog Energy 33(22):6749
Silvaggi J (2016) Integration of microbial fuel cell with in algal bioreactor. UWM Research Foundation
Singhvi P, Chhabra M (2013) Simultaneous chromium removal and power generation using algal biomass in a dual chambered salt bridge microbial fuel cell. J Bioremed Biodeg 4:5
Strik DPBTB, Terlouw H, HVM H, CJN B (2008) Renewable sustainable biocatalyzed electricity production in a photosynthetic algal microbial fuel cell (PAMFC). Appl Microbiol Biotechnol 81:659–668
Subhadra BG, Edwards M (2011) Co product market analysis and water footprint of simulated commercial algal biorefineries. Appl Energy 88:3515–3523
Subhash GV, Chandra R, Mohan SV (2013) Micro algae mediated bio-electro catalytic fuel cell facilitates bioelectricity generation through oxygenic photo mixotrophic mechanism. Bioresour Technol 136:644–653
Thorne R, Hu H, Schneider K, Bombelli P, Fisher A, Peter LM, Dent A, Cameron PJ (2011) Porous ceramic anode materials for photo-microbial fuel cells. J Mater Chem 21(44):18055–18060
Velasquez-Orta SB, Curtis TP, Logan BE (2009) Energy from algae using microbial fuel cells. Biotechnol Bioeng 103(6):1068–1076
Walter XA, Greenman J, Ieropoulos IA (2013) Oxygenic phototrophic biofilms for improved cathode performance in microbial fuel cells. Algal Res 2:183–187
Walter XA, Greenman J, Taylor B, Ieropoulos IA (2015) Microbial fuel cells continuously fuelled by untreated fresh algal biomass. Algal Res 11:103–107
Wang X, Feng Y, Liu J, Lee H, Li C, Li N et al (2010) Sequestration of CO2 discharged from anode by algal cathode in microbial carbon capture cells (MCCs). Biosens Bioelectron 25:2639–2643
Wang HY, Bernarda A, Huang CY, Lee DJ, Chang JS (2011) Micro-sized microbial fuel cell: a mini-review. Bioresour Technol 102(1):235
Wang H, Liu D, Lu L, Zhao Z, Xu Y, Cui F (2012) Degradation of algal organic matter using microbial fuel cells and its association with trihalomethane precursor removal. Bioresour Technol 116:80–85
Ward AJ, Lewis DM, Green FB (2014) Anaerobic digestion of algae biomass: a review. Algal Res:2014 https://doi.org/10.1016/j.algal.2014.02.001

Wei J, Liang P, Huang X (2011) Recent progress in electrodes for microbial fuel cells. Bioresour Technol 102(20):9335

Wu XY, Song TS, Zhu XJ, Wei P, Zhou CC (2013a) Construction and operation of microbial fuel cell with Chlorella vulgaris biocathode for electricity generation. Appl Biochem Biotechnol 171:2082–2092

Wu XY, Song TS, Zhu XJ, Wei P, Zhou CC (2013b) Construction and operation of microbial fuel cell with Chlorella vulgaris biocathode for electricity generation. Appl Biochem Biotechnol. https://doi.org/10.1007/s12013-013-0476-8

Winfield J et al (2012) Investigating a cascade of seven hydraulically connected microbial fuel cells. Bioresour Technol 110:245–250

Winfield J et al (2013) Comparing terracotta and earthenware for multiple functionalities in microbial fuel cells. Bioprocess Biosyst Eng 36:1913–1921

Xiao Z (2012) Renew Sustain Energy Rev 37:550–559

Xiao L, Young EB, Berges JA, He Z (2014) Integrated photo-bio electrochemical system for contaminants removal and bioenergy production. Environ Sci Technol 46:11459–11466

Xing DF, Zuo Y, Cheng SA, Regan JM, Logan BE (2008) Electricity generation by Rhodopseudomonas palustris DX-1. Environ Sci Technol 42:4146–4151

Xu C, Poon K, Choi MMF, Wang R (2015) Using live algae at the anode of a microbial fuel cell to generate electricity. Environ Sci Pollut Res 22(20):15621–15635

Yadav AK, Panda P, Rout P, Behara S, Patra AK, Nayak SK, Bag BP (2009) Entrapment of algae for wastewater treatment and bioelectricity generation in microbial fuel cell. XVIIth International Conference on Bioencapsulation, Groningen, Netherlands : 24–26,

Yagishita T, Sawayama S, Tsukahara K, Ogi T (1997) Effects of intensity of incident light and concentrations of Synechococcus sp. and 2-hydroxy-1,4-naphthoquinone on the current output of photosynthetic electrochemical cell. Sol Energy 61(5):347–353

Yang C, Hua Q, Shimizu K (2000) Energetics and carbon metabolism during growth of microalgal cells under photo autotrophic, mixotrophic and cyclic light-autotrophic/dark-heterotrophic conditions. Biochem Eng J 6:87–102

You SJ, Zhao QL, Jiang JQ, Zhang JN (2006) Treatment of domestic wastewater with simultaneous electricity generation in microbial fuel cell under continuous operation. Chem Biochem Eng 20:407–412

Yuan Y, Chen Q, Zhou S, Zhuang L, Hu P (2011) Bioelectricity generation and microcystins removal in a blue-green algae powered microbial fuel cell. J Hazard Mater 187(1–3):591–595

Zhang F, Brastad KS, He Z (2011) Integrating forward osmosis into microbial fuel cells for wastewater treatment, water extraction and bioelectricity generation. Environ Sci Technol 45(15):6690–6696

Zhou M, Chi M, Luo J, He H, Jin T (2011) An overview of electrode materials in microbial fuel cells. J Power Sources 196(10):4427

Zhou M, He H, Jin T, Wang H (2012) Power generation enhancement in novel microbial carbon capture cells with immobilized Chlorella vulgaris. J Power Sources 214:216–219

Zou Y, Pisciotta J, Billmyre RB, Baskakov IV (2009) Photosynthetic microbial fuel cells with positive light response. Biotechnol Bioprocess Eng 104:939–946

Chapter 6
Fungal Fuel Cells: Nature's Perpetual Energy Resource

Lavanyasri Rathinavel, Deepika Jothinathan, Santos D. Chicas, and Prabhakaran Mylsamy

6.1 Microbial Fuel Cell: Brief Introduction

Microbial fuel cells are considered to be the bioreactor that utilises the organic materials and converts the chemical energy into electricity with the help of microorganisms that undergo catalysis (Potter 1911). The general structural design of microbial fuel cell has an anode chamber and a cathode chamber separated by a membrane called proton-exchange membrane (Wilkinson 2000) which is actively involved in transporting the protons and besides the membrane prevents the transfer of oxygen and other compounds through it. The degradation of organic matters is initiated by microbes at the anode chamber to produce electrons and protons. As a result of degradation, carbon dioxide is also released in the anode chamber. The protons and electrons migrate to the cathode chamber through the external circuit that is connected externally. The protons and electron from the anode react with the oxygen at the cathode to form water molecules (Du et al. 2007). The above reactions take place within the MFC and enhance the generation of electricity. Microbial fuel cells have many advantages and greater application compared to the traditional technologies that were used to produce energy by utilising the organic substances. Pant et al. (2010) reported that the waste water that is used as substrates constitutes

L. Rathinavel · P. Mylsamy (✉)
Post Graduate and Research Department of Botany, Pachaiyappa's College, Chennai, 600030, India

D. Jothinathan
Department of Life Sciences, Central University of Tamil Nadu, Thiruvarur, 610005, India

S. D. Chicas
Faculty of Sciences and Technology, University of Belize, Belmopan Central Campus, Cayo District CA, Belize

V. Sivasankar et al. (eds.), *Microbial Fuel Cell Technology for Bioelectricity*,
https://doi.org/10.1007/978-3-319-92904-0_6

different kinds of organic substance as well as protein, lipid and carbohydrate. Besides MFCs utilise the substrate sources and convert the chemical energy into electricity. Verstraete et al. (2005) used a single-chambered MFC which had no mechanical aeration by which the cost of operation was reduced. Microbial fuel cells are most popular for their extensive application extended in treating waste waters: BOD (biological oxygen demand) sensor, production of hydrogen and generation of electricity (Logan and Regan 2006).

6.2 Introduction to Fungal Microbial Fuel Cell

Fungal fuel cell is a modified form of biofuel cell that has enormous application in the different kinds of industries. Most of the common research in biofuel cells has been carried out using bacteria and algae. Extensive applications of fungi have gained attention by the researchers of biofuels in recent times. The fungal cultures are used as pure cultures in anode and cathode compartment. Some research has proven the enzymes of fungi have been the best catalyst for oxidative reduction. Fungi species have been used to form biofilm around the anode or cathode by which the performance of the MFC has been efficiently proved. Here few fungi species are mentioned with their applications in industries. *Aspergillus awamori*, *Trichoderma viride and Trichoderma atroviride* are used in MFCs to treat waste wasters from the distillery industries and help in bioremediation process were as, *Trametes versicolor* was grown with *Shewanella oneidensis* which is a fungal-bacterial combination incorporated in MFC to decolourise the textile dye effluents, and *Pleurotus ostreatus* is also used in dye decolourisation. Pharmaceutical industries have many metabolite wastes in the waste waters where APAP is one common pollutant that is degraded by *Scedosporium dehoogii*. When using fungi in MFC, some of the by-products such as chitosan, laccase and manganese peroxide are recovered in between the process that have high commercial value; such kinds of fungi species used for research are *Aspergillus awamori* and *ligninolytic fungus*. Apart from these functions of the fungi species, all the abovementioned species are also involved in bioenergy production. Besides *Rhizopus* sp., *Aspergillus* sp., *Penicillium* spp., *Gloeophyllum* and *Rhizopus*, *Coriolus versicolor*, *Pleurotus ostreatus*, *Trichoderma viride* and *Trichoderma atroviride* are evaluated in research specifically for the production of electricity by incorporating the fungi species in MFC. Technologies and principles differ from the species and the application. Yet, many findings have to be optimised to better understand the biological principles behind the fungal enzymes and the electrochemical reactions. The research conducted using the abovementioned species of fungi is discussed in detail. The following chapter will explore the experimental ideas and give a clear vision and knowledge about implying fungi in MFC for bioenergy production.

6.3 Microbial Fuel Cell with Fungal Biofilm as Bio-anode

Scedosporium dehoogii is a filamentous fungi, where its biofilm was successfully transformed into a carbon felt which acts as a bio-anode for microbial fuel cell. Cathode electrode was designed by replacing the Pt by modifying the carbon felt using Poly-NiTSPc (polyNi(II)tetrasulfophthalocyanine). Under controlled conditions, this model of MFC has a stable power generation with 6.5mWm^{-2} (Mbokou et al. 2017).

APAP contaminant Different kinds of contaminants are emerging in the field of cosmetic and medicinal product manufacturing. Among them the products of pharmaceutical industries and metabolites obtained from them are very dangerous when they are mixed in the river or any water resources (Basha et al. 2015). Introducing these types of waste contaminants brings out drastic changes by accumulating in the tissues of aquatic animals further leading to different kinds of changes within the cell (Escher et al. 2011). The common contaminants found in the pharmaceutical waste waters are found to be acetanilide along with the APAP (acetaminophen) and their metabolites (Mazloum-Ardakani et al. 2015). APAP is the most popular drug known for its therapeutic values. It is used as an analgesic medicine for pain and it also has antipyretic activity. In recent times, water contaminants were identified, and one among them was APAP (Yoon et al. 2010). APAP was also identified as contaminant in different kinds of resources because of improper disposal of hospital wastes and domestic wastes resulting in severe contamination of APAP in groundwater (Ternes 1998), rivers (Ternes 1998), surface water (Bannwarth 2006), drinking waters (Benotti et al. 2008) and treated sewage plant waste waters. Westerhoff et al. (2005) and Basha et al. (2015) reported that chemical oxidation is the process that is carried out to treat APAP found in waste waters. The methodologies used for treating the pollutants seem to be very costly, and hence the use of this method has been limited (Waterson et al. 2006).

Another cost-effective and environment-friendly technique to degrade APAP is biodegradation using microorganisms (Kim et al. 2007; Wu et al. 2012b). Microbial fuel cells are used for treating waste waters and they produce energy (Heidrich et al. 2014; Wang et al. 2013; Ivanov et al. 2010; Willner 2002). This method seems to be the best alternative for the treatment of waste water due to its reliability and its special properties.

Fungi for bioremediation *Scedosporium apiospermum* is a species that has five kinds of different species. All these five species are having the potency to metabolise the aromatic and aliphatic compounds. These species utilise these compounds as a source of carbon and energy (Clauben and Schmidt 1998; April et al. 1998; Bond and Lovely 2003; Mbokou et al. 2016). Very recently it has been scientifically proved that the four different kinds of species are the precursor agents of various different kinds of disease and many different kinds of disease that result in death. But one species *Scedosporium dehoogii* is the one which has not been observed in any patients till date. This species has its own intrinsic properties for which it was applied in bioremediation (Lackner et al. 2012; Gilgado et al. 2009). A research was conducted for the first time using *Scedosporium dehoogii* for expanding a biofilm that is used as bio-anode in APAP biofuel cell. This method was simple to generate power with a microbial fuel cell to produce current as well as biodegrade the polluted water expelled out of pharma industries.

Preparation of biofilm of fungi For this research APAP was used in the form of powder. Solution of phosphate buffer with 0.1 M was used as electrolyte with pH of 7.4. *Scedosporium dehoogii* was isolated from the soil collected from France. The fungal sample was continuously cultured and maintained in YEPD (yeast extract peptone dextrose) agar medium. The efficacy of the fungus that utilises APAP as a source of carbon was evaluated by a subculturing method with APAP replacing the glucose. Two weeks later the conidia was harvested from the YEPD plates by pouring 15 ml of pure water on the agar surface, and the collected sample was filtered using nylon filter with pore size ranging 40 μm and the collected sample was centrifuged under 4 °C at 4000 g for 5 min. Later the sample was resuspended in 10 ml of pure water, and finally this was used for the biofilm elaboration where the suspension consisted of 106 spores/ml.

Anode and cathode expansion As a substrate for anode, carbon felt measuring 8 cm^2 which was cleaned using 1 M hydrogen chloride followed by a wash with pure water is used for fungal deposition. Carbon felt was immersed in a solution containing ethanol and water in the ratio of 1:1; later sonication process is carried out. The carbon felt immersed in the solution containing the fungal suspension is elaborated under sterile condition by polarisation. Electrochemical analyser was used to connect the three different electrodes such as working electrode (carbon felt), reference electrode (saturated calomel) and counter electrode (platinum wire). Seven days later the biofilm is obtained and utilised as biofilm anode in microbial fuel cell.

Cathode comprised a carbon felt which was pretreated electrochemically in 0.1 M sodium hydroxide for 10 cycles. This is performed mainly to improve the deposition of the poly-NiTSPc film. This is the one that helps to get the impurities apart, and the surface gets prepared for the oxygen to get grafted on it. The carbon felt modification is done to improve the reduction of oxygen and increase the performance of the microbial fuel cell.

Fungal bio-anode-assisted MFC Microbial fuel cell consisted of two different compartments separated by a membrane called Nafion which is a proton-exchange membrane and helps in transporting the protons from the anode chamber to the cathode chamber (Fig. 6.1). Anode compartment has a solution comprising 100 mgL^{-1} APAP mixed in 0.1 M PBS medium with a pH measuring 7.4. The cathode consists the same as the anode but additionally air is provided. The electrodes of both the compartment are connected with an external resistance to exchange the electrons between the anode and cathode. Generation of electrons by the fungus utilising APAP substrate is transferred to anode and then to cathode via resistor. The

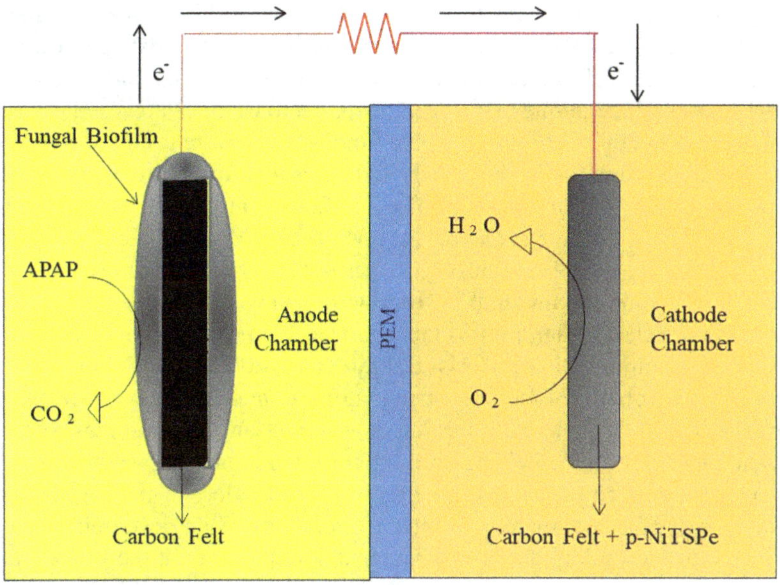

Fig. 6.1 Fungal biofilm based modified microbial fuel cell

electrons transported from the anode via resistor enters the cathode compartment to combine with the proton, which diffused through the PEM and also combines with the oxygen supplied by the air inlet resulting in water as a product. During this process the electron transfers from anode to cathode through an external resistance to generate electricity.

In this research *Scedosporium dehoogii* the filamentous fungi help in the bioremediation process where the catalytic activity was best found in the bio-anode. Besides oxygen reduction took place as a result of catalytic activity in the cathode. The microbial fuel cell constructed for this research has generated a power density of about 6.5Wm^{-2}. Finally, biodegradation of the APAP was achieved along with the power generation by the novel development of bio-anode using the fungal biofilm incorporated in microbial fuel cell for better performance.

6.4 Biodegradation Using Fungal MFC Yielding By-Products

Aspergillus awamori is used to remove organic matters produced from the cereal distillery units. This was achieved by the fermentation process in two stages and consecutively employed in MFC. There was a very decent percentile of reduction in the COD, and the suspended solids were greatly fermented by the fungi during the pretreatment. As a result fermentation by-product chitosan was obtained from the mycelia (Ray and Ghangrekar 2015).

Waste water treatment using fungal MFC Different kinds of techniques have been developed to treat the distillery waste water in the recent days; they are physical methods, chemical-related process, and biological organism-based and integrated or coupled techniques (Fig. 6.2). Filamentous fungi are used to treat the molasses of sugarcane. Different kinds of fungi species were used for pretreatment, and it resulted in reduction of COD with 10 days as their incubation period (Fig. 6.3) (González et al. 2000). Similary, Beltrán et al. (2001), introduced a treatment named footprint Technology which is one of the most feasible disposal technology. Nataraj et al. (2006) and Visvanathan et al. (2000) research were a guideline to remove dissolved inorganic and organic matter from waste using different kind of filtrations. Satyawali and Balakrishnan (2008) incorporated filamentous fungi pure strain for melanoidins decolourization process. Beltrán et al. (2001) and Wilkie et al. (2000) proposed several chemical methods using chlorine, hydrogen per-oxide and ozone to separate organic matters from the waste waters. Many technologies were developed, and they are into the field of research for the treatment of waste water, where the results weren't convincing since they were not safe and the disposal wasn't economically viable. So, there were many limitations while developing new treatment methods and finding the best permanent solution to treat the waste waters of distillery industries. *Aspergillus* spp. have a specific advantage to break the substrates such as complex starch and cellulose with the help of their own extracellular enzyme. They have a greater vitality to withstand high temperatures and varied pH and grow even when the rate of aeration differs along with the concentration of the substrate. The fungi species are selected on the basis of the ability to reduce the solids that are suspended in the waste waters and remove the COD. As a result of the fermentation process, the biomass is yielded with chitosan, organic acids and different amino acids (Van Leeuwen et al. 2013). Chitosan is said to be the natural polymer that is biodegradable present in the cell wall of fungi that are filamentous

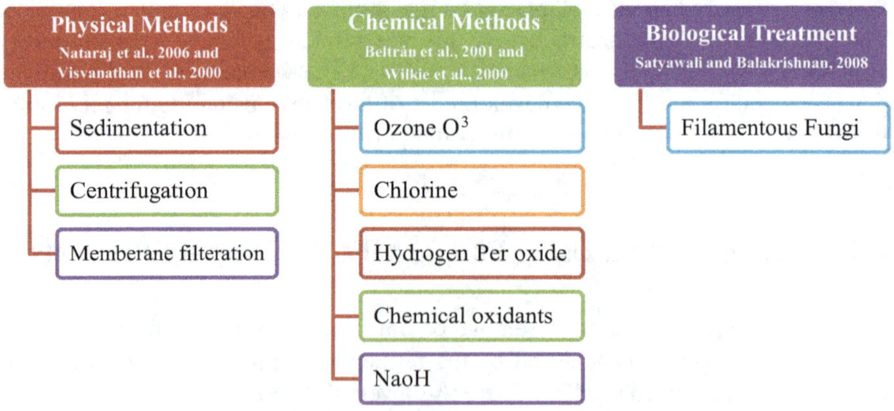

Fig. 6.2 Techniques to treat the distillery waste water

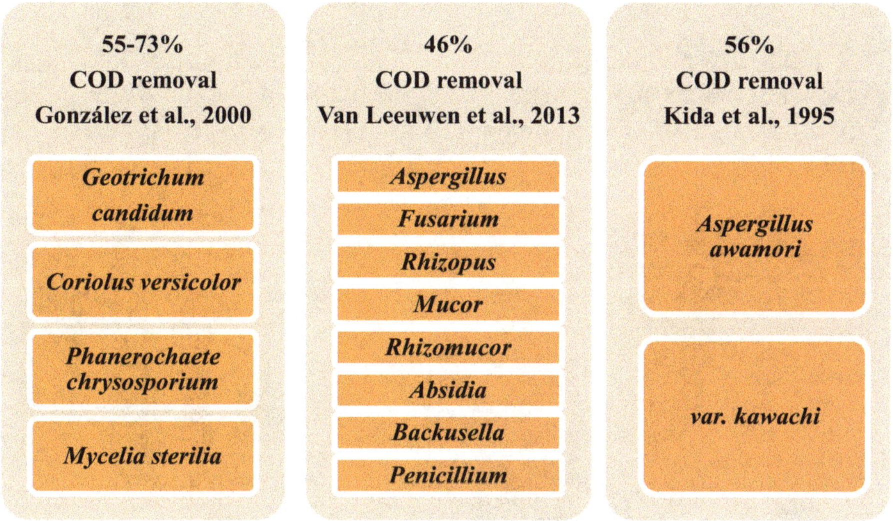

Fig. 6.3 Fungal strains treated to remove COD from the distillery stillage

in nature. Chitosan is highly potential and used in the biotech field to inhibit enzymes and used as coagulant in water engineering and additives in food as thickening and emulsifying agents. They are also used as antioxidant, found in dietary supplements and in the cosmetic product manufacturing (Kannan et al. 2011). Research prior to fermentation of the waste water from the industries is carried out in microbial fuel cells for oxidising the complex substance and to generate electricity.

Extraction of chitosan Lyophilised culture of *A. awamori* was cultivated in yeast extract agar at 25 °C for 5 days. Followed by the dense formation of fungi, it is sporulated and can be seen visualised as pallets. These pallets are used as inoculate in the culture medium used for the pretreatment. The pretreatment of the waste waters is processed in batch mode along with the sporulated fungal pallet culture. A fermenter jar with culture medium is inoculated with ten pallets under sterile conditions, and they are attached to the aeration unit (Fig. 6.4). The nutrient medium has no supplements such as nutrients or metals. The temperature is maintained at 25 °C, and the aeration rate is maintained with 1.75 V of inflow air/volume during the entire process for 10 days. The optimal pH for the fungal growth is 3.5, and the fermenter bottom is attached with the air diffuser which helps in the reduction of solid particles that settle at the bottom. This also facilitates the growth of fungi that are suspended in the medium. In between the process, the samples are collected to evaluate the reduction rate of the organic matter, and the pH is not adjusted during the fermentation process. On the 5th day of the fermentation process initiation, chitosan is extracted from the biomass produced. Later, at the interval of every 24 h, the chitosan is quantified till the fermentation process ends. At the end the fungi are recovered and repeatedly washed to remove the organic matters attached to the

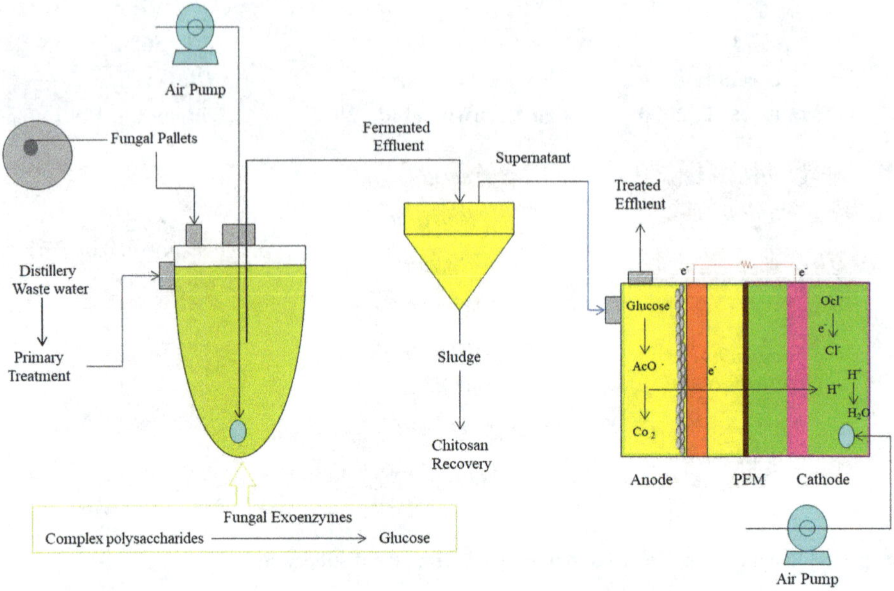

Fig. 6.4 Power production using fungal biofilm MFC and extraction of chitosan from distillery waste water

mycelia. The washed wet mycelia are treated with 1 M sodium hydroxide and sterilised by autoclaving. Insoluble alkali is collected at the end of repeated wash using distilled water to neutralise the pH. Later they are treated with 2% acetic acid with the ratio of 1:30 with the temperature maintained at 95 °C for a period of 8 h. The obtained slurry is centrifuged, and the pH is adjusted using sodium hydroxide solution to enhance the precipitation of chitosan. The precipitate is washed with distilled water and the pH is neutralised, and later they are dried at the temperature of 60 °C and the net weight is calculated (Maghsoodi et al. 2009).

Microbial fuel cell for electricity generation Dual-chambered MFC assisted at cathode was built with ceramic measuring 500 ml. A 4-mm-thick wall of clayware ceramic was used as PEM (proton-exchange membrane) (Jana et al. 2010). Carbon felt measuring 100cm^2 was used in both cathode and anode. The internal and external surface of the clayware chamber is fixed with the electrodes with the help of stainless steel. Insulated copper wire is used to connect the electrodes externally. The anode chamber was submerged within the cathode with sodium hypochlorite. The anode chamber is free from the exposure of air throughout the experiment (Fig. 6.3). With the help of an air sparger, the air flow rate is directed to the cathode solution at 3.51/min. The anode chamber is fed with the pretreated waste water, and raw waste materials for the treatment with 120 ml of anaerobic sludge mixed consortia which were obtained from septic tank are used. Fed-batch mode of operation is carried out in MFC under the temperature of 32 ± 2 °C for 91 days. The MFC performance is evaluated on the basis of COD removal and the stable voltage at which the operation of MFC is carried out. The process of treating the distillery waste waters using filamentous fungi strain has enhanced the removal of organic matter and simultaneously enhanced the generation of power with the help of cathode-assisted microbial fuel cell. This research has reported 99% of COD reduction and the solid suspended involves in improving the power generation. The COD values obtained at the end of multiple stages of biological treatment seem to be very less. Therefore the pretreatment of the waste water with the fungal culture and later implementing in the MFC could be beneficial and be the best solution to solve the water pollution problems raised by distillery waste water.

6.5 Fungi as Biocatalyst for Air-Cathode MFC

***Rhizopus* sp., *Aspergillus* sp. and *Penicillium* sp.** from the soil of Caatinga were isolated and applied for air cathodic in situ reduction of oxygen in MFC. This research has proved that these species could be used as biocatalyst for the reduction of oxygen in MFCs as well as improve the efficiency of generation of electricity (C.E. La Rotta Hernández et al. 2014). The evaluation of fungal cultures was done by visualising the formation of colour and the substrate obtained as a result of degradation. This was obtained by culturing the fungi samples in the plate that was incubated for 48 h with a temperature maintained at 28 °C. Submerged culture was prepared using the fungi sample disc that was obtained from the 2-day-old solid plates. The fungi sample in the disc was disrupted using test tubes, and later they were suspended and grown in the medium at 28 °C for a period of 48 h. Later, these cultures were used to inoculate the compartments of MFC and maintain the same condition. By filtering the culture, the mycelia were separated from the broth medium using nylon cloth. Further, they were centrifuged at 4 °C for 20 min at 4500 rpm. Later the collected mycelia were dried at 60 °C. The supernatant collected after centrifugation was quantified to know the enzyme activities. Microbial fuel cell was constructed using an air-cathode compartment with 100 ml of the fungal culture with electrodes immersed in the culture medium (Fig. 6.5). A modified type of carbon felt was coated with 0.5% platinum including carbon black, and the other electrode was used without platinum and black carbon-coated carbon felt as anode. Agar cation-exchange membrane was used as the bridge for transportation of electrons. A multimeter working based on acquisition software was used to monitor and record the data. The strain with oxidative activity was selected and used as the catalyst in cathodes of MFC transformation, and utilisation of glucose was observed in the strains of *Aspergillus* sp. and *Rhizopus* sp. Low oxidation was observed in all strains when the glycerol presence was confirmed. *Penicillium* sp. shows the highest production of energy by converting the glucose into energy with a higher efficiency of Coulombic effect. This seems to be a main fact helping to identify the potential strain to isolate and characterise to use as a biocatalyst in the cathodes Morant et al. (2014).

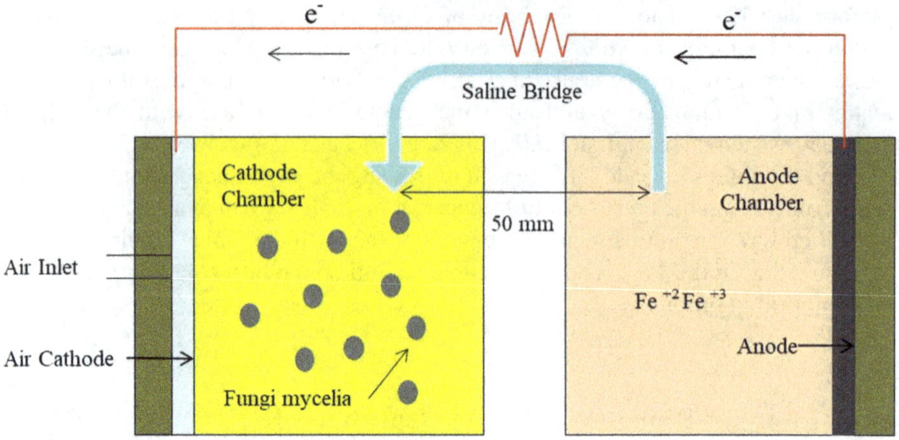

Fig. 6.5 Fungal MFC with air cathode

The results interpreted show that the oxidase enzymes are solely responsible for the electrochemical activity. The substrate produced by the above mentioned fungi species shows that they could catalyse reduce the four electrons and couple into one electron. Phenols and other substances are responsible for the supply of electrons and ions of inorganic compounds. Hence the biocatalyst using MFC could help in degrading heavy phenolic compound and waste water that contain dye. This research shows that the isolated fungi from the regions of casting could serve as an excellent biocatalyst supporting in the reduction of oxygen and generate electricity. This system offers cathodes composed of pure enzymes which are very low in cost for production, and the maintenance of MFC seems to be very easy and simple and it's an added advantage.

6.6 Fungal Enzyme-Based MFC

Trametes versicolor is a microorganism that secretes an enzyme called laccase, which is used at the cathode of MFC directly without any treatment, and this enzymatic biofuel cell helps in the conversion of chemical energy obtained from the biological resources into bioenergy. This methodology scientifically proves to be cost-effective and very simple. Moreover, this method doesn't require the mediators, and the cost of enzyme purification is highly reduced (Elena Kipf et al 2013; Sabine Sané et al. 2013). *Trametes versicolor* pure culture was subcultured in YPD (yeast extract peptone dextrose) agar plates, which were subjected to dark for 5 days, and the grown fungi were cultured in liquid laccase medium and maintained in different batches at 20 °C for the study. To evaluate the efficacy of the stability, the fungal cultures were prepared and monitored for a period of 74 days, and in between regular intervals, the supernatants of the medium were evaluated with the pH and the activity of the enzyme. These supernatants were also subjected to the electrophoresis and mass spec analysis.

The microbial fuel cell reactor was set up using the buckypaper as cathode electrodes which were from the nanotube dispersed carbon. Nanotube was loaded with nylon filter and used a cathode. The efficiency of the cathode was tested using configuration of a half cell. Following the procedure of (Kolke et al. 2010), the reactor was constructed using frames of polycarbonate where the silicone gasket is stacked alternatively. Cavities were created by creating vents and channels. The cation-exchange membrane is used to separate the compartments between the references and counter electrode. The culture medium sterility was maintained, and the syringe filters were used to separate the gas channels. The results were very extensively applicable where the enzymatic cathodes were self-generating and hence their lifetime was extended. The culture supernatant with the enzyme can be collected in a bioreactor for large-scale industrial applications. These enzymatic cathodes can be used in waste water treatment plants.

6.7 Microbial Fuel Cell with Fungal Biofilm as Bio-cathode

Coriolus versicolor is a white-rot fungus which hones the capacity to reduce the oxygen to water, and this has been highly used in enzyme-based microbial fuel cell. To improve the efficiency of electricity production, the fungi were employed in the cathode chamber where redox mediators facilitate the transfer of electrons between the laccase and the electrode. This is the very first research to report the bio-cathode consisting white-rot fungus in microbial fuel cell for the generation of the electricity (Wu et al. 2012a). For this research the pure culture strain of *Coriolus versicolor* was purchased from China's Institute of Microbiology, and subcultures were prepared. The MFC was designed in H shape with two chambers separated by a PEM (proton-exchange membrane); the sterilisation of the cultures and MFC chambers was done individually, whereas the PEM was sterilised with HNO_3 for half an hour. Each container of anode and cathode was measured a capacity of 100 ml. K4(Fe(CN)6) solution was fed up in anode and the fungi culture medium in cathode. Cathode chamber was supplied with the saturated air and stirred continuously; besides the anode was maintained in an air-tight condition. The temperature was maintained at 28 °C to operate the MFC. The output voltage of the MFC was calculated using the online data collection software (Fig. 6.6). The voltage drop was maintained by adding ABTS, and they were evaluated for 96 h for 3 cycles. The results were interpreted as the white-rot fungus functioned as an excellent bio-cathode helping to improvise the generation of electricity. The generation of power can be improved and can be commercialised in the future. The bio-cathode efficiency seems to be low, whereas there are many advantages with respect to the application and the cost. The MFCs' limiting factor has been the cathode, and hence the efficiency of the catalytic activity improvement is very essential. Till date the mechanism behind the electron transfer from the microbe group is limited. But in

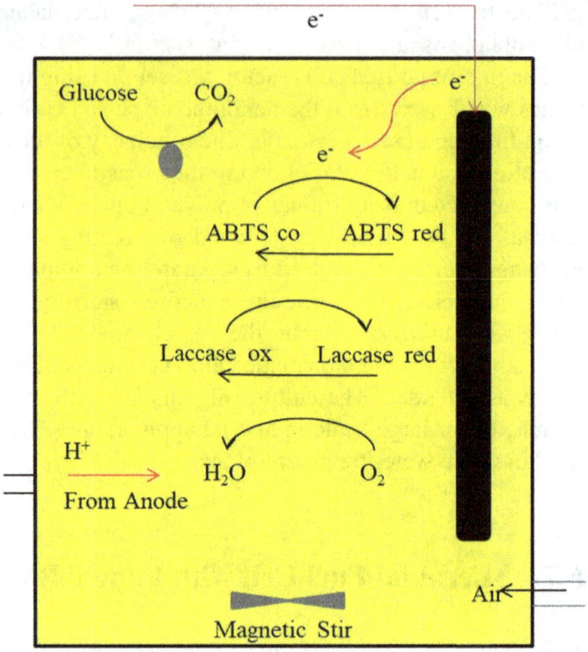

Fig. 6.6 White rot fungus bio cathode MFC

this research, the bio-cathode is convincing due to the great efficiency of the microorganism to catalyse the reduction of oxygen which secretes laccase. Hence, this model is supposed to be a promising model that could be standardised and improved. The white-rot fungus could biodegrade different kinds of pollutants, and hence this technology can be applied for a dual purpose by producing energy and treating the waste waters.

Another kind of research was carried out using *ligninolytic* fungus to find the effect on the performance of MFC when using fungal enzymes laccase and manganese peroxide as catalyst in the cathode chamber. The enzymatic electrode generated a higher power density than the graphite electrode. Manganese-per-oxide was produced by the white-rot fungus through adsorption and it get immobilised on the graphite electrode yielding maximum power density (Bakhshian and Kariminia 2011).

The MFC was constructed using two chambers with a short tube, and the chambers were separated using Nafion membrane. The membrane was pretreated. The electrodes of anode and cathode are composed of graphite. Both electrodes were connected using copper wire as the external resistance. Sludge waste was inoculated in the anode chamber, and with the help of magnetic stirrer, both the chambers were mixed without getting the sludge to get settled. Ligninolytic fungi produce manganese peroxide which is isolated using the fungal culture. The mycelia fungi were used as inoculum in potato-dextrose medium and incubated at the temperature of 32 °C. The fungal strain was collected after 7 days and inoculated in liquid medium maintaining the temperature of 32 °C, and it was continuously rotated at 160 rpm using rotary shaker. Fourteen days later the activity of manganese peroxide was observed in the supernatant of the centrifuged culture broth which was later used to immobilise the electrode made of graphite. The activity of the laccase was measured by using spectrophotometer where ABTS was used as substrate. The electrode comprised of graphite was immobilised with the white-rot fungi enzymes laccase and manganese peroxide, and this electrode was used in the compartment of cathode which increased the performance of MFC. The power density generated with the enzymatic cathode was twice than that of the cathode that is non-enzymatic. When the hydrogen peroxide was reduced by oxygen, it was due to the enzymes' catalytic effect by which the potential of MFC was decreased. In the future the laccase and manganese peroxide can be introduced as catalyst to obtain higher efficiency of the MFC.

6.8 Fungi-Bacteria-Assisted MFC for Bioenergy Production

Efficient bioenergy was produced by degrading the waste by employing *Trametes versicolor* with *Shewanella oneidensis* in microbial fuel cell as well as the electro-Fenton technology. This research was designed with dual benefits by achieving decolourisation of the dye and producing bioenergy. Biodegradation of the pollutant also generates energy, and the generated bioenergy was efficiently used to operate batch and continuous mode process in in situ electro-Fenton technique which generates a higher rate of bioenergy (María Ángeles Sanromán et al. 2013).

Trametes versicolor was cultured and maintained in malt agar medium plates at 4 °C, whereas the *Shewanella oneidensis* was cultured and maintained in plates of TSA agar medium at 20 °C. The microbial fuel cell was designed in the form of H type with two chambers separated by cation-exchange membrane (Fig. 6.7). Inside the chamber, the electrodes were placed in parallel maintaining a gap of 12 cm between them. The anode and cathode electrodes were made of graphite material. The generated power was recorded using Autolab (PGSTAT302N). Before initialising the MFC, the electrodes of anode were inoculated in submerged cultures of *T. versicolor* and *S. oneidensis* that were sterilised for 20 min at 121 °C. Anode electrode was fixed after 4 days in anode chamber. The anode compartment was continuously fed with the medium maintained with a pH of 7 and autoclaved at 121 °C for a period of 20 min. To maintain the aerobic or anaerobic condition, both the anode and cathode compartments were continuously supplied with air and nitrogen with an external pump. To maintain the homogeneous conditions in both the chambers was achieved by fixing magnetic stirring bars in both compartments. Electro-Fenton reactor of the MFC was researched using fungi-bacteria MFC. Graphite is used as electrodes in cathode where oxygen gets reduced to produce hydrogen peroxide. To attain a high rate of dissolved oxygen, the flow of air was controlled at the rate of 2 L/min which was inlet near to the cathode. The electro-Fenton in situ reaction is carried out in the cathode chamber to which the supplement of iron was given

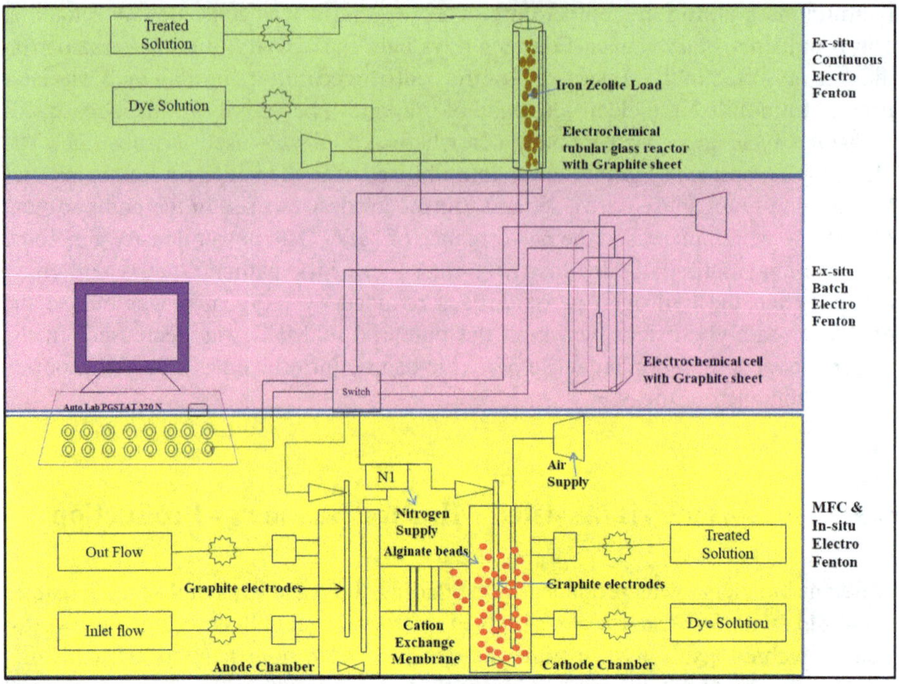

Fig. 6.7 Experimental set up of MFC, In-situ Electro Fenton, Ex-situ batch and continuous Electro Fenton: Diagram Ref: María Ángeles Fernández de Dios et al. (2013)

in the form of alginate beads. In continuous mode, the iron gets fixed into the matrix which enhances the operation. Lissamine Green colour dye solution was pumped to the cathode compartment with a period interval of 9 h, and the pH was maintained to 2. The concentration of dye residues was estimated from the reaction sample mixtures at room temperature. In each chamber the graphite electrodes in the form of rods were parallel to each other by maintaining a gap of 12 cm in between. The external resistance measuring 1000X was used as electrode connectors. The produced energy was calculated using the Autolab, whereas the electro-Fenton ex situ process was carried out in batch mode. For this experiment, a small electrochemical cell with a capacity of 4 ml was designed using graphite sheet as electrode linked to the MFC graphite electrode sheets was placed opposite to one another with a gap maintaining 1 cm and the air inlet was near the cathode.

In this research two dyes were evaluated. The dye was removed in the cathode chamber using a continuous mode for 9 h. Through this research 94% of Lissamine Green B and 83% of crystal violet were successfully removed using an electro-Fenton H-type MFC. Discolouration of the dye was achieved. Power generated through this experiment was used for electrochemical process. This experiment is evidential that under optimised conditions, the necessary electricity was driven using a small cell which is very stable. This electro-Fenton technique will be an alternate to treat the pollutants, and it is cost-effective.

6.9 Liquid Fungal Cultures as Anolyte and Catholyte in MFC

Two different strains *Gloeophyllum* and *Rhizopus* liquid cultures were used in the MFC. *Gloeophyllum* culture was used as anolyte, and *Rhizopus* culture was used as catholyte in microbial fuel cell to obtain bioenergy. The energy produced was directly used to operate remote sensor devices. Besides as a result of catalytic oxidation, laccase was produced by *Rhizopus* (Bakar et al. 2012). *Gloeophyllum trabeum* and *Rhizopus microsporus var. rhizopodiformis* fungal strains were utilised for this research. PDA plates were used to grow the strains and to prepare the broth culture. The fungal inoculums were utilised on the 4th day and 8th day following the incubation time. The liquid culture medium of *Gloeophyllum* acts as anolyte, and the liquid culture medium of *Rhizopus* acts as catholyte. MFC was constructed with two cylindrical containers. They were designed with two compartments, inner and outer. The liquid culture of *Gloeophyllum* was filled to the inner, and the liquid culture of *Rhizopus* was filled to the outer compartment. The opening of the inner compartment was attached to a cellulose separator. The power storage of anode chamber was done using the nickel mesh, and the air electrode was used in the cathode which permits a good amount of oxygen to diffuse into the chamber. Incubation of the anolyte in glucose has a greater influence for MFC performance; this case was not observed in the anolyte that was used without the glucose. In this case glucose is a donor of electrons, and they are likely to initialise the polysaccharide oxidation. Hence this research has given a promising result stating that the liquid cultures of fungal strain enhance the power production in the MFC which is enough to operate the remote sensor device.

6.10 Fungal Microbial Fuel Cell for Bioenergy Production

Pleurotus ostreatus is used as a microbial anode which oxidises the organic substances, and the enzymes act as cathode which reduces the oxygen. This method is an approach that degrades the waste water and generates electricity in return. For this research they have employed three different laccase-producing fungal strains. They have used textile dye waste waters in cathode, and anode was employed with urban waste water. Bioenergy was produced; besides, the toxicity levels were reduced in the treated ones compared to the original textile dye effluents. Today high interest has been shown towards the use of MFC to treat the waste waters from the industries and produce energy. The research carried out by C Ottoni et al. (2014) evaluated the performance of MFC by incorporating three different fungal strains that produce the laccase enzyme which was immobilised in the cathode and simultaneously filled with the dye effluents of textiles where as, the urban waste was implemented in the anode. Within the period of 72 h from the process initiation, more than 86% of the textile dye effluents were decolourised, and the power generated was more than 35 mW/m^2. The COD initial values drop down to 90% in 20 days. The toxicity of the textile effluents after treatment was having very low impact compared to the untreated textile dye effluents. Hence this research has laid the foundation for the MFC to be applied in the field of textile industries for waste water treatment and bioenergy production. Similar research was carried out by Ghosh Ray et al. (2017) using *Trichoderma viride*- and *Trichoderma atroviride*-derived peptaibiotics which are active compounds that help in biodegradation of sewage sludge effectively. Trichotoxin and alamethicin are from *Trichoderma viride* whereas neoatroviridin from *Trichoderma atroviride*. Efficient power was generated as a result of treating the inoculum with the mixed sewage sludge. Even after 15 cycles, the inhibition effects on methanogens were maintained, reporting with no effects and hence proving the availability of noncompetitive environment and enhancing the energy production from MFC.

6.11 Future Perspectives and Challenges

The biofuel developments and the applications of the fuel cells are yet in the early stages. Many researches have been carried out to improvise the power production using MFC, and recently the application of fungi is also joining the end list of biofuel cells for power production. High power density is achieved with the heavy biomass of wastes or sludge used in MFC where the catalytic reactions take place and hence the power optimisation is achieved greatly. Combination of the enzymes in biofuel cells has shown the versatility of bacterial- and fungal-based microbial fuel cells. Unfortunately these biofuel cells could be only a temporary source for the demands with respect towards the energy which can be applied only for a short period. Standardisation of the enzyme activities in MFC is a state of the art, and it's very hard to meet such type of devices. Incorporating genetic engineering to modify the fungal strains so as to stabilise the enzymes can be attempted. The compatibility of the device has been another problem. Many researches have to be carried in the future to stabilise and standardise the protocols and construct devices that are very simple and cost-effective. Generally, production of enzymes are too costly. By using the fungal strains in fuel cell, different enzymes are obtained as by-product in between the process hence, upcoming researches have to focus and try different fungal strains and develop novel technologies to produce enzymes in a cheaper way which can also be standardised in the future will mark milestones for the industries producing various kinds of enzymes.

6.12 Conclusion

So as to meet the demands of energy resources, fungal fuel cells will be a new horizon for the researchers working to find an alternative technology that could produce or be a source of energy. Different kinds of fungi species incorporated with the fuel cells in different modes have been discussed in the above chapter giving a clear vision about the fungi biofilm applications and the fungal pure cultures applied to the anode and cathode. The conventional modified biofuel cells with fungi have been distinctively used for bioremediation, biodegradation and bioenergy production. Besides efforts have to be initialised to apply these fungal fuel cells in large-scale industries. The most common problems that are faced by most of the researchers are the stability and the density of the power produced. Researchers must have an in-depth and detailed knowledge about the process of bio catalysis and the principle behind the process that takes place in the electrode surface by the electrons. Researching on the basic prinicples of working would bring emerging changes to develop new technologies using Fungal fuel Cells.

References

April MT, Abbott PS, Foght MJ, Currah RS (1998) Degradation of hydrocarbons in crude-oil by the ascomycete *Pseudallescheria boydii* (microascaceae). Can J Microbiol 44:270–278

Bakar ASA, Othman R, Yahya MZA, Othman R, Nik Din NMS (2012) Bioenergy from *Gloeophyllum-Rhizopus* fungal biofuel cell. Adv Mater Res 512-515:1461–1465

Bakhshian S, Kariminia HR (2011) Development of laccase and manganese peroxidase biocathodes for microbial fuel cell

Bannwarth B (2006) Acetaminophen or NSAIDs for the treatment of osteoarthritis. Res Clin Rheumatol 20:117–129

Basha S, Keane D, Nolan K, Oelgemo¨ller M, Lawler J, Tobin JM, Morrissey A (2015) UV-induced photo catalytic degradation of aqueous acetaminophen: the role of adsorption and reaction kinetics. Environ Sci Pollut Res 22:2219–2230

Beltrán FJ, Álvarez PM, Rodríguez EM, García-Araya JF, Rivas J (2001) Treatment of high strength distillery wastewater (cherry stillage) by integrated aerobic biological oxidation and ozonation. Biotechnol Prog 17(3):462–467

Benotti MJ, Trenholm RA, Vanderford BJ, Holady JC, Stanford BD, Snyder SA (2008) Pharmaceuticals and endocrine disrupting compounds in U.S. drinking water. Environ Sci Technol 43:597–603

Bond DR, Lovely DR (2003) Electricity production by Geobacter sulfurreducens attached to electrodes. Appl Environ Microbiol 69:1548–1555

Clauben M, Schmidt S (1998) Biodegradation of phenol and pcresol by the hyphomycete Scedosporium apiospermum. Res Microbiol 149:399–406

de Dios MÁF, del Campo AG, Fernández FJ (2013) b,Manuel Rodrigo, Marta Pazos, María Ángeles Sanromán. Bio Resour Technol 148:39–46

Du Z, Li H, Gu T (2007) A state of the art review on microbial fuel cells: a promising technology for wastewater treatment and bioenergy. Biotechnol Adv 25:464–482

Escher BI, Baumgartner R, Koller M, Treyer K, Lienert J, Mc Ardell CS (2011) Environmental toxicology and risk assessment of pharmaceuticals from hospital wastewater. Water Res 45:75–92

Ghosh Ray S, Ghangrekar MM (2015) Enhancing organic matter removal, biopolymer recovery and electricity generation from distillery wastewater by combining fungal fermentation and microbial fuel cell. Bioresour Technol 176:8–14

Gilgado F, Cano J, Gene´ J, Serena C, Guarro J (2009) Different virulence of the species of the Pseudallescheria boydii complex. Med Mycol 47(4):371–374

González T, Terrón MC, Yagüe S, Zapico E, Galletti GC, González AE (2000) Pyrolysis/gas chromatography/mass spectrometry monitoring of fungal bio treated distillery wastewater using Trametes sp. I-62 (CECT 20197). Rapid Commun Mass Spectrom 14(15):1417–1424

Heidrich ES, Edwards SR, Dolfing J, Cotterill SE, Curtis TP (2014) Performance of a pilot scale microbial electrolysis cell fed on domestic wastewater at ambient temperatures for a 12 month period. Bioresour Technol 173:87–95

Ivanov I, Vidakovic´-Koch T, Sundmacher K (2010) Recent advances in enzymatic fuel cells: experiments and modeling. Energies 3:803–846

Jana PS, Behera M, Ghangrekar MM (2010) Performance comparison of up-flow microbial fuel cells fabricated using proton exchange membrane and earthen cylinder. Int J Hydrog Energy 35(11):5681–5686

Kannan KP, George TS, Guru KSS, Vasanthi NS (2011) Extraction, purification and characterization of chitosan from endophytic fungi isolated from medicinal plants. World J Sci Technol 1(4):43

Kim SD, Cho J, Kim IS, Vanderford BJ, Snyder SA (2007) Occurrence and removal of pharmaceuticals and endocrine disruptors in south Korean surface, drinking, and waste waters. Water Res 41:1013–1021

Kipf E, Koch J, Geiger B, Erben J, Richter K, Gescher J, Zengerle R, Kerzenmacher S (2013) Systematic screening of carbon-based anode materials for microbial fuel cells with Shewanella oneidensis MR-1. Bio Resour Technol 146:386–392

Kloke A, Rubenwolf S, Bcking C, Gescher J, Kerzenmacher S, Zengerle R, von Stetten F (2010) Biosens Bioelectron 25:2559–2565

Lackner M, Najafzadeh MJ, Sun J, Lu Q, Sybren de Hoog G (2012) Rapid identification of Pseudallescheria and Scedosporium strains by using rolling circle amplification. Appl Environ Microbiol 78(1):126–133

Logan BE, Regan JM (2006) Electricity-producing bacterial communities in microbial fuel cells. Trends Microbiol 14:512–518

Maghsoodi et al (2009) Production of chitosan by submerged fermentation from Aspergillus niger. Sci Iran 16(2):145–148

Mazloum-Ardakani M, Sabaghian F, Khoshroo A, Abolhasani M, Naeimi H (2015) Electrochemical determination of captopril in the presence of acetaminophen, tryptophan, folic acid, and Lcysteine at the surface of modified carbon nanotube paste electrode. Ionics 21:239–250

Mbokou FS, Pontie´ M, Razafimandimby B, Bouchara J-P, Njanja E, Tonle KI (2016) Evaluation of the degradation of acetaminophen by the filamentous fungus Scedosporium dehoogii using carbon base modified electrodes. Anal Bioanal Chem 408:5895–5903

Mbokou SF, Tonle IK, Pontie M (2017) Development of a novel hybrid biofuel cell type APAP/O2 based on a fungal bioanode with a *Scedosporium dehoogii* biofilm. J Appl Electrochem 47(2):273 (8)

Morant KV, da Silva PH, de Campos-Takaki GM, La Rotta Hernández CE (2014) Isolation and bioelectrochemical characterization of novel fungal sources with oxidasic activity applied in situ for the cathodic oxygen reduction in microbial fuel cells. Enzym Microb Technol 66:20–27

Nataraj SK, Hosamani KM, Aminabhavi TM (2006) Distillery wastewater treatment by the membrane-based nano filtration and reverse osmosis processes. Water Res 40(12):2349–2356

Ottoni C, Peixoto L, Maiorano AE, Brito AG (2014) Use of filamentous fungi for improving electricity production and textile dye treatment in a microbial fuel cell. Center for Biological Engineering

Pant D, Bogaert GV, Diels L, Vanbroekhoven K (2010) A review of the substrates used in microbial fuel cells (MFCs) for sustainable energy production. Bioresour Technol 101:1533–1543

Potter MC (1911) Electrical effects accompanying the decomposition of organic compounds. Proc R Soc Lond 84:260–276
Ray G, Noorib MDT, Ghangrekar MM (2017) Novel application of peptaibiotics derived from Trichoderma sp for methanogenic suppression and enhanced power generation in microbial fuel cells. RSC Adv 7:10707–10717
Sané S, Jolivalt C, Mittler G, Nielsen PJ, Rubenwolf S, Zengerle R, Kerzenmacher S (2013) Overcoming bottlenecks of enzymatic biofuel cell cathodes: crude fungal culture supernatant can help to extend lifetime and reduce cost. ChemSusChem 6(7):1209–1215
Sanromán MÁ et al (2013) Bacterial–fungal interactions enhance power generation in microbial fuel cells and drive dye decolourisation by an ex situ and in situ electro-Fenton process. Bioresour Technol 148:39–46
Satyawali Y, Balakrishnan M (2008) Wastewater treatment in molasses-based alcohol distilleries for COD and color removal: a review. J Environ Manag 86(3):481–497
Ternes TA (1998) Occurrence of drugs in German sewage treatment plants and rivers. Water Res 32:3245–3260
Van Leeuwen J, Khanal SK, Pometto AL, Rasmussen ML, Mitra D (2013) Fungi cultivation on alcohol fermentation stillage for useful products and energy savings. Google Patents
Verstraete W, Morgan-Sagastume F, Aiyuk S, Rabaey K, Waweru M, Lissens G (2005) Anaerobic digestion as a core technology in sustainable management of organic matter. Water Sci Technol 52:59–66
Visvanathan C, Aim RB, Parameshwaran K (2000) Membrane separation bioreactors for wastewater treatment. Crit Rev Environ Sci Technol 30(1):1–48
Wang M, Yan Z, Huang B, Zhao J, Liu R (2013) Electricity generation by microbial fuel cells fuelled with Enteromorpha prolifera hydrolysis. Int J Electrochem Sci 8:2104–2111
Waterston K, Weijun Wang J, Bejan D, Bunce NJ (2006) Electrochemical waste water treatment: electrooxidation of acetaminophen. J Appl Electrochem 36:227–232
Westerhoff P, Yoon Y, Snyder S, Wert E (2005) Fate of endocrine-disruptor, pharmaceutical, and personal care product chemicals during simulated drinking water treatment processes. Environ Sci Technol 39(17):6649–6663
Wilkie AC, Riedesel KJ, Owens JM (2000) Stillage characterization and anaerobic treatment of ethanol stillage from conventional and cellulosic feedstocks. Biomass Bioenergy 19(2):63–102
Wilkinson S (2000) "Gastrobots"—benefits and challenges of microbial fuel cells in food powered robotapplications. Auton Robot 9:99–111
Willner I (2002) Biomaterials for sensors, fuel cells, and circuitry. Science 298:2407–2408
Wu C, Liu X-W, Li W-W, Sheng G-P, Zang G-L, Cheng Y-Y, Shen N, Yang Y-P, Yu H-Q (2012a) Applications white-rot fungus is used as a bio cathode to improve electricity production of a microbial fuel cell. Appl Energy 98:594–596
Wu S, Zhang L, Chen J (2012b) Paracetamol in the environment and its degradation by microorganism. Appl Microbiol Biotechnol 96:875–884
Yoon Y, Ryu J, Oh J, Choi BG, Snyder SA (2010) Occurrence of endocrine disrupting compounds, pharmaceuticals, and personal care products in the Han River (Seoul, South Korea). Sci Total Environ 408:636–646

Chapter 7
Bioelectricity Generation in Soil Microbial Fuel Cells Using Organic Waste

Kiyoshi Omine, Venkataraman Sivasankar, and Santos D. Chicas

7.1 Introduction

Recently, the amount of waste materials has been decreased due to recycling. However a large amount of organic wastes has been disposed at final landfill site by incineration process. To resolve resource and environmental issues such as global warming and depletion of fossil fuels, suppressing as much as possible the dependence on fossil fuels is important. The annual organic waste generated from the food industries and kitchen garbage in Japan is about 20 million tons per year (Koike et al. 2009). Most of these wastes are directly incinerated with other combustible waste, and the residual ash is disposed of in landfills. However, incineration of this water-containing waste is energy consuming. Microbial fuel cell (MFC) functions to primarily harness bioelectricity through microbial redox reactions and is gaining prominence due to its sustainable applications in multiple domains (Swathi et al. 2018).

Microbial fuel cells (MFCs) are devices that exploit microorganisms to generate electricity from a variety of reduced materials, including organic matter (Logan et al. 2006). MFCs have the prime function of harnessing the bioelectricity through microbial redox reactions and exhibit the multifaceted applications with environmental sustainability. There are several researches on electricity generation of MFCs from organic wastes or wastewaters that are conducted all over the world in the era of green energy generation for sustainable environment and future generation (Bennetto 1990; Miyahara et al. 2016). Researchers have also used MFCs to recover electricity from marine sediments (Reimers et al. 2001) and rice paddy

K. Omine (✉) · V. Sivasankar
Geo-tech Laboratory, Department of Civil Engineering, Graduate School of Engineering, Nagasaki University, Nagasaki, 8528521, Japan
e-mail: omine@nagasaki-u.ac.jp

S. D. Chicas
Faculty of Sciences and Technology, University of Belize, Belmopan Central Campus, Cayo District CA, Belize

V. Sivasankar et al. (eds.), *Microbial Fuel Cell Technology for Bioelectricity*,
https://doi.org/10.1007/978-3-319-92904-0_7

fields (Kaku et al. 2008). MFC in hybrid composting method by reusing the kitchen garbage as a raw material is also proposed (Moqsud et al. 2010, 2013).

In this study, a soil microbial fuel cell (SMFC) that generates electricity by the biodegradation of organic matter is developed. Influences of mixing materials and conditions of electrodes in the SMFC are investigated. A performance of the soil microbial fuel cell by composting under anaerobic condition is discussed based on the experimental results.

7.2 Test Materials and Methods

Generally microbial fuel cell is used under conditions of aerobic cathode with air and anaerobic anode in wastewater. Proton exchange membrane is also used as a separator between the cathode and anode in the microbial fuel cell. In this study, a new type of MFC with compost of organic wastes is developed. Cutting grass (organic waste), leaf mould and rice bran were mixed together with water and photosynthetic bacteria for promoting the fermentation process.

Activated bamboo carbon was used for anode and cathode. The schematic diagram of the experimental device of the SMFC is shown in Fig. 7.1. A plastic container (100 × 70 × 50 mm) is used as a cell. Then cutting grass, leaf mould and rice bran together with water and photosynthetic bacteria are blended properly and filled in the container and shown in Fig. 7.2. Test conditions of SMFC in different mixing ratio of the samples are shown in Table 7.1. For a purpose of increasing a performance of SMFC, anode with iron wire is used. SMFC test in a different electrode distance is also performed under the condition (Case E-1, E-2 and E-6). The anode is inserted into the sample, and the cathode is placed on a surface of the sample. Both the anode and cathode are related to a data logger. A filter paper is used to separate the anode and cathode. SMFC test is performed by wrapping the container up in plastic film. Aerobic or anaerobic condition was applied in the test. In aerobic condition, small holes on the surface of plastic film were opened using pin. In anaerobic condition, the container was sealed up without hole. The data logger is set to

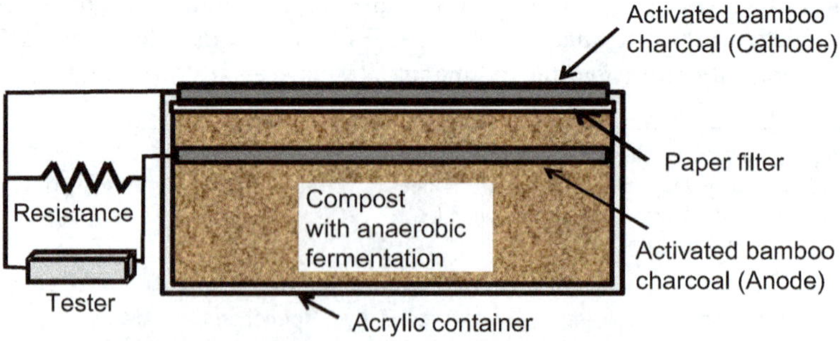

Fig. 7.1 Experimental device of the SMFC with compost of organic wastes

Fig. 7.2 Mixed constituents and electrode materials used in the SMFC study

measure the voltage in every 10 min' interval. The laboratory test is conducted in a constant room temperature of 30 °C.

Electrode output is measured in volts (V) against time. The current I in amperes (A) is calculated using Ohm's law, $I = V/R$, where V is the measured voltage in volts (V) and R is the known value of the external load resistor in ohms. From this it is possible to calculate the electric power output P in watts (W) of the SMFC by taking the product of the voltage and current, i.e. $P = I \times V$. For obtaining a maximum power of SMFC, values of voltage are measured using three different resistances (10, 100 and 1 kΩ).

7.3 Results and Discussion

7.3.1 Influence of Leaf Mould

In the mixture blend of organic waste, rice bran, photosynthetic bacteria and leaf mould, each constituent plays a vital role towards the generation of electricity as a consequence of biochemical process. With the cutting grass as organic waste, the mixed rice bran functions as a fermenting and nutritious material towards the multiplication of microorganism. The photosynthetic bacteria make its participation to promote the fermentation process under anaerobic condition. Leaf mould is a product of slow decomposition of deciduous shrub and tree leaves. It is also a form of compost produced primarily by fungal breakdown and is retentive of water and fertiliser. In order to investigate the influence of leaf mould from our experimental cases (A-1 to A-6), the amount of leaf mould was varied from 20 g to 120 g by keeping the amount of the other materials constant with the addition of adequate volume of water.

The relationship between voltage and elapsed time on the SMFCs at different mixed amounts of leaf mould during the time period of 48 h is represented in Fig. 7.3. The graphs illustrated that the SMFCs in cases A-1, A-2 and A-6 keep relatively high voltage. On the other hand, the voltage of other SMFCs in cases A-3, A-4 and A-5 was rather unstable and discrete. The recorded voltage of more than 0.4 V

Table 7.1 Test conditions of SMFC at different mixing ratio of samples

Case	Cuttinggrass (g)	Ricebran (g)	Leafmould (g)	Photosyntheticbacteria (g)	Water (g)	Anaerobic/ aerobic condition	Electrode distance (mm)	Iron wire
A-1	20	10	20	100	0	Anaerobic	30	Not used
A-2	20	10	40	100	20	Anaerobic	30	Not used
A-3	20	10	60	100	40	Anaerobic	30	Not used
A-4	20	10	80	100	60	Anaerobic	30	Not used
A-5	20	10	100	100	80	Anaerobic	30	Not used
A-6	20	10	120	100	100	Anaerobic	30	Not used
B-1	20	10	20	0	100	Anaerobic	30	Not used
B-2	20	10	40	0	120	Anaerobic	30	Not used
B-6	20	10	120	0	200	Anaerobic	30	Not used
C-1	20	0	20	100	0	Anaerobic	30	Not used
C-2	20	0	40	100	20	Anaerobic	30	Not used
C-6	20	0	120	100	100	Anaerobic	30	Not used
D-1	20	10	20	100	0	Aerobic	30	Not used
D-2	20	10	40	100	20	Aerobic	30	Not used
D-6	20	10	120	100	100	Aerobic	30	Not used
E-1	20	10	40	100	20	Anaerobic	10	Not used
E-2	20	10	40	100	20	Anaerobic	30	Not used
E-6	20	10	40	100	20	Anaerobic	50	Not used
F-1	20	10	20	100	0	Anaerobic	30	Used
F-2	20	10	40	100	20	Anaerobic	30	Used
F-6	20	10	120	100	100	Anaerobic	30	Used

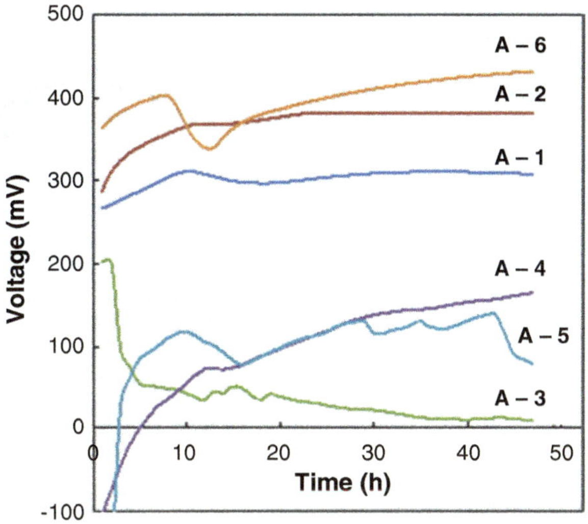

Fig. 7.3 Voltage: time graph of SMFCs at different leaf mould proportions

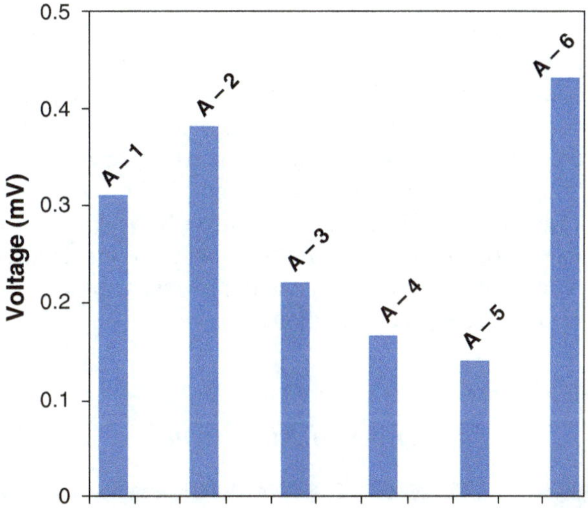

Fig. 7.4 Maximum voltage of the SMFCs at different leaf mould proportions for the time of 48 h

in 48 h for A-6 was the highest among the other cases (Fig. 7.4) with 120 g of leaf mould. However, the influence of leaf mould in voltage generation remains unclear.

7.3.2 *Influence of Photosynthetic Bacteria*

Photosynthetic bacteria have been used for the treatment of various wastewater and biodegradable solids (Choi et al. 2002). These bacteria have a relatively simple nutritional requirement and can grow actively, regardless of the oxygen diffusion

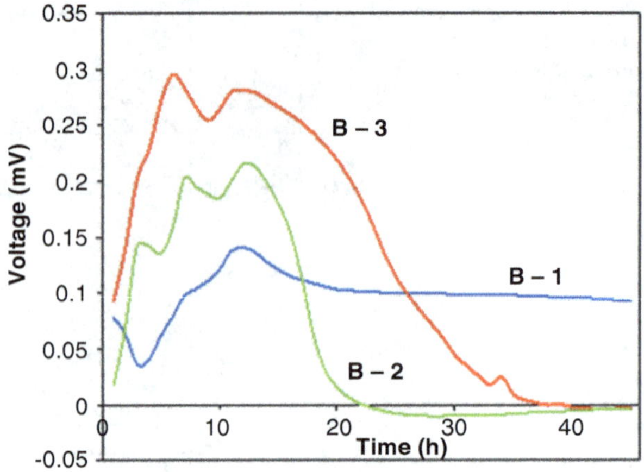

Fig. 7.5 Voltage: time graph for the SMFCs without photosynthetic bacteria

rate under aerobic/anaerobic conditions in the light or aerobic conditions in the dark.

In order to investigate the influence of photosynthetic bacteria, samples without photosynthetic bacteria in different mixing proportions of leaf mould were prepared as shown in cases B-1, B-2 and B-6. Figure 7.5 depicts the relationship between voltage and elapsed time for the SMFCs without photosynthetic bacteria for 45 h. In the case of B-2 and B-6, the maximum voltage generation in the range of 0.2–0.3 V could be studied up to 12 h of time which later decreased at a faster rate of 0.0116–0.0204 Vh^{-1} and declines at the end of 37 h and 23 h, respectively. Unlike the above cases, B-1 could be recorded with 0.14 V as the maximum initially and gradually falls to 0.1 V at the end of 20 h and then remained consistent. The significance of photosynthetic bacteria in the improved performance of SMFCs could be envisaged from the present observations.

7.3.3 Influences of Rice Bran

Many types of raw materials are effectively used as organic fertiliser (Sethuraman and Naidu 2008), and rice bran is among them. As a by-product of the rice milling industry, rice bran constitutes about 10% of the rough rice by its weight. It has the primary composition of aleurone, pericarp and subaleurone layer and germ and is a rich source of vitamins, minerals, essential fatty acids, dietary fibre and other sterols. Based on the above, it has been considered as a good fermentation material under aerobic and anaerobic conditions. Due to the sufficiently contained nutrients in rice bran, it enables the microorganisms for the electricity generation both in pure and mineral water as studied by Takahashi et al. (2016).

The influential characteristics of rice bran could be studied on comparing the SMFC conditions in the absence and presence of rice bran as denoted in cases C-1,

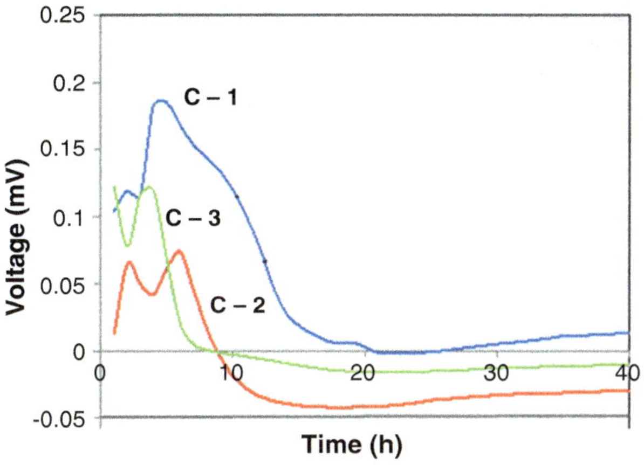

Fig. 7.6 Voltage: time graph for the SMFCs without rice bran

C-2 and C-6. It could be well illustrated from Fig. 7.6 that the SMFC system in absentia of rice bran was measured with the potential raised to 0.18 V, 0.07 V and 0.12 V for C-1, C-2 and C-6, respectively, during the initial hours. Later the cell voltage began to fall very quickly to zero at the end of 8 h for C-2 and C-6, but a comparatively lesser declining rate in C-1 was observed which became zero after 20 h with a residual amount of leaf mould. It is quite explicable that the output potential of SMFCs in the presence of rice bran was higher and consistent throughout the time period of 48 h as compared to the insubstantiality of voltage rate and the prolonged output during its absence. Hence the significance of rice bran towards the influence of potential in SMFCs becomes evident on the basis of the recorded results.

7.3.4 Influences of Aerobic Condition

The aerobic and anaerobic conditional environments are quite significant to drive the performance of SMFCs. It is rather decisive to maintain the aerobic and anaerobic maintenance of cathode and anode, respectively. Many types of membranes such as cation-exchange membranes, anion-exchange membranes, polymer/composite membranes and porous membranes have usually been applied in MFC systems which further extended to membraneless technology (Leong et al. 2013). In the present MFC setup, the separation of electrodes has been attempted with filter paper as it is deemed to be cost-effective. But at the same time, the appropriate ambience of the electrodes was failed from its adoption. The probable factor was that the anaerobically conditioned anode becomes oxidized feasibly on separating the aerobically set cathode with the filter paper.

On investigating the preponderance of aerobic condition, MFCs resembling the cases D-1, D-2 and D-6 were constructed with cathode set at defined leaf mould proportions.

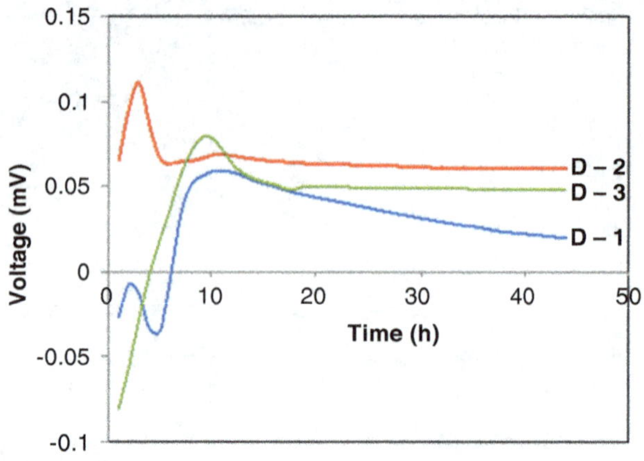

Fig. 7.7 Voltage–time graph for the SMFCs with cathode in the aerobic condition

Although the voltage–time graph (Fig. 7.7) for cases D1, D2 and D6 illustrated with curve dips and peaks initially, the potential was consistent with 0.07 V and 0.05 V for the prolonged time of 44 h for D-2 and D-3 cases, respectively. However case D-1 was observed with very gradual decrease in potential down to 0.02 V during the time period of 44 h. Evidently, the potential generated under anaerobic condition was several times higher than the potential generated under aerobic condition. The anaerobically conditioned cases such as A-1, A-2 and A-6 were greater in the consistent potential by 16, 3.4 and 8.4 times than the aerobically conditioned D-1, D-2 and D-6 cases during the time period of 44 h. It could be considered that the aerobic condition of cathode is suitable for this type of SMFCs with the priority of wrapping the container up using a plastic film to facilitate a perfect anaerobic (anode) and aerobic (cathode) conditions in SMFCs.

7.3.5 Influence Due to the Distance Between the Electrodes

The distance between anode and cathode, it plays a significant role in deciding the performance of SMFC as it affects the diffusion of protons from anode to cathode.

The anode/cathode distance is known to influence MFC performance, since it affects proton diffusion from anode to cathode (Cheng et al. 2006). Accordingly, cases E-1, E-2 and E-6 with reference to Table 7.1 were executed by mimicking the conditions of case A-2 with the distance between the electrodes as a variable. The voltage–time relationship is represented in Fig. 7.8a where the measurement of open-circuit voltage (equivalent to electromotive force) was carried out under no external load condition against time for the SMFC performance for 48 h. In the case of open-circuit MFCs, the potential of an anode becomes more negative, but on reconnecting the circuit, it tends to become less negative which ultimately results in greater power output and lower energy capture by bacterial organisms (Logan 2009).

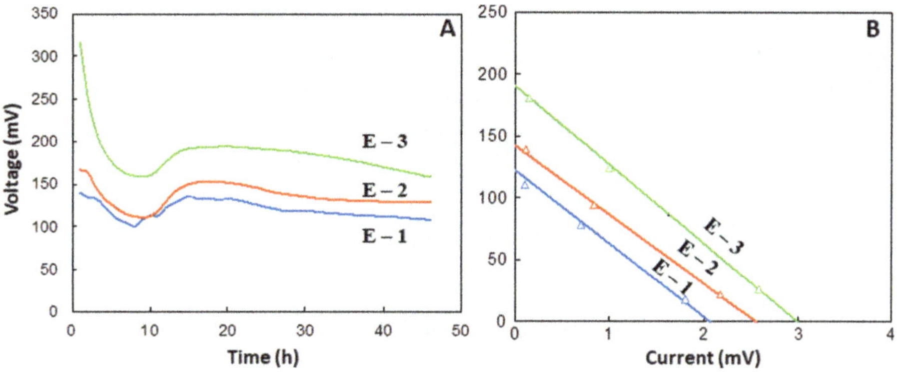

Fig. 7.8 Influence of the distance between electrodes: open-circuit voltage–time graph (**a**) polarization profile of SMFC (**b**)

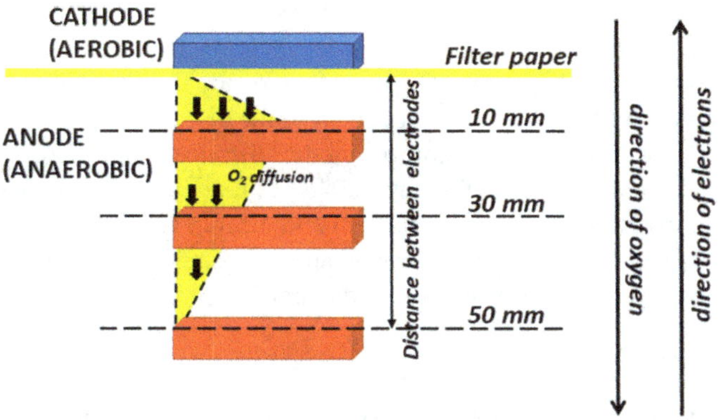

Fig. 7.9 Pictorial representation of the distance between electrodes and oxygen diffusion

The voltage–time graphs were observed to be identical in patterns by exhibiting a voltage drop from a maximum to a minimum followed by an increase which later attains constancy. The attainment of voltage constancy of E-1, E-2 and E-6 was measured with 0.12 V, 0.13 V and 0.17 V, respectively, and hence the cases were in the order: E-6 > E-2 > E-1. These observations revealed that the voltage was directly proportional to the distance between the electrodes and associated probably with the minimum oxygen diffusion towards the anode. Due to the diffusion of oxygen towards the anode, the possible reduction of oxygen molecule into oxide takes place (Eq. 7.1). Oxygen tends to undergo reduction by accepting electrons generated from the anode as its standard electrode potential is positive (Eq. 7.2). As the depth of anode is progressing higher, a highly anaerobic and oxygen-restricted condition prevails which facilitates the flow of electrons towards increasing the power rather than their consumption due to reduction by oxygen species (Fig. 7.9):

Table 7.2 Performance of SMFCs for varied distance between electrodes at 24 h

Electrical parameters	E-1	E-2	E-6
Electromotive force (mV)	122	142	191
Internal resistance (Ω)	58.76	55.47	63.66
Maximum electric power per area of anode (mW/m^2)	14.1	20.2	31.8

$$mO_2 + (4n+4)e^- \rightarrow 2mO^{n-} \; (m = 1,2,3,\ldots \text{and}\, n = 0,1,2,\ldots) \quad (7.1)$$

$$O_2(g) + 4H^+(aq) + 4e^- \rightleftharpoons 2H_2O(l); E^0 = +1.23\,V \quad (7.2)$$

Figure 7.8b shows the polarization profiles plotted between the voltage and current obtained for SMFCs E-1, E-2 and E-6 with 24 h of elapsed time. The curves exhibited a linearity in all the three cases where the slope and the intercept represent the internal resistance and electromotive force, respectively. The electromotive force was found to ascend from 122 mV to 191 mV proportionally to the increasing distance between the electrodes. On the other hand, the internal resistance was almost the same and found to be independent to the distance as shown in Table 7.2. It can be inferred that the extension of anodic distance multiplied the maximum electric power per anodic area of about 0.0045 m^2. It could also be conceivable with the fact that the depth variations of 20 mm and 40 mm from the cathode decrease the rate of oxygen diffusion towards anode and hence resulted in an increased voltage and maximum electric power per anodic area.

7.3.6 *Influence of Anode Modified with Iron Winding*

The role of anodic component is noteworthy as it is the primary source for electrons in MFCs. Recently, researchers intend to undertake experiments on MFCs with modified anodes with certain organic and inorganic compounds. The results established the enhanced performance of MFCs and proved that the modifications could improve the potential with economically feasible route (Hindatu et al. 2017; Sonawane et al. 2017).

The voltage–time profiles in Fig. 7.10a and b depict the prolonged voltage consistency between cases A and F. These two assembled SMFC cases are one and the same but with modified anodic parts, i.e. with (F cases) and without iron winding (A cases). Unlike the linear profiles of A (1, 2 and 6) cases, F (1, 2 and 6) cases appeared with initial voltage fluctuations. In cases F-1 and F-6, the consistency of output voltage was recorded after 12 h whereas in F-2 it was attained early after 5 h. The SMFC obtained voltage of F-1, F-2 and F-6 was 1.81, 1.84 and 1.88 times higher than the corresponding A-1, A-2 and A-6 cases. The remarkable raise in the voltage of SMFCs as a cause of synergistic effect in F cases could be substantiated with the modification by iron wire winding in the anode. The contribution of Fe oxidation

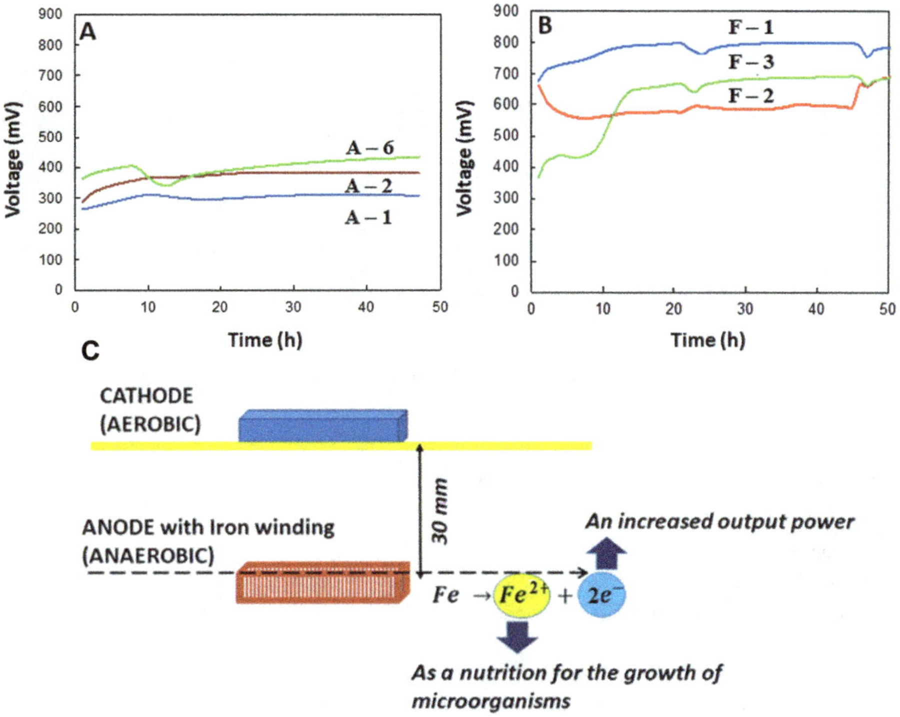

Fig. 7.10 Voltage – time graph for SMFCs: with iron – winding (**a**) without iron – winding (**b**) synergistic effect: modified anode (with iron winding) in voltage generation (**c**)

also played a prominent role in the generation of voltage from the cell. The standard reduction potential (E^0) for iron is 0.44 V (Eq. 7.3):

$$Fe^{2+}(aq) + 2e^- \rightleftharpoons Fe(s) \tag{7.3}$$

The oxidation of Fe to Fe (II) could be feasible as the standard potential value is more negative to drive the process. During the process, the loss of electron contributes for the potential increase along with the release of Fe (II) which facilitates the growth of microorganisms as a nutrient as shown in Fig. 7.10c.

The polarization profiles for F-1, F-2 and F-3 cases were linear for the time of 24 h as shown in Fig. 7.11. Among the three curves, F-1 and F-2 appear with a meagre difference of 12.82 Ω in internal resistance, but F-6 recorded with a lower value of 108.23 Ω.

Even though the electromotive force was higher for F-1 (800 mV), the role of internal resistance decreased the maximum power to 146.7 $mW.m^{-2}$. On the other hand, in F-6, the lower electromotive force of 671 mV was able to generate the power of 231.1 $mW.m^{-2}$ due to lesser internal resistance of 108.23 Ω. In F-2, the paradoxical influence due to the lower electromotive force and higher internal

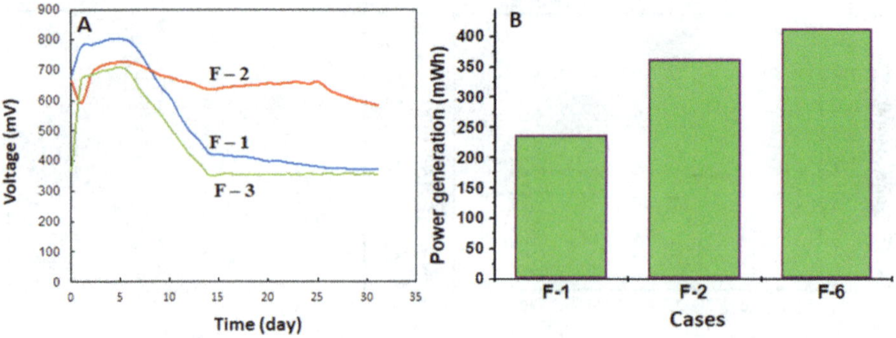

Fig. 7.11 SMFCs with iron wound anode for 31 days: voltage–time graph (**a**) maximum power generation (**b**)

Table 7.3 Test results of the SMFCs in F cases for 24 h

Electrical parameters	F-1	F-2	F-6
Electromotive force (mV)	800	574	671
Internal resistance (Ω)	242.42	229.60	108.23
Maximum electric power per area of anode (mW/m^2)	146.7	79.7	231.1

resistance, could generate the lowest power of 79.7 $mW.m^{-2}$. Hence the impact of multiplied addition (six times of F-1 and three times of F-2) of leaf mould ought to be the reason behind the decreased resistance to drive more output power from the SMFC system as given in Table 7.3.

7.3.7 Power Generation

The low-power output of SMFCs seems to be one of the major issues towards upscaling and practicable applications. Although the electric power of the SMFC is small, it is expected that the power generation will continue for a long period. Also it would be possible to increase the power by connecting several SMFCs in series and parallel.

Figure 7.11a illustrated the voltage variation every day for F cases. The voltage was measured under no external load condition (open-circuit), equivalent to an electromotive force. Remarkably, high voltage in these cases of SMFC continued for a long term more than a month of time period.

Case F-2 with the leaf mould quantity of 40 g was measured with higher voltage in the range of 560–670 mV from the 12th day of the total time period (31 days). On the other hand, in cases F-1 and F-6, the voltage was dropped down from the 14th day around 350 mV. The maximum recorded voltages for F-2 and F-6 cases are about 800 mV (3–7 days) and 700 mV (1–5 days), respectively. When a resistance is connected in the circuit as an external load, the electric power is calculated as a product of voltage and current. Assuming the values of internal resistance in Table 7.3 to be constant, the maximum power generation of respective SMFCs was calculated. Electric energy of the SMFCs was estimated by the integration of the power generation and elapsed time. The maximum electrical energy of the SMFCs generated in cases F-1, F-2 and F-6 during 31 days is represented in Fig. 7.11b. The electric power generated from SMFC of F-6 case was higher, and the value of 412.6 mWh equals the electric capacity of a small-sized dry battery. The electric power generated is of the following order: F-6 > F-3 > F-1. As the electric power of SMFC continued to generate more than a month, the ultimate solid residue can be used as a compost once the power generation is exhausted.

7.4 Conclusions

Soil microbial fuel cell that generates electricity through organic biodegradation was developed. The mixture of cutting grass, leaf mould and rice bran together is added with photosynthetic bacteria for promoting fermentation. Main conclusions drawn from the experimental cases are as follows:

1. The electromotive force of SMFCs assembled using anode made from activated bamboo charcoal was almost doubled due to modification of anode winding by iron wire. The maximum electric power density was recorded with 231.1 mW. m^{-2} for the modified anode.
2. The performance of SMFC was proportional to the distance between the electrodes. It was ascertained that the restricted oxygen diffusion at greater anodic depth was favourable for electronic ejection rather from anode than reduction of oxygen.
3. The SMFC with Fe wire-wound anode continued to generate the output power of 412.6 mWh during 31 days of time period that was equivalent to that of an electric capacity of a small-sized dry battery.
4. Apart from the bioenergy output, the residual solid remaining in the exhausted SMFC can be utilized as a compost in an agricultural field. Future studies would be aimed at the maximization of power by adopting SMFC modifications with respect to anode, different organic waste and proportions and series and parallel SMFC connections.

Acknowledgement The authors thank Nagasaki University, Nagasaki, Japan.

References

Bennetto HP (1990) Electricity generation by micro-organisms. Biotechnol Educ 1(4):163–168
Cheng S, Liu H, Logan BE (2006) Increased power generation in a continuous flow MFC with advective flow through the porous anode and reduced electrode spacing. Environ Sci Technol 40:2426–2432

Choi H-P, Kang H-J, Seo H-C, Sung H-C (2002) Isolation and identification of photosynthetic bacterium useful for wastewater treatment. J Microbiol Biotechnol 12(4):643–648

Hindatu Y, Annuar MSM, Gumel AM (2017) Mini-review: anode modification for improved performance of microbial fuel cell. Renew Sust Energ Rev 73:236–248

Koike Y, An M, Tang Y, Syo T, Osaka N, Morimura S, Kida K (2009) Production of fuel ethanol and methane from garbage by high-efficiency two-stage fermentation process. J Biosci Bioeng 108:508–512

Kaku N, Yonezawa N, Kodama Y, Watanabe K (2008). Plant/microbe cooperation for electricity generation in a rice paddy field. Appl Microbiol Biotechnol 79:43–49. https://doi.org/10.1007/s00253-008-1410-9.

Leong JX, Dauda WRW, Ghasemi M, Liew KB, Ismail M (2013) Ion exchange membranes as separators in microbial fuel cells for bioenergy conversion: a comprehensive review. Renew Sust Energ Rev 28:575–587

Logan BE, Hamelers B, Rozendal R, Schröder U, Keller J, Freguia S, Aelterman P, Verstraete W, Rabaey K (2006) Microbial fuel cells: methodology and technology. Environ Sci Technol 40:5181–5192

Logan BE (2009) Exoelectrogenic bacteria that power microbial fuel cells. Nat Rev Microbiol 7:375–381

Moqsud MA., Omine K, Hayashi S (2010) Microbial fuel cells by using kitchen garbage in hybrid composting method. Proceedings of the 5th civil engineering conference in the Asian region and Australasian structural engineering conference, Sydney

Moqsud MA, Omine K, Yasufuku N, Hyodo M, Nakata Y (2013) Microbial fuel cell (MFC) for bioelectricity generation from organic wastes. Waste Manag 33:2465–2469

Miyahara STM, Kouzuma A, Watanabe K (2016) Electricity generation from rice bran in microbial fuel cells. Bioresour Bioprocess 3:50. https://doi.org/10.1186/s40643-016-0129-1

Reimers CE, Tender LM, Fertig S, Wang W (2001) Harvesting energy from the marine sediment-water interface. Environ Sci Technol 35:192–195

Sethuraman G, Naidu S (2008) Soils and composts. In: International encyclopaedia of agricultural science and technology, vol 4. Mittal Publications, New Delhi

Takahashi S, Miyahara M, Kouzuma A, Watanabe K (2016) Electricity generation from rice bran in microbial fuel cells. Bioresour Bioprocess 3: 1–5

Sonawane JM, Yadav A, Ghosh PC, Adeloju SB (2017) Recent advances in the development and utilization of modern anode materials for high performance microbial fuel cells. Biosens Bioelectron 90:558–576

Swathi K, Sarkar O, Butt SK, Velvizhi G, Venkat Mohan S (2018) Stacking of microbial fuel cells with continuous mode operation for higher bioelectrogenic activity. Bioresour Technol 257:210–216

Chapter 8
Microbial Fuel Cell Research Using Animal Waste: A Feebly-Explored Area to Others

Deepika Jothinathan, Nasrin Fathima A. H., Prabhakaran Mylsamy, L. Benedict Bruno, and Venkatraman Sivasankar

8.1 Introduction

A fuel cell is an electrochemical cell that converts chemical energy from the fuel supplied to it into electricity, through an electrochemical reaction of hydrogen fuel with oxygen or another oxidizing agent. In contrast to batteries, fuel cells need a constant supply of fuel in order to continue the electrochemical reaction that generates electricity. Batteries produce electricity only from the chemicals already present in them while fuel cells can produce electricity continuously for as long as fuel and oxygen are supplied. Fuel cells are also similar to batteries in that they have an anode, a cathode and an electrolyte.

Many researchers in the field of biology and environmental technology have always demonstrated the potentiality of microbial fuel cell (MFC) technology. Annually, huge volumes of waste are produced from industries, animals and agricultural field. Microbial fuel cell has an ability to convert the waste food materials to energetic form. The MFCs fed with food waste (FWs) are affected by organic loading rate.

D. Jothinathan (✉)
Department of Life Sciences, Central University of Tamil Nadu, Thiruvarur, 610005, India

A. H. Nasrin Fathima
Department of Plant Biology and Plant Biotechnology, Presidency College, Chennai, 600005, India

P. Mylsamy
PG & Research Department of Botany, Pachaiyappa's College, Chennai, 600030, India

L. B. Bruno
Department of Environmental Sciences, Bharathiar University, Coimbatore, 641046, India

V. Sivasankar
Department of Civil Engineering, Nagasaki University, Nagasaki, 8528521, Japan

V. Sivasankar et al. (eds.), *Microbial Fuel Cell Technology for Bioelectricity*,
https://doi.org/10.1007/978-3-319-92904-0_8

Higher amounts of food waste are mainly produced from our own domestic and from commercial cookeries. Approximately 60 million loads of FWs are generated per year in China alone and have exceeded 1.0 thousand tons in Beijing and Shanghai of China (Yan et al. 2012). The 34 million tons of FWs that were generated in the USA in 2010 was reported by the US Environmental Protection Agency (EPA) of which only less than 3% were recycled. The rest of the obtained FWs were thrown away. These FWs are considered as one of the valuable resources and tend to process higher energy, corrosion and ubiquity. Therefore, the problem produced by these FWs has been a demanding field of research due to the awareness on environmental conservation and energy revival (Goud et al. 2011).

In current years, researchers simultaneously attain waste treatment and energy creation of anaerobic digestion of FWs (Zhang et al. 2012). Microbial fuel cell (MFC) is used as an important and promising technique for anaerobic waste treatment in which bacteria are used as a catalyst and the bacteria help in generating bioelectricity from organic wastes (Min et al. 2005). The open access energies also innovated that bacteria has an ability to produce electricity from waste and renewable sources. Recently, bacteria named *Geobacter sulfurreducens* KN400 have the capacity to produce high electric current and were considered as the most important innovation for the year 2009 as announced by Time Magazine. Through this it was finalized that anaerobic digestion was found to be beneficial for the environment and also for treating the commercial waste (Levis and Barlaz 2011). Up to our knowledge, the communication between the microbes in MFCs fed with FWs is uncertain.

The animal wastewater discharged in the environment should be avoided to prevent (Suzuki et al. 2002) water contamination and odour issues (Luo et al. 2002). The water pollution is the major cause nowadays. The presence of higher concentration of nitrate and phosphate in wastewater confers to water pollution through eutrophication of surface water (Luo et al. 2002; Ra et al. 2000).

The present chapter discusses microbial fuel cells (MFCs) using different waste including animal waste and their application as an affordable and reliable anaerobic treatment technology. It has also discussed how rumen waste has bestowed its part for the energy production.

8.1.1 Microbial Fuel Cells in Waste Management

Microbial fuel cells (MFCs) have been proved experimentally that it has an ability to produce electricity (Oliveira et al. 2013; Poggi-Varaldo et al. 2014). Whereas in recent research, it has been proved that organic bioenergy is obtained from biofuels, biodiesel from algae, hydrogen from microbial electrolysis cells and electricity from MFCs. The current and power density (PD) can be affected by operational conditions, such as pH, temperature, substrate concentration, organic loading rate, hydraulic retention time (HRT), microorganisms' activity, parallel or serial connection and static magnetic field (Akman et al. 2013; Jadhav and Ghangrekar 2009; Jafary et al. 2013; Li et al. 2011).

In order to enhance the PD from MFCs, various nano-engineered electrode materials, electrode architectures and cost-effective electrodes have been considered (Gadhamshetty and Koratkar 2012; Kumar et al. 2013; Lefebvre et al. 2013). MFCs have an ability to convert biomass to electricity while leaving the anaerobic macerate to be used as soil improvement, etc. MFCs treatment has an ability to produce electricity from wastewater or biological waste products by converting the waste products to soil reformation. The antagonistic electron acceptor produced by them may be detrimental in the wastewater or biological waste treatment (Chiu et al. 2016).

The coupling of MFC with electricity results in the degradation of organic compounds, solid and liquid wastes. The waste management in MFC attains vital results in the removal of COD and power output. MFC never wanted a chemical catalyst or a temperature. The author confirms that MFC to waste treatment is much more reliable, and it has shown positive results from the last 20 years (Nastro et al. 2013). As most of the researchers perceive that MFC is a system which is efficient enough to convert the organic biodegradable substances to chemical energy into electrical energy, the first METs (microbial electrochemical technologies) was detected. METs are technologies which gain various energy and resources from the wastewater which had various subsystems, targeted to various objectives, and they also include MFCs (Logan and Rabaey 2012). Mainly, MFC has been studied for treating civil and industrial wastewater with a gain of feasible energy that could balance waste treatment cost (Puig et al. 2011; Capodaglio et al. 2013; Cercado-Quezada et al. 2010). Urban wastewater is generally said to have more than nine times of energy than the energy produced from wastewater treatment processes (WWTPs) (Shizas and Bagley, 2004). Recovery of that segment of energy will reduce the cost of both economical and environmental waste. Practical application of MFC in WWTPs construction is considered to be a fluctuated system because of low voltage and low power density (Kaur et al. 2014; Capodaglio et al. 2015). The achievement of MFC is found to be increasing depending upon the OLR (organic loading rate), pH, HRT and electricity resistance (Aelterman et al. 2008; Ieropoulos et al. 2010; Molognoni et al. 2016).

8.2 Energy Production from Various Sources

The fuel in an MFC varies depending on various factors such as the nature of the microorganism that is being used, the amount of energy to be produced by the MFC, the cost of the fuel that is being used and so on. The sources of MFC have been depicted in Fig. 8.1. Some commonly used fuels are:

1. Sewage sludge (Jiang et al. 2009; Zhang et al. 2012)
2. Domestic waste (Fornero et al. 2010)

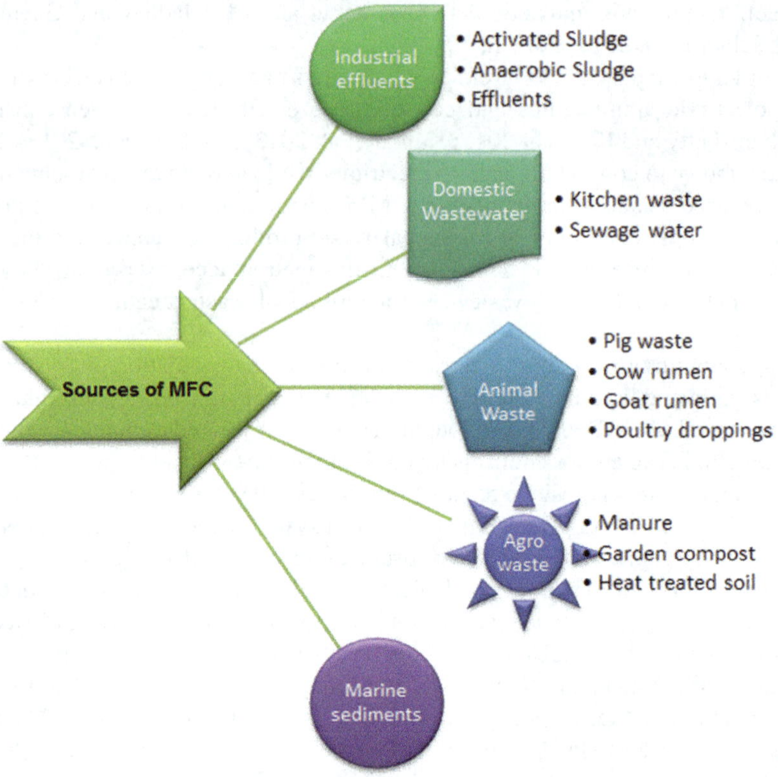

Fig. 8.1 Sources of MFC

3. Industrial waste (Angenent et al. 2004; Wang et al. 2008)
4. Animal waste (Min et al. 2005; Yokoyama et al. 2006; Kim et al., 2008b)
5. Marine sediments (Dumas et al. 2007; Donovan et al. 2011)

8.2.1 Sewage Sludge

Industrial and urban areas are said to pollute the water. There are tremendous wastewater plants which produce huge quantity of sludge per year. Many countries are involved in the production of sludge. In USA, 7.6 million tons of sludge were recovered in the year 2005, and this rate is expected to increase to 8.2 million tons by 2010 (Lee et al. 2005). In the year 2005, EU produced 8.2 million tons of sludge, and China produced approximately 1 million tons of sludge in the same year (Strünkmann et al. 2006). This kind of sludge can be disposed only when it undergoes a proper treatment under appropriate conditions (Burke et al. 2003; Cusido et al. 2003). The sludge treatment is quite expensive, and China still carries out with sludge treatment which costs 25–60% of the total expense.

Biological fuel cell is one of the promising and innovative technologies to solve this issue. When compared to conventional fuel cells, the biological fuels cells have a mild reaction temperature like ambient temperature, normal pressure and neutral pH. The authors Rodrigo et al. studied that the generated power density was found to be in organic matters but not on the wastewater flow. The maximum power density obtained in the system is 25 mW m^{-2}. The presence of oxygen occurred in both anodic chamber and cathodic chamber. The oxygen in cathodic chamber is found to be low. A couple reaction takes place between oxidation-reduction and COD oxidation in anodic chamber. The removal of 0.25% of COD was obtained during the generation of electricity. Occasionally, the anodic chamber worsens the MFC performance because of the presence of oxygen. For a better performance of MFC, the algal growth should be controlled. His work concluded that from wastewater, he was able to generate electricity by means of biological cultures (Rodrigo et al. 2007).

Later, the anaerobic digestion method was identified for sewage sludge treatment because it consumes very less amount of energy, generated very limited amount of solids and requires less nutrition and recovery of energy from biogas. During anaerobic digestion, the sewage sludge maintains its stability by converting the organic matter into biogas (Hwang et al. 2004). There are two anaerobic digestion processes, mesophilic and thermophilic processes. Mesophilic process requires a long retention time, whereas thermophilic needs less retention time and desires to have extreme heating (Zupancic and Ros 2003). The biogas produced from the digested sludge is now considered as bioenergy source.

MFC can convert organic substances into electricity. The organic compounds are simple carbohydrates such as glucose, acetate and butyrate (Liu and Logan, 2004) and complex organic wastes that include swine waste, domestic waste and sludge waste which produce maximum electricity through MFC (Scott and Murano 2007).

8.2.2 Domestic Waste

Domestic wastewater is treated in wastewater treatment plants which produce large amounts of excess sludge. The treatment and disposal of this sludge are becoming more and more of a challenge due to economic, environmental and regulatory considerations (Wei et al. 2003; Aelterman et al. 2006).

In this situation, MFCs that use exoelectrogenic bacteria to produce electricity while also treating the domestic wastewater become attractive options. The electricity produced can be cycled back to the same wastewater treatment plant. The other property of MFCs that make them attractive as wastewater treatment plants is that they are carbon neutral. MFCs oxidize organic matter which releases recently fixed carbon back into the atmosphere (Lovley 2006). Because MFC-based wastewater treatment plants can be installed far from existing treatment plant networks, such decentralized wastewater treatment can also be financially attractive (Wilderer and Schreff 2000).

One of the goals of present studies is to make MFCs net electricity producers. The high energy content of domestic wastewater provides the possibility of higher energy production. Cusick et al. (2010) examined the procedures and processes that can lead to higher energy production using MFCs where the fuel/substrate is domestic wastewater or wastewater from winery. But all domestic wastewater treatments produce nitrogen which needs to be treated. Especially, for decentralized treatment plants that use MFCs, the processing of nitrogen is an issue that still needs an optimal solution that is financially viable.

In another study, the wastewater treatments at two different temperatures using batch as well as continuous flow systems were examined. The treatment was dependent upon the calculation of the removal of COD, power generation, recovery of energy and removal of nitrogen. These are the certain factors which determined the potentiality of MFC to produce power and reduce the solid production from the treatment systems compared to aerobic conventional process. Under mesophilic conditions and continuous flow, the highest power density achieved was 422 mW/m^2 and 12.8 W/m^3 at an organic loading rate of 54 g COD/L-d, and the COD removal was only 25.8%. Energy recovery was obtained by proper operational conditions of flow mode, temperature HRT and organic loading (Ahn and Logan 2010).

However, in recent years, researchers have been using single-chambered MFCs. Adeniran et al. (2016) designed a sandwich domestic wastewater-fed dual-chambered microbial fuel cell for the generation of energy and wastewater treatment. The power density generated by MFC seems to increase COD of domestic wastewater. When the COD was 3400 mg L^{-1} at a current density of 0.054 mA cm^{-2}, the maximum power density was 251 mW m^{-2} and external resistance of 200 Ω. These records were declined when they used 91% diluted wastewater. The domestic wastewater reduces cost, and it might be the bright future for large-scale industries (Adeniran et al. 2016).

There are many studies carried on treating long-term operation of MFCs. In a municipal wastewater treatment facility, two 4 L microbial fuel cells (MFCs) were installed on primary effluent for more than 400 days. Nearly 65–70% of COD at hydraulic retention time of 11 h were removed by both the MFCs, and it reduced about 50% of the solids. Some fluctuation occurred like discharge of anode for 1–3 days or different HRTs. The groundwork analysis of production of energy and consumption indicates that the two MFCs can hypothetically achieve energy consumption, and positive energy balance can be reduced by using large tubing connectors. By denitrifying MFC, the MFC system enhances the removal of total nitrogen from 27.1% to 76.2%. However, the production of energy gradually declines because of the conception of organic substances in the denitrifying MFC. Establishing a carbon balance discloses that sulphate reduction was found to be a major scavenger and methane production plays a very minimal role in the distribution of electrons. These results determine the technical capability of MFC technology, and the advantages are recovery of energy from waste, low conception of energy and low production of sludge (Zhang and He 2013).

8.2.2.1 Kitchen and Bamboo Waste

Moqsud et al. (2014) studied the generation of bioelectricity through kitchen waste and bamboo waste by a microbial fuel cell (MFC) method. The beneficial nutriments like nitrogen, phosphorus and potassium (NPK) were determined to use them as soil amendments. By using both kitchen and bamboo wastes, one-chambered MFC was used for the generation of bioelectricity. The room temperature is maintained for 25 °C for approximately 45 days, and the data logger was recorded. The voltage generation of kitchen and bamboo waste showed different peaks. In kitchen waste, the voltage was found be more in the initial stage and then reached to a peak of 620 mV, whereas the bamboo's waste reached the voltage of 540 mV. This result concluded that MFC shows productive and environment-friendly results for organic waste management which is very useful for less unaware countries and can contribute safe electricity from organic waste materials.

Five two-chambered MFCs were connected to the MFC stack which was further incorporated into the sink pipe connected with the kitchen wastewater treatment. The performance of the MFC stack functioning with real and artificial wastewater was reviewed. Practically, the voltage was checked at different flow rate and temperature. The results detected were with an average open circuit voltage of 3.44 ± 0.02 V, a coulombic efficiency of 78.2 ± 3.6% and a peak power of 45.74 ± 1.39 mW. The performance of MFC is agitated by a process called the flushing process. The MFC stack can be operated by flushing the substrate at 50 °C, and beyond that, irreversible performance corrosion was observed. The suggested MFC stack is likely to function as a potential power source for light and low power devices, specifically in off-grid rural areas (Yang et al. 2016).

Apart from wastewater, solid organic waste has also been looked into as a possible substrate/fuel for commercial production of electricity. Solid food waste from commercial kitchens is anaerobically digested by bacteria, and this kind of anaerobic digestion has been concluded to be the most environmentally beneficial treatment option for commercial FWs (Levis and Barlaz 2011). Now, studies are considering to not only treat the solid food waste but also research on how to use the microbes that digest the food waste as part of viable microbial fuel cells.

8.2.3 Industrial Waste

Of the large variety of industrial waste products, the waste from organic matter and organic matter-related products is the most promising candidate to be used as substrate/fuel for MFCs. Organic matter is used as a substrate by an innumerable variety of microorganisms, and using some of these microorganisms as electrogenic bacteria is only a matter of research.

Some of the industries whose wastewaters are used for electricity generation include brewery/winery, chocolate making, food processing, meat packing and paper recycling industries. In the case of industrial waste, the primary purpose is to reduce the cost of treating the wastewater so that the water itself becomes recycled. The resultant electricity is considered more a bonus rather than a useful commodity that can be commercialized.

8.2.3.1 Winery Wastewater

One of the reasons for this being that enough research has not been done to make electricity generation from the treatment of industrial wastewaters treatment commercially viable. For example, Cusick et al. (2010) compared the amount of electricity generated by using winery wastewater versus domestic wastewater and found that winery wastewater is a better substrate for MFCs while domestic wastewater processing are better processed by so-called microbial electrolysis cells (MECs) which need a supplemental voltage supply in order to function.

8.2.3.2 Brewery Wastewater

The need for energy has increased worldwide. Researchers have also focused on the production of electricity and MFC modelling from beer brewery wastewater. According to Wen et al. (2009), a single air cathode MFC was constructed in which carbon fibre was used as anode and brewery wastewater allowed to dilute is then used as substrate. The open circuit voltage displayed by the MFC is 0.578 V, and the maximum powered density was found to be 9.52 W/m^2 (264 mW/m^2). Kinetic loss and transport loss are the most important factors which manipulates the implementation of MFC. There are also many other factors which decrease these losses, for example, increase in reactant concentration, retaining much effective and cheaper electrode catalysts, retaining irregular electrode, increase in reaction temperature, improvement of flow structure and many more.

8.2.3.3 Food Industry

There are large amount of fats and oils used in food industries. These comprise of organic compounds and vegetable oils. The organic compounds include fatty acid, glycerine, waxes and hydrocarbons. A considerable amount of compounds are discharged as waste. Many investigations have been undergone in food waste industries, and it was recorded that nearly 500,000 tons per year of cooking oil is wasted in Japan.

Large amounts of fats and oils are discharged into wastewater from food industries. In a recent study the researcher estimated the risk of using MFC for the generation of electricity from wastewater containing vegetable oils. Single-chambered MFCs were used and were furnished with artificially created wastewater-bearing soybean oil, removed oil and examination of electric output at many different terms. It was also found that MFC functioning can be upgraded by inoculation of oil-contaminated soil; enhancing wastewater with emulsifier and graphite-coated anode with carbon nanotubes resulting in 2 Wm^{-2} power output. The bacterium "*Burkholderia*" that helps in degrading oil in oil-contaminated soil and anode biofilms was detected by means of PCR amplification of 16 S rRNA fragment. Through these results, it has been concluded that MFCs can be used for energy recovery from food industrials wastewater containing oils and fats (Hamamoto et al. 2016).

8.2.3.4 Potato-Processing Wastewater

MFCs appear as a new chance to deal with organic waste (Logan and Regan 2006; Rabaey and Verstraete 2005). The MFCs play an important role in the potato-processing industries, which undergo certain sequences that include the production of methane as the anaerobic one. In most countries, potato processing plays an important role. The wastewater compost mainly consists of debris and chipping of potato peeling. In order to reduce the organic materials from the waste, both aerobic (Lasik et al. 2010) and anaerobic way of processing has been done (Linke 2006). The researcher investigates the possibility of methanogenesis with many new technologies of MFCs. They have also studied the production of electrified biofilms from real anaerobic sludge and renovation of potato-processing wastewater into electricity. The MFCs had an ability to process the wastewater with prohibitive amount of COD removal but with low energetic conversion productivity. The methanogenesis helps to improvise conversion productivity and gradually degrades the organic matter from the final collected effluent. The author described the production of methanogenesis and a removal of electricity from better quality COD as optimal achievement.

8.2.3.5 Dairy Industry

MFC is the bioreactor which has an ability to convert chemical energy to electrical energy in an anaerobic condition with the help of catalytic reacting microorganism (Du et al. 2007). In MFCs the substrate of cheese whey from dairy industry has been tried out (Antonopoulo et al. 2010; Nasirahmadi and Safekordi 2011; Kassongo and Togo 2010; Dalvi et al. 2011). During the cheesemaking process, the precipitation and removal of milk casein ostracize liquid fraction which is called cheese whey which has a significant amount of carbohydrates, lactose, protein, lactic acid, fat and salt (Gelegenis et al. 2007).

According to Tremouli et al. (2013), MFC produced electricity from different organic loads of sterilized cheese whey. The investigation was further carried on two-chambered MFC. The enactment of the cell was detected at the highest concentration of the pretreated cheese whey (6.7 g COD/L) corresponding to the maximum power density of about 46 mW/m^2. For comparative reasons, the experiment was carried out using glucose (0.35 g COD/L). In this study, the open circuit impedance of MFC varied virtually to the same magnitude on both ohmic resistance between anode and cathode in the complete polarization resistance. This overpotential of ohmic resistance is the main purpose for the energy loss in two-chambered MFC.

A country like India has always raised the need of processing milk, and people and the mechanization have activated the latest trend of relocation of rural people to urban areas for employment. In such countries, process of milk is approximately 94.6 million tons every year. 2–2.5 L of wastewater is produced for every litre of milk (Ramasamy et al. 2004). Therefore, a large amount of dairy wastewater is wasted without utilizing it and contaminated the environment when released without treatment. The power production of a two-chambered MFC with dairy wastewater was employed with two different metabolism: aerobic and anaerobic. The initial COD concentration was 1600 mg/L and pH 7 was maintained in the anode chamber. Comparatively, anaerobic metabolism favoured the MFC performance by producing better columbic efficiency. Conversely, high power production was evident in MFC with aerobic metabolism. (Elakkiya and Matheswaran 2013).

In another study, the author reported the fabrication of novel annular single-chambered microbial fuel cells (ASCMFC) with spiral anode. Anode has a graphite coating with stainless steel. Dairy wastewater with organic matter was used as a substrate. By operating ASCMFC for 450 h, the outcome indicates a high open circuit voltage of about 810 mV. The maximum power density obtained was 20.2 W/m^3, and 91% of COD removal was achieved. Thus, he proved that ASCMFC is a promising alternative to predictable MFCs for wastewater treatment and power generation (Mardanpour et al. 2012).

8.2.4 *Animal Waste*

A large volume of wastewater has been produced annually from industries and agricultures. For example, in each year, the USA generates approximately 5.8×10^7 tons of animal manures (Dentel et al. 2004). The animal waste should be treated in order to protect the environment from getting polluted and to avoid water contamination and odour problem (Luo et al. 2002). Many treatment techniques are there in order remove organic and inorganic compounds from water. Now, MFC plays an important role in the treatment of wastewater by generating electricity directly from marine sediments, anaerobically digested sludge domestic and food wastewater.

8.2.4.1 Slaughterhouse Wastewater

Though slaughterhouse wastewater is classified as industrial waste, the high concentration of contaminants such as fats, blood, manure and other organic compounds make slaughterhouse wastewater a strong pollutant of the environment in general and a major degrader of aquatic ecosystems, in particular. The same contaminants that make slaughterhouse wastewater a strong pollutant also make it a high-strength wastewater – the fats, blood, manure and other organic compounds are all biodegradable organic compounds that can be broken down by bacterial treatment. For instance, more than 30,000 m^3/day of slaughterhouse wastewater is produced in Ireland alone.

Katuri et al. (2012) studied direct electricity production from slaughterhouse wastewater using MFCs. The conclusion was that the generation of electricity from slaughterhouse wastewater using two-chambered MFCs is feasible and should be considered for large-scale application. Deepika et al. (2015) investigated the usage of goat rumen fluid as an MFC substrate/fuel for the production of bioelectricity. The rumen fluid is normally drained from the slaughterhouses and has no use. By using MFCs to produce electricity using rumen fluid, this fluid instead can serve a useful purpose instead of being discharged to pollute the environment.

In another study done by Oladejo et al. (2015), poultry droppings were used as a substrate/fuel for an MFC arrangement to determine if electricity can be produced from such a system. They demonstrated energy production from poultry droppings, but further research is needed to establish the feasibility of such a system. As with other systems, the electricity generation is dependent on the concentration of the substrate (poultry droppings), and these act as a limiting factor generating a sustainable power supply.

8.2.4.2 Swine Wastewater Treatment

Swine and wastewater treatment and control of odour are important constituents for ecological animal production. Harsh systems of heifer industry require to control the swine wastewater and odour (Zahn et al. 1997). The researcher carried out anaerobic treatment which can generate methane gas, but ammonia and other odour stuffs are not completely removed. Whereas aerobic treatment of animal wastewater is bit costly and do not produce any useful products, so it was found that anaerobic treatment can be energy demanding (Kim et al. 2004; Chen et al. 2004).

The removal of several distinctive aliphatic and aromatic hydrocarbon, phenolic compounds and indoles is known to contribute to nuisance chemical odours from swine wastewater. The chemical removals were compared with the MFC operated in an open circuit mode and with a sealed reactor over the same period of time. Single-chambered MFCs were used to reduce ten chemicals which created the odours by 99.76% and acetate, butyrate and propionate, the three volatile organic acids, by >99%. The MFC seems to produce 228 mW/m^2, and approximately 84% of organic compounds were removed in 260 h. These results showed that it is possible to accelerate the odour removal using MFCs, though at the same time removing the organic compounds and electricity production (Kim et al. 2008a).

Bioelectrochemical systems (BES) are operated in MFC or MEC (microbial electrolysis cell) mode, when the electricity is produced or supply of energy to non-spontaneous reaction can be combined to anaerobic digestion in order to expand its implementation and effluent quality (Hamelers et al. 2010). These systems are considered as one of the most effective systems for the combinations of wastewater treatments to the productions of chemical matters and energy carriers (Pant et al. 2012).

The raw and anaerobically digested pig slurry was inspected in batch say in two-chambered BES run in MFC and MEC mode. Recovery of nitrogen, cation transport, COD and microbial population evolution were evaluated. The anaerobic digestion-MEC integrated system achieved utmost of 60% in 48 h removal of COD, while the maximum removal of ammonium was achieved in MFC mode fed with digested pig slurry in 24 h. High pH in MEC mode could favour recovery of ammonium in a successive shedding and absorption process. *Thermoplasmatales* is favourable to acetotrophic *Methanosaetaceae* and has hydrogenotrophic and methylotrophic methanogen phylotypes while the known family of exoelectrogenic bacteria *Desulfuromonadaceae* was developed under MEC mode. These results show that intermingling of anaerobic digestion in BES seems to be an interesting substitution for the treatment of complex substrates. Recovery of ammonia can be used in the future as a fertilizer (Cerrillo et al. 2016).

8.2.5 Agrowaste Industries

The innovative hypothesis of circular economy claims many fresh scientific approaches for energy and resource regain. Agro-food industries yield great amounts of organic materials as a secondary stream and waste (Fava et al. 2013). Four sets of membrane-less single-chambered MFC were operated in a parallel form for more than 100 days and were immunized with anaerobic sludge from biogas production plant and fed with different organic substrates regularly. The substrates are kitchen waste, cheese whey from dairy industries, fish residual oil and pulpy waste of citrus fruits. All MFCs were able to produce electricity, and biodegradation rates were accomplished. Radical decrease of biodegradation proficiency in sequential batch cycle corresponded to weakening peaks of MFC potentials. The pH drop below 6.5 will affect the biodegradation and anodic exoelectrogenic activity. The bioreactor with complex organic matter delays electric signal and COD degradation. Low current was produced by anodic and cathodic polarization curves. Here, the author explains the electrical signal produced by MFCs during anaerobic biodegradation of four unusual types of agro-industrial residual materials of significance in Mediterranean agro-food sectors (Colombo et al. 2017).

8.2.6 Marine Sediments

MFC is also considered as a novel energy harvesting technology which provides reliability and power conception for a long time and has a strong lifetime of sensor and communication hardware. Ten years ago, MFC was organized in marine sediments (Reimers et al. 2001). In 2007, these MFCS were first established to feasible power sources for undersea sensors and communication systems (Tender et al. 2008). The progress of Trophos Energy which is positioned in these aquatic systems and made in the development of BMFC (benthic microbial fuel cells) technologies and further steps are taken into consideration for power generation. BMFC is a novel form of energy harvesting for marine environment. After a lot of experimentation, it was proved that Trophos Energy has developed novel BMFC that expands a generation of power density. Merging of BMFCs with Trophos offers robust and long-term power generations for various marine applications. The author clearly explains about the importance of BMFC systems as power sources for marine networks (Guzman et al. 2010).

8.3 Rumen Waste in MFC

8.3.1 Rumen Fluid as a Cheap Energy Source

Microbial fuel cell studies have been dealt with different wastewater from treatment plants to generate bioelectricity where there are various parameters to be maintained in a particular limit. However, there are very less reports in using animal waste in MFC research. In the rumen ecosystem, a wide range of bacterial community habitats with different temperature gradients shows complex metabolic function, such as physical, physiological and environmental factors. Ruminants feed on various cellulosic materials with different composition which leads to acquiring different microbial consortia. These microbes breakdown the feed into volatile fatty acids (VFAs): acetic acid, propionic acid and butyric acid.

Rumen microbial fuel cells (RMFCs) are a favourable technology for viable production of green energy and fibre waste treatment. The ruminal chamber of these animals possesses abundant microbes, consisting of bacteria, fungi and protozoa. The microorganisms residing inside the rumen produce active enzymes to commendably degrade the complex ingested plant material under anaerobic conditions (Hobson and Stewart 1997).

8.3.1.1 Pros and Cons of Using Animal Waste

Animal waste is generated at places such as slaughterhouses and meat factories, and so it is easier to procure. Since it is more uniform in composition than plant waste, it can also be studied more easily. In places where more meat is consumed, animal waste will be cheaper to obtain than plant waste. The fermentation of ruminants is usually complemented by the production of electrons and protons. This has made a pavement for the electricity production (Offner and Sauvant 2006).

Recent studies have confirmed that addition of plant fibres (Zang et al. 2010) and purified cellulose (Rismani-Yazdi et al. 2007) could be transformed into bioelectricity by rumen microbes in the anodic chamber without the aid of any external mediators. Thus, usage of rumen waste in MFC would definitely solve the piling of agricultural waste in an indirect manner.

The problem with using animal waste is the timeframe of the microbes that ingest the cellulose present in the rumen fluid. Since it is a complex carbohydrate, assimilation is quite tough and takes a longer time. One other problem is that animal rights, which already don't want animals to be used as food, will oppose the usage of animals and animal products for energy production.

8.4 Conclusion

The energy source has become so important, and its running out day by day is quite noticeable. In order to overcome this problem, renewable and environmental friendly sources are used as alternative tools. The microbial ecology of electrochemically active bacterial population is insufficient. The ecophysiology and microbial association within MFCs are just about to be explored. MFCs incorporated with microbial community are a major compound to be studied. Since the microbial colonies release electrons which helps in the generation of electricity are affected by some parameters such as MFCs types and substrate. The changes in microbial ecology occur based on the type of sources used in the anode chamber. For additional exhibition, the MFC technology is crucial to examine the long-term performance and firmness of large-sized MFCs with actual wastewater.

References

Adeniran JA, Huberts R, De-Koker JJ, Arotiba OA, Olorundare OF, Van-Zyl E, Du-Plessis SC (2016) Energy generation from domestic wastewater using sandwich dual-chamber microbial fuel cell with mesh current collector cathode. Int J Environ Sci Technol 13(9):2209–2218

Aelterman P, Rabaey K, Clauwaert P, Verstraete W (2006) Microbial fuel cells for wastewater treatment. Water Sci Technol 54(8):9–15

Aelterman P, Versichele M, Marzorati M, Boon N, Verstraete W (2008) Loading rate and external resistance control the electricity generation of microbial fuel cells with different three-dimensional anodes. Bioresour Technol 99:8895–8902

Ahn Y, Logan BE (2010) Effectiveness of domestic wastewater treatment using microbial fuel cells at ambient and mesophilic temperatures. Bioresour Technol 101(2):469–475

Akman D, Cirik K, Ozdemir S, Ozkaya B, Cinar O (2013) Bioelectricity generation in continuously-fed microbial fuel cell: effects of anode electrode material and hydraulic retention time. Bioresour Technol 149:459–464

Angenent LT, Karim K, Al-Dahhan MH, Wrenn BA, Domíguez-Espinosa R (2004) Production of bioenergy and biochemicals from industrial and agricultural wastewater. Trends Biotechnol 22(9):477–485

Antonopoulo G, Stamatelatou K, Beblis S, Lyberatos G (2010) Electricity generation from synthetic substrates and cheese whey using a two chamber microbial fuel cell. Biochem Eng J 50:10–15

Burke S, Heathwaite L, Quinn P, Merrett S, Whitehead P, Preedy N, Lerner D, Saul A (2003) Strategic management of non-point source pollution from sewage sludge. Water Sci Technol 47(8):305–310

Capodaglio AG, Molognoni D, Dallago E, Liberale A, Cella R, Longoni P et al (2013) Microbial fuel cells for direct electrical energy recovery from urban wastewaters. Sci World J https://doi.org/10.1155/2013/634738 2013:1–8
Capodaglio AG, Molognoni D, Puig S, Balaguer MD, Colprim J (2015) Role of operating conditions on energetic pathways in a microbial fuel cell. Energy Procedia 74:728–735
Cercado-Quezada B, Delia ML, Bergel A (2010) Testing various food-industry wastes for electricity production in microbial fuel cell. Bioresour Technol 101:2748–2754
Cerrillo M, Oliveras J, Viñas M, Bonmatí A (2016) Comparative assessment of raw and digested pig slurry treatment in bioelectrochemical systems. Bioelectrochemistry 110:69–78
Chen M, Kim JH, Kishida N, Nishimura O, Sudo R (2004) Enhanced nitrogen removal using C/N load adjustment and real-time control strategy in sequencing batch reactors for swine wastewater treatment. Water Sci Technol 49:309–314
Chiu HY, Pai TY, Liu MH, Chang CA, Lo FC, Chang TC, Lo HM, Chiang CF, Chao KP, Lo WY, Lo SW (2016) Electricity production from municipal solid waste using microbial fuel cells. Waste Manag Res 34(7):619–629.
Colombo A, Schievano A, Trasatti SP, Morrone R, D'Antona N, Cristiani P (2017) Signal trends of microbial fuel cells fed with different food-industry residues. Int J Hydrog Energy 19;42(3):1841–1852
Cusick RD, Kiely PD, Logan BE (2010) A monetary comparison of energy recovered from microbial fuel cells and microbial electrolysis cells fed winery or domestic wastewaters. Int J Hydrog Energy 35(17):8855–8861
Cusido JA, Cremades LV, Gonzalez M (2003) Gaseous emissions from ceramics manufactured with urban sewage sludge during firing processes. Waste Manag 23(3):273–280
Dalvi AD, Mohandas N, Shinde OA, Kininge PT (2011) Microbial fuel cell for production of bioelectricity from whey and biological wastewater. Int J Adv Biotechnol Res 2:263–268
Deepika J, Meignanalakshmi S, Thilagaraj RW (2015) The optimization of parameters for increased electricity production by a microbial fuel cell using rumen fluid. Int J Green Energy 12(4):333–338
Dentel SK, Strogen B, Sharma A, Chiu P (2004) Direct generation of electricity from sludges and other liquid wastes. In: Proceedings, IWA conference on resources from sludge, Singapore, March 2–3
Donovan C, Dewan A, Peng H, Heo D, Beyenal H (2011) Power management system for a 2.5 W remote sensor powered by a sediment microbial fuel cell. J Power Sources 196(3):1171–1177
Du ZW, Li HR, Gu T (2007) A state of the art review on microbial fuel cells: a promising technology for wastewater treatment and bioenergy. Biotechnol Adv 25:464–482
Dumas C, Mollica A, Féron D, Basséguy R, Etcheverry L, Bergel A (2007) Marine microbial fuel cell: use of stainless steel electrodes as anode and cathode materials. Electrochim Acta 53(2):468–473
Elakkiya E, Matheswaran M (2013) Comparison of anodic metabolisms in bioelectricity production during treatment of dairy wastewater in microbial fuel cell. Bioresour Technol 136:407–412
Fava F, Totaro G, Diels L, Reis M, Duarte J, Carioca OB, Poggi-Varaldo HM, Ferreira BS (2013) Biowaste biorefinery in Europe: opportunities and research & development needs. New Biotechnol 32(1):100–108
Fornero JJ, Rosenbaum M, Angenent LT (2010) Electric power generation from municipal, food, and animal wastewaters using microbial fuel cells. Electroanalysis 22(7–8):832–843
Gadhamshetty V, Koratkar N (2012) Nano-engineered biocatalyst-electrode structures for next generation microbial fuel cells. Nano Energy 1(1):35
Gelegenis J, Georgakakis D, Angelidaki I, Mavris V (2007) Optimization of biogas production by co-digesting whey with diluted poultry manure. Renew Energy 32:2147–2160
Goud RK, Babu PS, Mohan SV (2011) Canteen based composite food waste as potential anodic fuel for bioelectricity generation in single chambered microbial fuel cell (MFC): bio-electrochemical evaluation under increasing substrate loading condition. Int J Hydrog Energy 36:6210e8

Guzman JJ, Cooke KG, Gay MO, Radachowsky SE, Girguis PR, Chiu MA (2010) Benthic microbial fuel cells: Long-term power sources for wireless marine sensor networks. In: Sensors, and command, control, communications, and intelligence (C3I) technologies for homeland security and homeland defense IX, vol. 7666. International Society for Optics and Photonics, p 76662M

Hamamoto K, Miyahara M, Kouzuma A, Matsumoto A, Yoda M, Ishiguro T, Watanabe K (2016) Evaluation of microbial fuel cells for electricity generation from oil-contaminated wastewater. J Biosci Bioeng 122(5):589–593

Hamelers HM, Heijne A, Sleutels TJA, Jeremiasse A, Strik DBTB, Buisman CN (2010) New applications and performance of bioelectrochemical systems. Appl Microbiol Biotechnol 85:1673–1685

Hobson PN, Stewart CS (1997) The rumen microbial ecosystem. Blackie Academic & Professional, New York, pp 10–72

Hwang MH, Jang NJ, Hyun SH, Kim IS (2004) Anaerobic bio-hydrogen production from ethanol fermentation: the role of pH. J Biotechnol 111:297–309

Ieropoulos I, Winfield J, Greenman J (2010) Effects of flow-rate, inoculum and time on the internal resistance of microbial fuel cells. Bioresour Technol 101:3520–3525

Jadhav GS, Ghangrekar MM (2009) Performance of microbial fuel cell subjected to variation in pH, temperature, external load and substrate concentration. Bioresour Technol 100(2):717–723

Jafary T, Rahimnejad M, Ghoreyshi AA, Najafpour G, Hghparast F, Daud WRW (2013) Assessment of bioelectricity production in microbial fuel cells through series and parallel connections. Energy Convers Manag 75:256–262

Jiang J, Zhao Q, Zhang J, Zhang G, Lee DJ (2009) Electricity generation from bio-treatment of sewage sludge with microbial fuel cell. Bioresour Technol 100(23):5808–5812

Kassongo J, Togo CA (2010) The potential of whey in driving microbial fuel cells: a dual prospect of energy recovery and remediation. Afr J Biotechnol 9(46):7885–7890

Katuri KP, Enright AM, O'Flaherty V, Leech D (2012) Microbial analysis of anodic biofilm in a microbial fuel cell using slaughterhouse wastewater. Bioelectrochemistry 87:164–171

Kaur A, Boghani HC, Michie I, Dinsdale RM, Guwy AJ, Premier GC (2014) Inhibition of methane production in microbial fuel cells: operating strategies which select electrogens over methanogens. Bioresour Technol 173:75–81

Kim JH, Chen M, Kishida N, Sudo R (2004) Integrated real-time control strategy for nitrogen removal in swine wastewater treatment using sequencing batch reactors. Water Res 38:3340–3348

Kim JR, Dec J, Bruns MA, Logan BE (2008a) Removal of odors from swine wastewater by using microbial fuel cells. Appl Environ Microbiol 74(8):2540–2543

Kim JR, Zuo Y, Regan JM, Logan BE (2008b) Analysis of ammonia loss mechanisms in microbial fuel cells treating animal wastewater. Biotechnol Bioeng 99(5):1120–1127

Kumar GG, Sarathi VGS, Nahm KS (2013) Recent advances and challenges in the anode architecture and their modifications for the applications of microbial fuel cells. Biosens Bioelectron 43:461–475

Lasik M, Nowak J, Krzywonos M, Cibis E (2010) Impact of batch, repeated-batch (with cell recycle and medium replacement) and continuous processes on the course and efficiency of aerobic thermophilic biodegradation of potato processing wastewater. Bioresour Technol 101:3444–3451

Lee JW, Cha HY, Park KY, Song KG, Ahn KH (2005) Operational strategies for an activated sludge process in conjunction with ozone oxidation for zero excess sludge production during winter season. Water Res 39(7):1199–1204

Lefebvre O, Tan Z, Shen Y, Ng HY (2013) Optimization of a microbial fuel cell for wastewater treatment using recycled scrap metals as a cost-effective cathode material. Bioresour Technol 127:158–164

Levis JW, Barlaz MA (2011) What is the most environmentally beneficial way to treat commercial food waste? Environ Sci Technol 45(17):7438–7444

Li WW, Sheng GP, Liu XW, Cai PJ, Sun M, Xiao X, Wang YK, Tong ZH, Dong F, Yu HQ (2011) Impact of a static magnetic field on the electricity production of Shewanella-inoculated microbial fuel cells. Biosens Bioelectron 26(10):3987–3992

Linke B (2006) Kinetic study of thermophilic anaerobic digestion of solid wastes from potato processing. Biomass Bioenergy 30:892–896

Liu H, Logan BE (2004) Electricity generation using an air-cathode single chamber microbial fuel cell in the presence and absence of a proton exchange membrane. Environ Sci Technol 38:4040–4046

Logan BE, Rabaey K (2012) Conversion of wastes into bioelectricity and chemicals by using microbial electrochemical technologies. Science 337(6095):686–690

Logan BE, Regan JM (2006) Microbial fuel cells-challenges and applications. Environ Sci Technol 40(17):5172–5180

Lovley DR (2006) Bug juice: harvesting electricity with microorganisms. Nat Rev Microbiol 4(7):497–508

Luo A, Zhu J, Ndegwa PM (2002) Removal of carbon, nitrogen, and phosphorus in pig manure by continuous and intermittent aeration at low redox potentials. Biosyst Eng 82:209–215

Mardanpour MM, Esfahany MN, Behzad T, Sedaqatvand R (2012) Single chamber microbial fuel cell with spiral anode for dairy wastewater treatment. Biosens Bioelectron 38(1):264–269

Min B, Kim J, Oh S, Regan JM, Logan BE (2005) Electricity generation from swine wastewater using microbial fuel cells. Water Res 39(20):4961–4968

Molognoni D, Puig S, Balaguer MD, Capodaglio AG, Callegari A, Colprim J (2016) Multiparametric control for enhanced biofilm selection in microbial fuel cells. J Chem Technol Biotechnol 91(6):1720–1727

Moqsud MA, Omine K, Yasufuku N, Bushra QS, Hyodo M, Nakata Y (2014) Bioelectricity from kitchen and bamboo waste in a microbial fuel cell. Waste Manag Res 32(2):124–130

Nasirahmadi S, Safekordi AA (2011) Whey as a substrate for generation of bioelectricity in microbial fuel cell using *E. coli*. Int J Environ Sci Technol 8(4):823–830

Nastro RA, Dumontet S, Ulgiati S, Falcucci G, Vadursi M, Jannelli E, Minutillo M, Cozzolino R, Trifuoggi M, Erme G, De Santis E (2013) Microbial fuel cells fed by solid organic waste: a preliminar experimental study. In: European fuel cell. Piero Lunghi Conference & Exhibition – Rome, 11–13 December

Offner A, Sauvant D (2006) Thermodynamic modeling of ruminal fermentations. Anim Res 55(5):343–365

Oladejo D, Shoewu OO, Yussouff AA, Rapheal H (2015) Evaluation of electricity generation from animal based wastes in a microbial fuel cell. Int J Sci Technol Res 4(4):85–90

Oliveira VB, Simões M, Melo LF, Pinto AMFR (2013) Overview on the developments of microbial fuel cells. Biochem Eng J 73:53–64

Pant D, Singh A, Van Bogaert G, Singh Nigam P, Diels L, Vanbroekhoven K (2012) Bioelectrochemical systems (BES) for sustainable energy production and product recovery from organic wastes and industrial wastewaters. RSC Adv 2:1248–1263

Poggi-Varaldo HM, Munoz-Paez KM, Escamilla-Alvarado C, Robledo-Narváez PN, Ponce-Noyola MT, Calva-Calva G, Ríos-Leal E, Galíndez-Mayer J, Estrada-Vázquez C, Ortega-Clemente A, Rinderknecht-Seijas NF (2014) Biohydrogen, biomethane and bioelectricity as crucial components of biorefinery of organic wastes: a review. Waste Manag Res 32(5):353–365

Puig S, Serra M, Coma M, Balaguer MD, Colprim J (2011) Simultaneous domestic wastewater treatment and renewable energy production using microbial fuel cells (MFCs). Water Sci Technol 64:904–909

Ra CS, Lo KV, Shin JS, Oh JS, Hong BJ (2000) Biological nutrient removal with an internal organic carbon source in piggery wastewater treatment. Water Res 34:965–973

Rabaey K, Verstraete W (2005) Microbial fuel cells: novel biotechnology for energy generation. Trends Biotechnol 23:291–298

Reimers CE, Tender LM, Fertig S, Wang W (2001) Harvesting energy from the marine sediment-water interface. Environ Sci Technol 35:192–195

Rismani-Yazdi H, Christy AD, Dehority BA, Morrison M, Yu Z, Tuovinen OH (2007) Electricity generation from cellulose by rumen microorganisms in microbial fuel cells. Biotechnol Bioeng 97(6):1398–1407

Rodrigo MA, Cañizares P, Lobato J, Paz R, Sáez C, Linares JJ (2007) Production of electricity from the treatment of urban wastewater using a microbial fuel cell. J Power Sources 169:198–204

Ramasamy, EV, Gajalakshmi S, Sanjeevi R, Jithesh MN, Abbasi SA (2004) Feasibility studies on the treatment of dairy wastewaters with upflow anaerobic sludge blanket reactors. Bioresour. Technol. 93, 209–212.

Scott K, Murano C (2007) A study of a microbial fuel cell battery using manure sludge waste. J Chem Technol Biotechnol 82:809–817

Shizas I, Bagley DM (2004) Experimental determination of energy content of unknown organics in municipal wastewater streams. J Energy Eng 130:45–53

Strünkmann GW, Müller JA, Albert F, Schwedes J (2006) Reduction of excess sludge production using mechanical disintegration devices. Water Sci Technol 54(5):69–76

Suzuki K, Tanaka Y, Osada T, Waki M (2002) Removal of phosphate, magnesium and calcium from swine wastewater through crystallization enhanced by aeration. Water Res 36:2991–2998

Tender L, Gray S, Groveman E, Lowry D, Kauffman P (2008) The first demonstration of a microbial fuel cell as a viable power supply: powering a meteorological buoy. J Power Sources 179:571–575

Tremouli A, Antonopoulou G, Bebelis S, Lyberatos G (2013) Operation and characterization of a microbial fuel cell fed with pretreated cheese whey at different organic loads. Bioresour Technol 131:380–389

Wang X, Feng YJ, Lee H (2008) Electricity production from beer brewery wastewater using single chamber microbial fuel cell. Water Sci Technol 57(7):1117–1121

Wei Y, Van Houten RT, Borger AR, Eikelboom DH, Fan Y (2003) Minimization of excess sludge production for biological wastewater treatment. Water Res 37(18):4453–4467

Wen Q, Wu Y, Cao D, Zhao L, Sun Q (2009) Electricity generation and modeling of microbial fuel cell from continuous beer brewery wastewater. Bioresour Technol 100(18):4171–4175

Wilderer PA, Schreff D (2000) Decentralized and centralized wastewater management: a challenge for technology developers. Water Sci Technol 41(1):1–8

Yan S, Chen X, Wu J, Wang P (2012) Ethanol production from concentrated food waste hydrolysates with yeast cells immobilized on corn stalk. Appl Microbiol Biotechnol 94(3):829–838

Yang Y, Liu T, Liao Q, Ye D, Zhu X, Li J, Li Y (2016) A three-dimensional nitrogen-doped graphene aerogel-activated carbon composite catalyst that enables low-cost microfluidic microbial fuel cells with superior performance. J Mater Chem A 4(41):15913–15919

Yokoyama H, Ohmori H, Ishida M, Waki M, Tanaka Y (2006) Treatment of cow-waste slurry by a microbial fuel cell and the properties of the treated slurry as a liquid manure. Anim Sci J 77(6):634–638

Zahn JA, Hatfield JL, Do YS, DiSpirito AA, Laird DA, Pfeiffer RL (1997) Characterization of volatile organic emissions and wastes from a swine production facility. J Environ Qual 26:1687–1696

Zang GL, Sheng GP, Tong ZH, Liu XW, Teng SX, Li WW, Yu HQ (2010) Direct electricity recovery from *Canna indica* by an air-cathode microbial fuel cell inoculated with rumen microorganisms. Environ Sci Technol 44(7):2715–2720

Zhang F, He Z (2013) A cooperative microbial fuel cell system for waste treatment and energy recovery. Environ Technol 34(13–14):1905–1913

Zhang G, Zhao Q, Jiao Y, Wang K, Lee DJ, Ren N (2012) Efficient electricity generation from sewage sludge using biocathode microbial fuel cell. Water Res 46(1):43–52

Zupancic GD, Ros M (2003) Heat and energy requirements in thermophilic anaerobic sludge digestion. Renew Energy 28:2255–2267

Chapter 9
Electricigens: Role and Prominence in Microbial Fuel Cell Performance

Deepika Jothinathan, Prabhakaran Mylsamy, and L. Benedict Bruno

9.1 Introduction

This chapter mainly focuses on the participation and importance of microbes in power generation in microbial fuel cell. MFC can be regarded truly as an alternative source of energy as it emerges from everlasting microbes. A wide range of microbes and their electron transfer mechanism have been discussed in detail. Most of the microbes that take part in the electron transfer in MFC are of anaerobic in nature (Reimers et al. 2001). However, few microbes are aerobic and facultative aerobic in nature inside the microbial fuel cell (Song et al. 2003; Fournier et al. 2005). A clear discussion about all the microorganisms contributed in MFC has been attempted in this chapter. A list of techniques has been used in characterizing the biofilm such as scanning electron microscopy, transmission electron microscopy, confocal scanning microscopy and PCR-DGGE profiling. Many mesophiles have been worked on in the anodic chamber of the MFC (Bélafi-Bako et al. 2011; Hussain et al. 2014). More recently, a wide range of extremophiles have been concentrated, and molecular techniques have been performed to identify the nature of the microbe. There are few reports on psychrophilic microorganisms such as *Pseudomonas fragi* DRR-2 that works efficiently in microbial fuel cell (Jothinathan and Wilson 2017). Metagenomics work has been a trend in this research.

D. Jothinathan (✉)
Department of Life Sciences, Central University of Tamil Nadu, Thiruvarur, 610005, India

P. Mylsamy
Post Graduate and Research Department of Botany, Pachaiyappa's College, Chennai, 600030, India

L. B. Bruno
Department of Environmental Sciences, Bharathiar University, Coimbatore, 641046, India

V. Sivasankar et al. (eds.), *Microbial Fuel Cell Technology for Bioelectricity*,
https://doi.org/10.1007/978-3-319-92904-0_9

9.2 Electricigens

Electricigens are the microbes which actively transport the electrons to the electrode. These microorganisms exist in various natural habitats such as sewage sludge, wastewater, aquatic sediments, anaerobic sludge, marine sediments, paddy soil, compost and submerged soil.

9.2.1 Electron Transport Mechanism

In MFCs there are two electron transfer mechanism, namely, direct electron transfer and indirect electron transfer (Yan-Ping 2008). The direct electron transfer mechanism usually needs an efficient binding between the microbial cell surface attachment and the surface of the electrode. The electrons are accepted from the cytochromes present in the external part of the cell. *Shewanella putrefaciens*, *Geobacter sulfurreducens*, *Rhodoferax ferrireducens*, etc. have the ability to exchange electrons from the cell to e^- acceptors via biofilms, highly conductive appendages, namely, pili and c-type cytochromes (Lovley 2008). In an indirect electron transfer, the mechanism either takes place by bacteria's own mediators or by some chemical mediators added in the anode chamber. Mediators assist the microbes to produce electrochemically effective output. The reduced form of the mediator is cell penetrable which receives electrons from the e^- transporter to the anode surface (Lovley 2006).

Electron transfer by mediators: The electron transfer in this mechanism either takes place by bacteria's own mediators, thereby promoting extracellular electron transfer or with the help of some chemical mediators added in the anode chamber. Mediators offer a dais for the microbes to produce electrochemically energetic products (Lovley 2006). Neutral red, thionine, methylene blue, anthraquinone-2, 6-disulfonate, phenzines and iron chelates are some of the redox mediators that are commonly used in MFC (Du et al. 2007). *E. coli*, *Pseudomonas* species and *Proteus vulgaris* require a mediator as they are unable to transfer electrons outside the cell. An active mediator must enter the cell membrane and capture the electrons from the carrier in the electron transport, must be stable even after long periods of redox cycling and should not harm the microbes (Osman et al. 2010; Du et al. 2007).

9.2.2 Etymology of Microbes in Microbial Fuel Cell

Till date, there are a plenty of research articles in this field of research that use different vocabularies for expressing microbe's activity inside the microbial fuel cell. These terms came into existence because of their nature, ability to transfer electrons, origin in the chamber, respiration mechanism, etc. Below is the list of

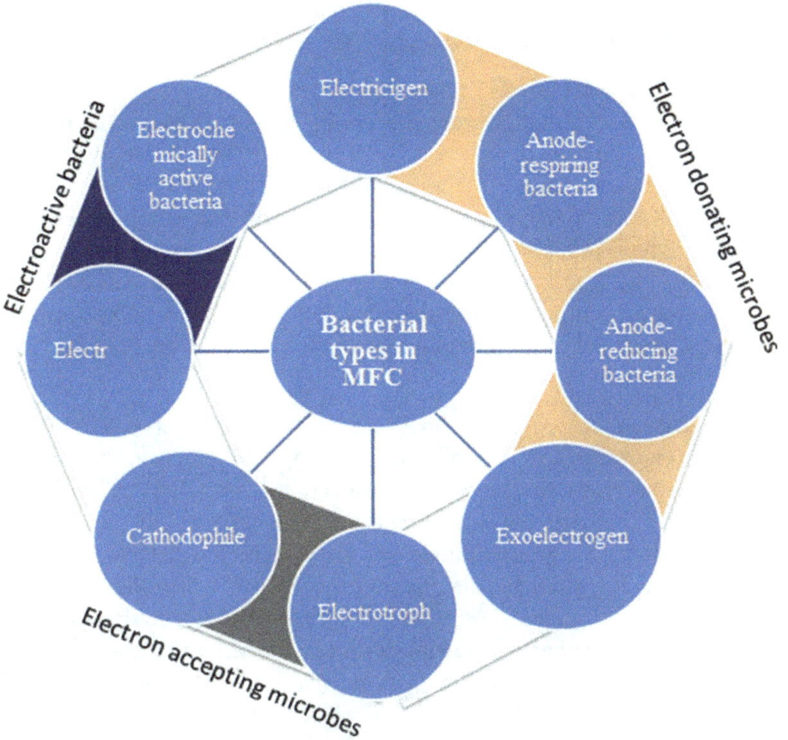

Fig. 9.1 Etymology of microbes in microbial fuel cell

different names of MFC microbes that have been cited in literature and their characteristic feature inside the chamber. A schematic representation of the microbial terms has been categorized in Fig. 9.1.

Anode-Respiring Bacteria

Anode-respiring bacteria (ARB) are the microorganisms that have the ability to conserve energy via respiration using anode as an electron acceptor. The anode potential plays a vital role in regulating the energy for ARB. The ARB community might increase energy and change the behaviour as per the anode potential having various kinds of respiratory pathways (Torres et al. 2007). It has been reported that the ARB cultures within *Shewanellaceae* family when added with iron have improved the current densities (Feng et al. 2013). This was tested experimentally by Zhang et al. (2014) where he used Fe(III) in the bioelectrochemical systems (BESs) solutions. At the end of the experiments, it was found that adding iron has (a) aided in selecting iron-reducing bacteria from the wastewater and (b) increased the current production thrice when compared to the solution without iron. These ARB have been tested against various anode potentials. A pure culture of *G. sulfurreducens* has shown two electrochemical responses one at 0.1 V and another at 0.4 V vs SHE indicating the adaptation to different anode potentials (Busalmen et al. 2008).

Another study indicated that there is rapid growth of ARB when exposed to high potential which in turn helped to start the MFC operation earlier (Wang et al. 2009).

Anode-Reducing Bacteria
These are the microbes that possess the ability to donate electrons to an anode (Logan and Regan 2006). Among these microbes, sulphur-reducing bacteria (SRB) and iron-reducing bacteria have been reported predominantly (Liu et al. 2012). SRB secrete enzymes which reduce the sulphate to hydrogen sulphide in anaerobic environment. Metal-reducing bacteria predominantly fall under the family *Geobacteraceae* that can transfer the electrons to the electrode via direct transfer mechanism using cytochromes. Other than this, *Pseudomonas* and *Desulfosporosinus* were also predominant in iron-reducing bacterial group (Yokoyama et al. 2016).

Electricigen
Electricigens are microorganisms that are well capable of oxidizing organic compounds completely to carbon dioxide. The only electron acceptor is the electrode, and the conserved energy is utilized for growth (Lovley 2006). The mechanism is based on the chemical compound being oxidized on the one hand, and the other is reduced. This group of microbes can transfer the electrons to the surface of the electrode, using them as sole electron acceptors for the effective working of MFCs. One of the advantages of using electricigens is the high coulombic efficiency. Another merit of electricigens is the long-term sustainability of the fuel cells holding them. Some of the electricigens are *Desulfuromonas acetoxidans* and *Geobacter metallireducens* (Bond et al. 2002), *Rhodoferax ferrireducens* (Finneran et al. 2003) and *Geothrix fermentans* (Bond and Lovley 2005).

Exoelectrogen
Exoelectrogens are basically described as organisms that have the potential to contribute electrons from the cell to an anode through the two different electron transfer mechanisms, namely, a direct contact mechanism or indirect mechanism, using self-secreted mediators (Oh and Logan 2007). Genetic engineering of exoelectrogens plays a major role in electron transfer by producing specific molecules that will further enhance the electricity production. For instance, a genetic manipulation was performed in *S. oneidensis* MR-1, in which a synthetic flavin biosynthesis pathway from *Bacillus subtilis* was expressed in the exoelectrogen (Yang et al. 2015).

Electrotroph
Electrotrophs are contradictory to electrogens where the microbes pull the electrons from the cathode through direct electron transfer or indirect electron mechanisms (Lovley 2011). Some electrotrophs reported in literature are *Ochrobactrum* sp. X1, *Ochrobactrum anthropi* X7, *Pseudomonas* sp. X3 and *Pseudomonas delhiensis* X5 which are sequestered from Cd(II)-removal biocathodes of MECs (Huang et al. 2018).

Cathodophile

These microorganisms favourably colonize the surface of the cathode. There is no specific explanation in literature about their electron transfer from the cathode (Rinaldi et al. 2008). They can also be described as cathode-oxidizing bacteria which preferentially attract electrons from a cathode (Martin et al. 2011).

Electrochemically Active Bacteria

Microorganisms which are capable of providing electrons or receiving electrons from an electrode through direct transfer mechanism or self-secreted mediators are known to be electrochemically active bacteria (Chang et al. 2006; Scott et al. 2007; Zhang et al. 2015). Electrochemically active bacteria (EAB) can be isolated by few techniques such as dilution-to-extinction method (Zuo et al. 2008), dilution-plate method (Wang et al. 2010) and biological laser printing method (Ringeisen et al. 2010). The dilution-to-extinction method is one of the most potent tools reported to isolate the EAB. Other techniques are quite sensitive and have their own drawbacks.

9.3 Pioneering Microbes

There is a wide range of microbes so far contributed in microbial fuel cell. Pure cultures and mixed cultures have been used in microbial fuel cells, and the power production has been monitored using high-end potentiostat. The most studied microorganisms in this field are *Geobacter* and *Shewanella* which gave better results when inoculated as pure cultures. *Geobacter* group and *Shewanella* genus are the utmost well-known bacterial group in microbial fuel cells (Coates et al. 1996). Lovely (2003) showed that *Geobacter sulfurreducens* offers 3000-fold rise in e^- movement in contrast to other microbes like *Shewanella putrefaciens*. *Shewanella* is a more legendary bacterium in microbial fuel cell, having a great usage in the biosensor field (Kim et al. 1999). It is previously known that *S. putrefaciens* and *Geobacter* group have the capacity to reduce Fe(III) with surface-active cytochromes. Other bacteria such as *Clostridium beijerinckii*, *Clostridium butyricium*, *Rhodobacter capsulatus*, *Desulfotomaculum reducens*, *Thiobacillus ferrooxidans* and *Geovibrio* genus are used in a mediatorless fuel cell (Park et al. 2001). *Klebsiella pneumoniae*, *Saccharomyces cerevisiae*, *Staphylococcus aureus* and sewage sample used in MFC produced a higher-voltage production (Dalvi et al. 2011). Certain gram-positive microbes including *Micrococcus luteus*, *Bacillus subtilis* and *Staphylococcus carnosus* were studied by cyclic voltammetry, an electrochemical method which shows that the bacteria can achieve prominent e^- transfer. This provides a prevalent ability found among bacteria. *Enterobacter aerogenes*, facultative anaerobes, are well known, strong and efficient producers of hydrogen. *Ochrobactrum anthropi* YZ-1, exoelectrogenic bacteria which were isolated from U-tube-shaped MFC, produced power using acetate as the electron donor (Zuo et al. 2008).

9.3.1 *Geobacter sp. and Shewanella sp.*

When compared to *Geobacter* species, *Shewanella* has the ability to partly oxidize a narrow range of organic acids into acetate. This process makes *Shewanella* less efficient since majority of the electrons existing in the original fuel persist in the acetate (Lovley 2006; Lanthier et al. 2008). On the other hand, *Geobacter* species have the ability to completely oxidize organic compounds to carbon dioxide, and the electron delivery is greater than 90% (Bond et al. 2002; Nevin et al. 2008). To compare them in terms of electron transfer mechanism, *Geobacter* species exhibit a direct contact to the anodic surface for the transferring electrons via redox-active proteins (Reguera et al. 2006; Holmes et al. 2006), and *Shewanella* species seems to transfer electrons to anodes through the self-mediators that are released in the medium (Lanthier et al. 2008).

Geobacter sulfurreducens is the highly power-producing species among *Geobacter* group. It has been reported that these bacteria transfer the electrons via c-type cytochromes present in the cell's outer surface (Holmes et al. 2006). *Desulfuromonas acetoxidans* is a marine bacterium that belongs to *Geobacteraceae* which could efficiently oxidize acetate in sediment MFC in the absence of mediators (Bond et al. 2002). Additionally, there have been reports on *Geobacter metallireducens*, which is a freshwater isolate that oxidizes toluene and benzoate to carbon dioxide.

9.3.2 *Pseudomonas sp.*

Among the gram-negative bacteria, *Pseudomonas* sp. has been astonishing in the power production in microbial fuel cell. Some bacterial strains such as *Shewanella* sp. and *Geobacter* sp. transfer electrons through direct electron transfer. However, certain exoelectrogens are believed to transfer electrons via self-produced metabolites which are commonly grouped under indirect electron transfer mechanism (Pham et al. 2008). *Pseudomonas* sp. is one such group of bacteria that exhibit this activity by transferring the produced electrons by its own mediators like pyocyanin (PYO) and 1-hydroxy-phenazine (OHPHZ) (Bellin et al. 2014; Chen et al. 2015). Among the various metabolites produced by the organism, pyocyanin is known to display the most electrochemical features (Vukomanovic et al. 1996). They also act as a cell-to-cell signal and play a vital role in quorum sensing (Logan 2009).

There is a recent report stating that the PYO production can be significantly increased by overexpressing phzM – one of the key genes responsible for PYO synthesis (Yong et al. (2014a, b)). This increased production of PYO by *Pseudomonas* sp. will obviously improvise the current generation (Shen et al. 2014). But the exact mechanism by which these bacteria produce the metabolite in MFC is still unclear.

9.3.3 Clostridium *sp.*

Clostridium sp. belongs to gram-positive group which is a wonderful microbe worthy of both fermentation and bioelectricity production. Hence, researchers showed a keen interest in studying this microbe's nature in bioelectrochemical systems (BES). Among this genus, *Clostridium acetobutylicum* has been concentrated in most of the studies recently. It is an obligate anaerobe, chemoorganotroph and mesophilic in nature with ideal temperature range of 10–45 °C and is capable of utilizing complex substrates (Cato et al. 1986; Wells and Wilkins 1996). In 2010, a study was done to compare the power production of *Saccharomyces cerevisiae* and *Clostridium acetobutylicum*, where a considerable current production of 10.89 mA and 10.5 mA, respectively, has been produced (Mathuriya and Sharma 2010). It is well known that this bacterium is popular in industrial fermentation exhibiting diauxic growth pattern, and thus the same was reflected when it is used in MFC producing two voltage peaks. This work was further explored by Amethist S. Finch in 2010, where it was found that *Clostridium acetobutylicum* produced acetate and butyrate in the first exponential phase corresponding to the first peak and acetone and butanol (solvents) in the second growth phase analogous to the second peak. Rather the power production in MFC, the metabolic pathway was focused by this group. It was also demonstrated that this bacterium does not require any mediators (Finch et al. 2011). Contradictory to this, another report says *Clostridium acetobutylicum* utilizes mediators such as resazurin and methylene blue (Mathuriya and Sharma 2009).

Similar to the above-mentioned bacterium, *Clostridium butyricum* was isolated from MFC fed with starch processing water and shown to reduce Fe(III) ion during the process in MFC (Park et al. 2001). In a very recent report, a consortium of sulphate-reducing bacteria and sulphate-oxidizing bacteria was inoculated in MFC fed with landfill leachate. The enriched community was dominated by *Clostridium* followed by *Desulfovibrio*, *Aeromonas* and *Tetrathiobacter.* Thus in a synchronous manner, these microbes worked efficiently to remove the pollutants from the effluent (Kumar et al. 2017).

9.3.4 Enterobacter *Species*

Among the *Proteobacteria* group, *Aeromonas* sp., *Enterobacter* sp. and *Klebsiella* sp. were reported to be the electricigens in the microbial fuel cell (Pham et al. 2003; Park et al. 2008; Rezaei et al. 2009). It has been demonstrated that *Enterobacter cloacae* was able to utilize cellulose and produce efficient electricity without any external mediators in a U-tube MFC. Initially a consortium was enriched in the MFC which could solely use cellulose as the electron donor. Later using techniques like denaturing gradient gel electrophoresis (DGGE) and band sequencing, it was revealed that the predominant bacterium was *Enterobacter cloacae* FR. It was found to be 100% identical with the strain ATCC 13047. The test strain *Enterobacter cloacae* FR was also efficient in power production of 4.8 ± 0.01 mW/m^2 comparatively to the ATCC strain with 5.4 ± 0.3 mW/m^2 for strain ATCC 13047 (Rezaei et al. 2009).

In another study, *Enterobacter cloacae* IIT-BT 08 was tested in MFC in the presence of the mediators methylene blue (MB) and methyl viologen (MV). In the presence of MV, the bacteria produced only 0.4 V and no current. However, with MB it showed 0.37 V and 56.7 μA. In 0.03 mM of MB, the power density was observed to be 9.3 mW/m^2. In comparison with the previous study, the power production was almost doubled in the presence of mediator like methylene blue. These kinds of studies insist that even though the bacteria can produce much power without the mediator, the efficiency has not been failed to improve when a mediator has been supplied in an adequate concentration (Mohan et al. 2008).

9.3.5 Aeromonas *Species*

A renowned microbe, *Aeromonas* species, has been recently concentrated for its power production and by-product formation. It has been well-known for its Fe(III)-reducing capacity. In a recent study, *Aeromonas hydrophila* YC 57 was able to produce high-power output when fed with wastewater containing crystal violet (CV). The maximum power generation was calculated as 240 ± 5.6 mW/m^2 in the single-chambered MFC. It seems that the CV metabolites were regarded as non-toxic (Cheng et al. 2014). There is an interesting report published in 2017 where *Aeromonas hydrophila* has been effectively used in chitin degradation during the course of electricity generation in MFC. Since this bacterium is worthy enough both as a fermentative organism and power-producing microbe, the chitin degradation was attempted, and to the surprise, 0.13 and 0.03 mM C/d/mM of chitin were degraded (Li et al. 2017). Based on the 16S rRNA sequencing, a novel strain ISO2–ISO3 was found to be electrochemically active in a glucose-fed microbial fuel cell. This strain was phylogenetically related to *Aeromonas sp.* The maximum current produced by this cell was 0.2 A/m^2 (Chung & Okabe 2009).

9.3.6 Saccharomyces cerevisiae

In yeast, *Saccharomyces cerevisiae* is the widely used microorganism that has been tried in microbial fuel cell for the bioelectricity production. *S. cerevisiae* can grow either in aerobic and anaerobic condition. There are strong reports stating that *S. cerevisiae* requires artificial mediators such as neutral red (NR), thionine, humic acid and methylene blue (MB) for the effective electron transfer (Rahimnejad et al. 2012b).

It has been analysed that ferricyanide reductase of *S. cerevisiae* has been characterized as a b-type cytochrome which is analogous to flavocytochrome b558 of human neutrophils (Shatwell et al. 1996). As we discussed earlier, the electricigens present on the anode surface directly transfer the electrons to the anode via specific cytochromes available on the surface of the membrane. In case of *S. cerevisiae*, the electron transfer to the electron acceptors is accompanied by metal-reductase receptors (Holmes et al. 2006; Gorby et al. 2006; Hubenova and Mitov 2015). *Saccharomyces cerevisiae* reduce external metals by using numerous metal-reductase enzymes and carry these metals into the cell (Eide 1998). Recently, Rossi et al. (2015) has exclusively investigated the performance of yeast in microbial fuel cell. He also studied the fermentation of glucose by yeast in the presence and absence of mediators and observed that there was an increase in ethanol production in the absence of mediator and decrease in ethanol production in the presence of mediators. This supports the statement said by Babanova et al. (2011) where methylene blue acted as an electron acceptor and aided in regenerating yeast's coenzymes. *S. cerevisiae* has also been applied in desalination microbial fuel cell to remove NaCl from marine water. After 30 days of operation, 64% of ions moved from the dilute compartment to the concentrate compartment (Mardiana et al. 2016).

Some yeast varieties such as *Candida melibiose* (Hubenova and Mitov 2010) and *Arxula adeninivorans* (Haslett et al. 2011) have been also investigated in the microbial fuel cell apart from *S. cerevisiae*.

9.3.7 Other Microbes

When coculture of *Pelobacter carbinolicus* and *Geobacter sulfurreducens* was used with ethanol as the fuel, current was generated. As a pure culture, *P. carbinolicus* could not able to produce current. Likewise, *G. sulfurreducens* is unable to break down ethanol which implies that in a coculture form, *P. carbinolicus* metabolize ethanol to hydrogen and acetate. This is taken by *Geobacter sulfurreducens* for the electron transfer to the anode (Richter et al. 2007). The list of various microbes that have been shown to produce power is provided in Table 9.1.

9.4 Characterization of Biofilm

9.4.1 Scanning Electron Microscopy

The bacterial associations on the anode surfaces were studied using the images from the scanning electron microscopy. It uses an electron beam on the specified sample and scans them to receive an image of high magnification. These images give detailed information about the morphology of the sample or microbe, 3-D profile of the biofilm. This is not a direct technique to confirm the electrochemical activity but can provide supporting information of the bacterial adhesion towards the electrode surface. For instance, the presence of conductive pili and nanowires can be clearly observed using the scanning tunneling electron microscopy. Nanowires can

Table 9.1 List of electrogens that are employed in microbial fuel cell

S. No.	Microorganism	Characteristic features	Power production in MFC	References
1	*Arcobacter butzleri*	Gram-negative, curved rod, aerotolerant	296 mW/L	Fedorovich et al. (2009)
2	*Aeromonas hydrophila*	Gram-negative rod, heterotrophic	240 ± 5.6 mW/m^2	Cheng et al. (2014)
3	*Aeromonas jandaei*	Gram-negative, short rods, facultative anaerobic bacteria, motile	Not mentioned	Sharma et al. (2016)
4	*Klebsiella pneumoniae* L17	Gram-negative rods, facultative anaerobic bacteria, nonmotile, encapsulated	409.71 mW/m^2	Zhang et al. (2008)
5	*Klebsiella oxytoca* ADR13	Gram-negative rods, diazotroph	4.87 mW/m^2	Kingsly et al. (2017)
6	*Corynebacterium humireducens* sp. nov.	Gram-positive, facultative anaerobic, nonmotile, halotolerant, humic acid-reducing bacterium	Not mentioned	Wu et al. (2011)
7	*Citrobacter* sp.	Gram-negative coliform bacteria, facultative anaerobic, motile	88.1 mW/m^2	Xu & Liu. (2011)
8	*Cupriavidus basilensis*	Gram-negative rod, facultative aerobe, motile	44 mW/m^2	Friman et al. (2013)
9	*Shewanella oneidensis* DSP10	Gram-negative, facultative anaerobic, metal-reducing bacterium	24 mW/m^2	Ringeisen et al. (2006)
10	*Pseudomonas* sp. C27	Gram-negative curved rod, motile, autotrophic denitrifier	40 mW/m^2	Lee et al. (2012)
11	*Ochrobactrum anthropi* YZ-1	Gram-negative rod, aerobic, motile and non-pigmented	89 mW/m^2	Zuo et al. (2008)

facilitate electron transfer from bacterial cells to the surface of the electrode (Gorby et al. 2006). The understanding of the microbial attraction towards electrode is still in the early stage, and the mechanism of the bacterial proteins that are involved in the process has to be studied in the near future. This might help in concentrating of using less chemical mediators which are toxic to the environment.

The major disadvantage of SEM is their low resolution in comparison with transmission electron microscope (TEM). Samples that are handled in SEM must resist the low pressure inside the vacuum chamber (Niemantsverdriet 2007).

9.4.2 Atomic Force Spectroscopy

Atomic force spectroscopy (AFM) is used to study the cell-electrode interface and highly conductive pili (Gorby et al. 2006; Reguera et al. 2005). The shape and surface characteristic of the anode electrode can be determined by AFM (Rahimnejad et al. 2012a). The AFM tip can be either used in direct interaction mode, where the tip actually comes in contact with the surface or in indirect mode where vander Waals forces helps in the interface between the tip and the sample surface. Contradictory to SEM, AFM does not require sample processing, and there is no modification of the sample. It can be operated in an ambient air or in liquid conditions (Niemantsverdriet 2007). Since there is no special procedure required for sample preparation, the biological samples, bacterial biofilm and macromolecules can be examined easily by AFM. This technique can be used to analyse properties such as surface charges, molecular interactions, mechanical properties and electrochemical characters (Dufrêne 2010). There are few reports where AFM has been widely used in studying the EPS formation in the biofilm (Beech et al. 2002). The drawback of AFM is the small area of the scanned image, and it is extremely sensitive to superficially produced electrical noises (Niemantsverdriet 2007).

9.4.3 Confocal Scanning Laser Microscopy

Confocal scanning laser microscopy (CSLM) has been used to screen the live and dead bacterial cells in the biofilm. Apart from this, the bacterial profiling and architecture can be studied by CSLM. It enables to take an array of optical sections of complete biological samples as thin as 0.3 μm. CSLM is used in microbial fuel cell systems that involve pure and cocultures and more commonly used in as an ex situ method (Astner 2010; Murphy and Davidson 2012). The electrodes containing biofilm were removed from the system and washed fluorescently stained with the LIVE/DEAD BacLight bacterial viability kit (L7012, Molecular Probes, Inc., Eugene, OR) and examined with a Zeiss LSM700. The confocal microscope was equipped with an Argon laser and two HeNe lasers, a krypton-argon dual laser and a diode laser (Manz et al. 1999).

This technique will reveal the live microbial cells in green colour and dead cells in red colour. Ramasamy et al. (2008) reported that after 2 weeks of MFC operation, the biofilm forms thick aggregates. It was also reported that the live cells of *Geobacter sulfurreducens* was favourably colonized on the anodic surface directly and the dead cells were predominantly located in the topmost layer of the biofilm (Reguera et al. 2006).

9.4.4 Thermogravimetric Analysis

Thermogravimetric analysis is a method which is recently used in microbial fuel cell research to observe the differences in the weight loss of the modified electrodes during the course of operation with respect to the temperature (Kramer et al. 2012). CSLM is commonly used to study the biofilm thickness; however TGA is comparatively a cost-effective, rapid and simple method to characterize the material in MFC and will surely be used in the future for the semi-quantitative measurement of biofilms. However it lacks the accuracy as in the case of CSLM. A recent report suggests that between 100 and 200 °C, there is a 12% change of weight loss after 14 days of MFC operation and no change before the process (Baranitharan et al. 2015). In this temperature range, the organic matter has been suspected to decompose, thus creating a variation in microbial biofilm (Kramer et al. 2012).

9.4.5 DGGE and Sequence Analysis

Denaturing gradient gel electrophoresis studies are generally carried out to find the presence of efficient electrochemically active bacteria in the bacterial consortia. This technique is used in environmental microbiology to study the microbial community and structures (Chen and Kucernak 2004). Ha et al. (2008) conducted an experiment with MFCs fed with formate and acetate. The bacteria prevalent in the consortia were not given the proper environment. A majority of the population in the culture were yet present in the anodic biofilm of the microbial fuel cell. Nevertheless, there were some observable variances on the gels, especially the DNA fragments that possess a greater G + C content. The DGGE gel of DNA sampled from the MFC fed with acetate presented limited bands than the sample obtained from the formate-fed MFCs. *Acetobacterium* sp. was commonly observed in the final DNA sample from formate-fed MFCs based on the DGGE analysis. This demonstrated the normal presence of an *Acetobacterium* sp. in the formate-nourished MFCs. It was expected that these microaerophiles are mostly available in the inoculum. Furthermore, the anodic chamber of the fuel cells could possess micro-aerobic conditions due to the oxygen disseminating via cathodic chamber of the microbial fuel cells (Ha et al. 2008).

9.5 Summary and Conclusion

Electricigens are considered as a key factor for the bioelectricity production in the microbial fuel cell. The microorganisms tend to grow in various sources such as industrial effluents, marine water and domestic wastewater and remediate the pollutants during the course of MFC operation. Based on the electron transfer method, they are likely to derive different names. Electrochemically active bacteria are the microbes that can be able to produce and transfer the electrons on their own to the electrode surface without any added mediators. Researchers currently search for more EABs so that the need of mediators can be reduced. The biofilm formation on the electrode surface plays a critical role in the power production, and thus its characterization will certainly boost up the investigation in this area.

Acknowledgement Dr. J. Deepika is very much thankful to the Head of the Department, Department of Life Sciences, Central University of Tamil Nadu, for motivating her in writing the chapter. Dr. L. Benedict Bruno thankfully acknowledges the Science and Engineering Board, National Postdoctoral fellowship Scheme (Grant No.PDF/2017/001074), for financial support.

References

Astner S (2010) Ulrich M. Confocal laser scanning microscopy. Hautarzt 61:421–428

Babanova S, Hubenova Y, Mitov M (2011) Influence of artificial mediators on yeast-based fuel cell performance. J Biosci Bioeng 112(4):379–387

Baranitharan E, Khan MR, Prasad DM, Teo WF, Tan GY, Jose R (2015) Effect of biofilm formation on the performance of microbial fuel cell for the treatment of palm oil mill effluent. Bioprocess Biosyst Eng 38(1):15–24

Beech IB, Smith JR, Steele AA, Penegar I, Campbell SA (2002) The use of atomic force microscopy for studying interactions of bacterial biofilms with surfaces. Colloids Surf B: Biointerfaces 23(2):231–247

Bélafi-Bako K, Vajda B, Nemestothy N (2011) Study on operation of a microbial fuel cell using mesophilic anaerobic sludge. Desalination Water Treat 35(1–3):222–226

Bellin DL, Sakhtah H, Rosenstein JK, Levine PM, Thimot J, Emmett K, Dietrich LE, Shepard KL (2014) Integrated circuit-based electrochemical sensor for spatially resolved detection of redox-active metabolites in biofilms. Nat Commun 5:3256

Bond DR, Lovley DR (2005) Evidence for involvement of an electron shuttle in electricity generation by *Geothrix fermentans*. Appl Environ Microbiol 71:2186–2189

Bond DR, Holmes DE, Tender LM, Lovley DR (2002) Electrode-reducing microorganisms that harvest energy from marine sediments. Science 295(5554):483–485

Busalmen JP, Esteve-Nunez A, Feliu JM (2008) Whole cell electrochemistry of electricity-producing microorganisms evidence an adaptation for optimal exocellular electron transport. Environ Sci Technol 42(7):2445–2450

Cato EP, George WL, Finegold SM (1986) Genus *Clostridium*. In: Sneath PHA et al (eds) Bergey's manual of systematic bacteriology, vol 2. Williams and Wilkins, Baltimore, MD, pp 1141–1200

Chang IS, Moon H, Bretschger O, Jang JK, Park HI, Nealson KH, Kim BH (2006) Electrochemically active bacteria (EAB) and mediator-less microbial fuel cells. J Microbiol Biotechnol 16:163–177

Chen SL, Kucernak A (2004) Electrocatalysis under conditions of high mass transport: investigation of hydrogen oxidation on single submicron Pt particles supported on carbon. J Phys Chem B 108(37):13984–13994

Chen W, Liu X-Y, Qian C, Song X-N, Li W-W, Yu H-Q (2015) An UV–vis spectroelectrochemical approach for rapid detection of phenazines and exploration of their redox characteristics. Biosens Bioelectron 64:25–29

Cheng CY, Liang FY, Chung YC (2014) Electricity generation from crystal violet using a single-chambered microbial fuel cell inoculated *Aeromonas hydrophila* YC 57. Adv Mater Res 860:466–471

Chung K, Okabe S (2009) Characterization of electrochemical activity of a strain ISO2-3 phylogenetically related to *Aeromonas sp.* isolated from a glucose-fed microbial fuel cell. Biotechnol Bioeng 104(5):901–910

Coates JD, Phillips EJP, Lonergan DJ, Jenter H, Lovely DR (1996) Isolation of *Geobacter species* from diverse sedimentary environments. Appl Environ Microbiol 62:1531–1536

Dalvi AD, Mohandas N, Shinde OA, Kininge PT (2011) Microbial fuel cell for production of bioelectricity from whey and biological waste treatment. Int J Adv Biotechnol Res 2(2):263–268

Du Z, Li H, Gu T (2007) A state of the art review on microbial fuel cells: a promising technology for wastewater treatment and bioenergy. Biotechnol Adv 25:464–482

Dufrêne YF (2010) Atomic force microscopy of fungal cell walls: an update. Yeast 27(8):465–471

Eide DJ (1998) The molecular biology of metal ion transport in *Saccharomyces cerevisiae*. Annu Rev Nutr 18:1–469

Fedorovich V, Knighton MC, Pagaling E, Ward FB, Free A, Goryanin I (2009) Novel electrochemically active bacterium phylogenetically related to *Arcobacter butzleri*, isolated from a microbial fuel cell. Appl Environ Microbiol 75(23):7326–7334

Feng C, Yue X, Li F, Weia C (2013) Bio-current as an indicator for biogenic Fe(II) generation driven by dissimilatory iron reducing bacteria. Biosens Bioelectron 39(1):51–56

Finch AS, Mackie TD, Sund CJ, Sumner JJ (2011) Metabolite analysis of *Clostridium acetobutylicum*: fermentation in a microbial fuel cell. Bioresour Technol 102(1):312–315

Finneran K, Johnsen CV, Lovley DR (2003) *Rhodoferax ferrireducens* sp. nov., a psychrotolerant, facultatively anaerobic bacterium that oxidizes acetate with the reduction of Fe(III). Int J Syst Evol Microbiol 53:669–673

Fournier D, Trott S, Hawari J, Spain J (2005) Metabolism of the aliphatic nitramine 4-nitro-2,4-diazabutanal by *Methylobacterium sp.* strain JS178. Appl Environ Microbiol 71(8):4199–4202

Friman H, Schechter A, Ioffe Y, Nitzan Y, Cahan R (2013) Current production in a microbial fuel cell using a pure culture of *Cupriavidus basilensis* growing in acetate or phenol as a carbon source. Microb Biotechnol 6(4):425–434

Gorby YA, Yanina S, McLean JS, Rosso KM, Moyles D, Dohnalkova A, Beveridge TJ, Chang IS, Kim BH, Kim KS, Culley DE, Reed SB, Romine MF, Saffarini DA, Hill EA, Shi L, Elias DA, Kennedy DW, Pinchuk G, Watanabe K, Ishii S, Logan B, Nealson KH, Fredrickson JK (2006) Electrically conductive bacterial nanowires produced by *Shewanella oneidensis* strain MR-1 and other microorganisms. Proc Natl Acad Sci U S A 103(30):11358–11363

Ha PT, Tae B, Chang IS (2008) Performance and bacterial consortium of microbial fuel cell fed with formate. Energy Fuel 22:164–168

Haslett ND, Rawson FJ, Barrière F, Kunze G, Pasco N, Gooneratne R, Baronian KHR (2011) Characterisation of yeast microbial fuel cell with the yeast *Arxula adeninivorans* as the biocatalyst. Biosens Bioelectron 26(9):3742–3747

Holmes DE, Chaudhuri SK, Nevin KP, Mehta T, Methe BA, Liu A, Ward JE, Woodard TL, Webster J, Lovley DR (2006) Microarray and genetic analysis of electron transfer to electrodes in *Geobacter sulfurreducens*. Environ Microbiol 8(10):1805–1815

Huang L, Xue H, Zhou Q, Zhou P, Quan X (2018) Imaging and distribution of Cd (II) ions in electrotrophs and its response to current and electron transfer inhibitor in microbial electrolysis cells. Sensors Actuators B Chem 255:244–254

Hubenova Y, Mitov M (2010) Bioelectrochemistry potential application of *Candida melibiosica* in biofuel cells. Bioelectrochemistry 78(1):57–61

Hubenova Y, Mitov M (2015) Extracellular electron transfer in yeast–based biofuel cells: a review. Bioelectrochemistry 106:177–185

Hussain A, Bruant G, Mehta P, Raghavan V, Tartakovsky B, Guiot SR (2014) Population analysis of mesophilic microbial fuel cells fed with carbon monoxide. Appl Biochem Biotechnol 172(2):713–726

Jothinathan D, Wilson RT (2017) Production of bioelectricity in MFC by *Pseudomonas fragi* DRR-2 (psychrophilic) isolated from goat rumen fluid. Energy Sources A 39(4):433–440

Kim HJ, Park DH, Hyun MS, Chang IS, Kim M, Kim BH (1999) Mediatorless fuel cell. US Patent 5,976,719

Kingsly A, Jothinathan D, Thilagaraj WR (2017) Degradation of oleic acid and simultaneous bioelectricity production by *Klebsiella oxytoca* ADR 13. Energy Sources A 39(9):874–882

Kramer J, Soukiazian S, Mahoney S, Hicks-Garner J (2012) Microbial fuel cell biofilm characterization with thermogravimetric analysis on bare and polyethyleneimine surface modified carbon foam anodes. J Power Sources 210:122–128

Kumar SS, Malyan SK, Basu S, Bishnoi NR (2017) Syntrophic association and performance of *Clostridium*, *Desulfovibrio*, *Aeromonas* and *Tetrathiobacter* as anodic biocatalysts for bioelectricity generation in dual chamber microbial fuel cell. Environ Sci Pollut Res 23:1–2

Lanthier M, Gregory KB, Lovley DR (2008) Growth with high planktonic biomass in *Shewanella oneidensis* fuel cells. FEMS Microbiol Lett 278(1):29–35

Lee CY, Ho KL, Lee DJ, Su A, Chang JS (2012) Electricity harvest from nitrate/sulfide-containing wastewaters using microbial fuel cell with autotrophic denitrifier, *Pseudomonas* sp. C27. Int J Hydrog Energy 37(20):15827–15832

Li SW, He H, Zeng RJ, Sheng GP (2017) Chitin degradation and electricity generation by *Aeromonas hydrophila* in microbial fuel cells. Chemosphere 168:293–299

Liu F, Zhang J, Zhang S, Li W, Duan J, Hou B (2012) Effect of sulphate reducing bacteria on corrosion of Al Zn In Sn sacrificial anodes in marine sediment. Mater Corros 63(5):431–437

Logan BE (2009) Exoelectrogenic bacteria that power microbial fuel cells. Nat Rev Microbiol 7:375–381

Logan BE, Regan JM (2006) Electricity-producing bacterial communities in microbial fuel cells. Trends Microbiol 14:512–518

Lovley DR (2003) Electricity production by *Geobacter sulfurreducens* attached to electrodes. Appl Environ Microbiol 69(3):1548–1555

Lovley DR (2006) Bug juice: harvesting electricity with microorganisms. Nat Rev Microbiol 4(7):497–508

Lovley DR (2008) The microbe electric: conversion of organic matter to electricity. Curr Opin Biotechnol 19:1–8

Lovley DR (2011) Powering microbes with electricity: direct electron transfer from electrodes to microbes. Environ Microbiol Rep 3:27–35

Manz W, Wendt-Potthoff K, Neu TR, Szewzyk U, Lawrence JR (1999) Phylogenetic composition, spatial structure, and dynamics of lotic bacterial biofilms investigated by fluorescent in situ hybridization and confocal laser scanning microscopy. Microb Ecol 37:225–237

Mardiana U, Innocent C, Cretin M, Buchari B, Gandasasmita S (2016) Yeast fuel cell: application for desalination. IOP Conf Ser Mater Sci Eng 107:012049

Martin E, Tartakovsky B, Savadogo O (2011) Cathode materials evaluation in microbial fuel cells: a comparison of carbon, Mn_2O_3, Fe_2O_3 and platinum materials. Electrochim Acta 58:58–66

Mathuriya AS, Sharma VN (2009) Bioelectricity production from paper industry waste using a microbial fuel cell by *Clostridium species*. J Biochem Technol 1(2):49–52

Mathuriya AS, Sharma VN (2010) Electricity generation by *Saccharomyces cerevisiae* and *Clostridium acetobutylicum* via microbial fuel cell technology: a comparative study. Adv Biol Res 4(4):217–223

Mohan Y, Kumar SM, Das D (2008) Electricity generation using microbial fuel cells. Int J Hydrog Energy 33(1):423–426

Murphy DB, Davidson MW (2012) Confocal laser scanning microscopy. In: Fundamentals of light microscopy and electronic imaging. Wiley, New York, pp 265–305

Nevin KP, Richter H, Covalla SF, Johnson JP, Woodard TL, Orloff AL, Jia H, Zhang M, Lovley DR (2008) Power output and columbic efficiencies from biofilms of *Geobacter sulfurreducens* comparable to mixed community microbial fuel cells. Environ Microbiol 10(10):2505–2514

Niemantsverdriet JW (2007) Spectroscopy in catalysis. Wiley-VCH Verlag GmbH & Co. KGaA, Weinheim

Oh SE, Logan BE (2007) Voltage reversal during microbial fuel cell stack operation. J Power Sources 167:11–17

Osman MH, Shah AA, Walsh FC (2010) Recent progress and continuing challenges in bio-fuel cells part II: microbial. Biosens Bioelectron 26:953–963

Park HS, Kim BH, Kim HS, Kim HJ, Kim GT, Kim M, Chang IS, Park YK, Chang HI (2001) A novel electrochemically active and Fe (III)-reducing bacterium phylogenetically related to *Clostridium butyricum* isolated from a microbial fuel cell. Anaerobe 7(6):297–306

Park HI, Sanchez D, Cho SK, Yun M (2008) Bacterial communities on electron-beam Pt-deposited electrodes in a mediator-less microbial fuel cell. Environ Sci Technol 42(16):6243–6249

Pham CA, Jung SJ, Phung NT, Lee J, Chang IS, Kim BH, Yi H, Chun J (2003) A novel electrochemically active and Fe (III)-reducing bacterium phylogenetically related to *Aeromonas hydrophila*, isolated from a microbial fuel cell. FEMS Microbiol Lett 223(1):129–134

Pham TH, Boon N, Aelterman P, Clauwaert P, De Schamphelaire L, Vanhaecke L, De Maeyer K, Hoefte M, Verstraete W, Rabaey K (2008) Metabolites produced by *Pseudomonas* sp. enable a gram-positive bacterium to achieve extracellular electron transfer. Appl Microbiol Biotechnol 77:1119–1129

Rahimnejad M, Ghoreyshi AA, Najafpour GD, Younesi H, Shakeri M (2012a) A novel microbial fuel cell stack for continuous production of clean energy. Int J Hydrog Energy 37:5992–6000

Rahimnejad M, Najafpour GD, Ghoreyshi AA, Talebnia F, Premier GC, Bakeri G, Kim JR, Oh SE (2012b) Thionine increases electricity generation from microbial fuel cell using *Saccharomyces cerevisiae* and exoelectrogenic mixed culture. J Microbiol 50:575–580

Ramasamy RP, Ren Z, Redcloud-Owen S, Mench MM, Regan JM (2008) Effect of biofilm properties on the electrochemical performance of microbial fuel cells. ECS Trans 13(21):11–17

Reguera G, McCarthy KD, Mehta T, Nicoll JS, Tuominen MT, Lovley DR (2005) Extracellular electron transfer via microbial nanowires. Nature 435:1098–1101

Reguera G, Nevin KP, Nicoll JS, Covalla SF, Woodard TL, Lovley DR (2006) Biofilm and nanowire production leads to increased current in *Geobacter sulfurreducens* fuel cells. Appl Environ Microbiol 72(11):7345–7348

Reimers CE, Tender LM, Fertig S, Wang W (2001) Harvesting energy from the marine sediment-water interface. Environ Sci Technol 35:192–195

Rezaei F, Xing D, Wagner R, Regan JM, Richard TL, Logan BE (2009) Simultaneous cellulose degradation and electricity production by *Enterobacter cloacae* in a microbial fuel cell. Appl Environ Microbiol 75(11):3673–3678

Richter H, Lanthier M, Nevin KP, Lovley DR (2007) Lack of electricity production by *Pelobacter carbinolicus* indicates that the capacity for Fe (III) oxide reduction does not necessarily confer electron transfer ability to fuel cell anodes. Appl Environ Microbiol 73(16):5347–5353

Rinaldi A, Mecheri B, Garavaglia V, Licoccia S, Di Nardo P, Traversa E (2008) Engineering materials and biology to boost performance of microbial fuel cells: a critical review. Energy Environ Sci 1:417–429

Ringeisen BR, Henderson E, Wu PK, Pietron J, Ray R, Little B, Biffinger JC, Jones-Meehan JM (2006) High power density from a miniature microbial fuel cell using *Shewanella oneidensis* DSP10. Environ Sci Technol 40(8):2629–2634

Ringeisen BR, Lizewski SE, Fitzgerald LA, Biffinger JC, Knight CL, Crookes-Goodson WJ, Wu PK (2010) Single cell isolation of bacteria from microbial fuel cells and Potomac River sediment. Electroanalysis 22(7–8):875–882

Rossi R, Fedrigucci A, Setti L (2015) Characterization of electron mediated microbial fuel cell by *Saccharomyces cerevisiae*. Chem Eng Trans 43:337–342

Scott K, Rimbu GA, Katuri KP, Prasad KK, Head IM (2007) Application of modified carbon anodes in microbial fuel cells. Process Saf Environ Prot 85:481–488

Sharma SC, Sun Q, Li J, Wang Y, Suanon F, Yang J, Yu CP (2016) Decolorization of azo dye methyl red by suspended and co-immobilized bacterial cells with mediators anthraquinone-2, 6-disulfonate and Fe_3O_4 nanoparticles. Int Biodeterior Biodegrad 112:88–97

Shatwell KP, Dancis A, Cross AR, Klausnert RD (1996) The FRE1 ferric reductase of *Saccharomyces cerevisiae* is a cytochrome b similar to that of NADPH oxidase. J Biol Chem 271(24):14240–14244

Shen HB, Yong XY, Chen YL, Liao ZH, Si RW, Zhou J, Wang SY, Yong YC, OuYang PK, Zheng T (2014) Enhanced bioelectricity generation by improving pyocyanin production and membrane permeability through sophorolipid addition in *Pseudomonas aeruginosa*-inoculated microbial fuel cells. Bioresour Technol 167:490–494

Song Z, Zhou J, Wang J, Yan B, Du C (2003) Decolorization of azo dyes by *Rhodobacter sphaeroides*. Biotechnol Lett 25(21):1815–1818

Torres CI, Marcus AK, Rittmann BE (2007) Kinetics of consumption of fermentation products by anode-respiring bacteria. Appl Microbiol Biotechnol 77:689–697

Vukomanovic DV, Zoutman DE, Marks GS, Brien JF, van Loon GW, Nakatsu K (1996) Analysis of pyocyanin from *Pseudomonas aeruginosa* by adsorptive stripping voltammetry. J Pharmacol Toxicol Methods 36(2):97–102

Wang X, Feng YJ, Ren NQ, Wang HM, Lee H, Li N, Zhao QL (2009) Accelerated start-up of two-chambered microbial fuel cells: effect of anodic positive poised potential. Electrochim Acta 54(3):1109–1114

Wang A, Sun D, Ren N, Liu C, Liu W, Logan BE, Wu WM (2010) A rapid selection strategy for an anodophilic consortium for microbial fuel cells. Bioresour Technol 101(14):5733–5735

Wells CL, Wilkins TD (1996) Clostridia: sporeforming anaerobic bacilli. In: Baron S et al (eds) Baron's medical microbiology, 4th edn. University of Texas Medical Branch, Galveston, TX

Wu CY, Zhuang L, Zhou SG, Li FB, He J (2011) *Corynebacterium humireducens* sp. nov., an alkaliphilic, humic acid-reducing bacterium isolated from a microbial fuel cell. Int J Syst Evol Microbiol 61(4):882–887

Xu S, Liu H (2011) New exoelectrogen *Citrobacter sp.* SX-1 isolated from a microbial fuel cell. J Appl Microbiol 111(5):1108–1115

Yang Y, Ding Y, Hu Y, Cao B, Rice SA, Kjelleberg S, Song H (2015) Enhancing bidirectional electron transfer of *Shewanella oneidensis* by a synthetic flavin pathway. ACS Synth Biol 4(7):815–823

Yan-Ping XMYS (2008) Preliminary study on *E.coli* microbial fuel cell and on-electrode taming of the biocatalyst. Chin J Process Eng 8:179–1184

Yokoyama H, Ishida M, Yamashita T (2016) Comparison of anodic community in microbial fuel cells with iron oxide-reducing community. J Microbiol Biotechnol 26(4):757–762

Yong XY, Feng J, Chen YL, Shi DY, Xu YS, Zhou J, Wang SY, Xu L, Yong YC, Sun YM (2014a) Enhancement of bioelectricity generation by cofactor manipulation in microbial fuel cell. Biosens Bioelectron 56:19–25

Yong XY, Shi DY, Chen YL, Jiao F, Lin X, Zhou J, Wang SY, Yong YC, Sun YM, OuYang PK (2014b) Enhancement of bioelectricity generation by manipulation of the electron shuttles synthesis pathway in microbial fuel cells. Bioresour Technol 152:220–224

Zhang L, Zhou S, Zhuang L, Li W, Zhang J, Lu N, Deng L (2008) Microbial fuel cell based on *Klebsiella pneumoniae* biofilm. Electrochem Commun 10(10):1641–1643

Zhang J, Zhang Y, Liu B, Dai Y, Quan X, Chen S (2014) A direct approach for enhancing the performances of a microbial electrolysis cell (MEC) combined anaerobic reactor dosing ferric iron: enrichment and isolation of Fe(III) reducing bacteria. Chem Eng J 248:223–229

Zhang YC, Jiang ZH, Ying LI (2015) Application of electrochemically active bacteria as anodic biocatalyst in microbial fuel cells. Chin J Anal Chem 43(1):155–163

Zuo Y, Xing D, Regan JM, Logan BE (2008) Isolation of the exoelectrogenic bacterium *Ochrobactrum anthropi* YZ-1 by using a U-tube microbial fuel cell. Appl Environ Microbiol 74(10):3130–3137

Chapter 10
Rumen Fluid Microbes for Bioelectricity Production: A Novel Approach

Deepika Jothinathan, Prabhakaran Mylsamy, and L. Benedict Bruno

10.1 Introduction

Microbial fuel cell (MFC) is a promising technology for the production of sustainable energy. Almost all wastewater and waste have been tried out as the raw material for the energy production under mesophilic conditions. In this chapter, power production of rumen microbes is being discussed. The utilization of slaughterhouse waste is compared to the other wastewater collected from industries. Slaughterhouse waste comprises of blood, skin, digestive contents, etc. Ruminants such as cow, sheep, camel, etc. have a four compartmental stomach comprising of rumen, omasum, abomasum, and reticulum. While slaughtering the ruminants, the rumen fluid is thrown away as a waste. Million tonnes of rumen fluid gets wasted in slaughter houses. The wastewater from slaughterhouse is heavy in pollution, and, therefore, it should not be allowed to mix with the municipal drain system without pretreatment meeting sewage standards as per the Bureau of Indian Standards (BIS).

In a large slaughterhouse per day, more than 200 large animals are slaughtered, and annually 40,000 animals are slaughtered approximately, which create a waste of 6–7 tonnes/day. To efficiently convert this waste into energy, microbial fuel cell can be employed.

One milliliter of rumen contains roughly 10–50 billion bacteria and 1 million protozoa, and certain yeast and anaerobic fungi also comprise the group. *Fibrobacter (Bacteroides) succinogenes*, *Ruminococcus albus*, *Ruminococcus flavefaciens*,

D. Jothinathan (✉)
Department of Life Sciences, Central University of Tamil Nadu, Thiruvarur, 610005, India

P. Mylsamy
Post Graduate and Research Department of Botany, Pachaiyappa's College, Chennai, 600030, India

L. Benedict Bruno
Department of Environmental Sciences, Bharathiar University, Coimbatore, 641046, India

V. Sivasankar et al. (eds.), *Microbial Fuel Cell Technology for Bioelectricity*,
https://doi.org/10.1007/978-3-319-92904-0_10

Butyrivibrio fibrisolvens, *Streptococcus bovis*, *Succinimonas amylolytica*, *Selenomonas ruminantium*, *Succinivibrio dextrinosolvens*, *Lactobacillus sp.*, *Anaerovibrio lipolytica*, *Eubacterium ruminantium*, *Oxalobacter formigenes*, *Methanosarcina barkeri*, *Wolinella succinogenes*, *Megasphaera elsdenii*, etc. are the common bacteria found in the rumen. These bacteria are anaerobic and carbohydrate fermenters.

Microbes in rumen exist either to the rumen epithelium or feed and free floating cells in rumen fluid portion (Chen et al. 2008). These microorganisms help in the degradation of ingested plant material. As per Cheng and Costerton (1980), rumen was considered as an ecosystem for studying the microbial behavior that is adhered to the biological surfaces. In 2007, Rismani-yazdi et al demonstrated that rumen microbes in MFC depend on the inoculum source and size and substrate composition. Cellulolytic bacterium is the most active species involved in the digestion of plant cell walls due to its high cellulase and hemicellulase activity. It produces hydrogen acetate and succinate as end products. Electrochemically active microorganisms are also present in the rumen. Similarly in other MFC reports, the physical and chemical parameters affected the performance of the microbes in the system (Reimers et al. 2007). Chen (2010) observed that, in the presence of protozoa, ruminal redox potential was more negative and produced a higher maximal voltage output of 595 mV (Chen 2010).

This chapter gives an insight of the power production by rumen microbes in MFC. The parameters that are favorable for the biofilm formation and electron transfer are tested. The bacterial strains isolated were checked for their efficiency in bioelectricity generation, and also their electrochemical activity was tested using cyclic voltammetry (CV) and electron impedance spectroscopy (EIS) techniques.

10.1.1 Optimization of Parameters for the Increased Electricity Production by the Microbial Fuel Cell Using Rumen Fluid

(a) The first parameter was electrodes where copper and zinc electrodes gave a maximum of 840 mV and 0.820 mA. However, the voltage dropped after the 4th day drastically and reached 100 mV. Carbon electrodes produced a stable voltage and current of 540 mV and 0.510 mA, respectively. Graphite and stainless steel produced 300 mV and 0.420 mA and 90 mV and 0.320 mA, respectively. Aluminum produced a negligible amount of voltage and current. Since carbon electrodes produced a considerable power, it was used for the further experiments. In an earlier report, carbon paper used as anode produced 14.92 mA with sugar industry wastewater as the anolyte (Mathuriya and Sharma 2009). MFC with carbon cloth utilizing beer brewery wastewater produced 63 mW/cm^2 as reported by Feng et al. (2008). In another report, plain graphite plates which were used as anode in a dual-chambered MFC produced 271.5 mV which has confirmed that carbon is the best electrode material in MFC (Venkatamohan et al. 2008).

(b) The second parameter being the pH which plays a vital role in many biological experiments was selected ranging from 5.0 to 9.0 in the anode chamber, and the results showed that pH 7.0 gave a maximum voltage of 590 mV and 0.420 mA. When the rumen fluid pH is changed to acidic, the voltage and current production is increased. When it is alkaline, the voltage production was stable. Various studies have focused mainly on the pH of the medium in anode chamber. For instance, the anodic sludge of pH 6.0–6.8 gave a power density of 10.4 W/cm^3, and when the pH was increased to 7.55, the power output increased to 11.8 W/m^3 (Jiang et al. 2009). Beer brewery wastewater produced 10.92 mA of current at pH – 6.4; municipal wastewater produced 9.01 mA of current at pH – 7.6 (Mathuriya and Sharma 2009).

(c) Substrates are the source of the bacteria during the process of bioelectricity generation and hence were selected as the third parameter. A variety of substrates in the final concentration of 2 g which contributed for the oxidation process were used in anode chamber. Among them, spinach gave the maximum voltage and current production of 600 mV and 0.300 mA, respectively. Cabbage peel produced 410 mV and 0.20 mA, respectively. All other substrates except paddy straw gave less amount of electricity. These results show that cheap substrates or agro-waste material can be used for the current production. In an earlier study, monosodium glutamate wastewater was used as a substrate in MFC inoculated with *Rhodoferax ferrireducens*, and it produced 0.18 V (Liu and Li (2007)). In another study, cheese whey was found to produce 29.1 W/m^2 (Kassongo and Togo 2011). Abattoir wastewater being an exceptional substrate used in a MFC produced 12.26 mW/cm^2 (Momoh and Neayor 2010). These reports support the finding implying that when an appropriate substrate is used based on the waste, a maximum power production can be achieved.

(d) Among the catholytes tested, acetic acid gave the maximum electricity production. Hydrogen peroxide produced the least current. Acetic acid gave the maximum voltage of 0.47 V and current of 0.05 mA. In the earlier reports, oxidizing agents like hexacyanoferrate (Rabaey et al. 2005) and acidic permanganate (You et al. 2006) have been used as catholytes in MFCs. The maximum power density of using a single-brush anode in a double-air cathode MFC was 154 ± 1 W/m^3, which is 108% more than the single-cathode MFC (Xiaoyuan Zhang et al. 2011). When calcium hypochlorite powder ($Ca(OCl)_2$) was used as a catholyte, it produced 12.26 and 20.71 mW/cm^2 for single- and dual-chambered MFC, respectively (Momoh and Neayor 2010). From our experimental results, the catholyte acetic acid produced 470 mV and 0.05 mA.

(e) Among the buffers ranging from pH 5 to 9 tested, acetate gave a maximum of 1.4 V and 0.140 mA. Phosphate buffer produced a maximum of 720 mV and 0.100 mA. Citrate phosphate buffer produced very less voltage among the buffers. From the previous literature, it is understood that phosphate-buffered saline of concentration 50/50 v/v was used for the efficiency (Cheng et al. 2009). A maximum power density of 1550 W/m^3 (2770 mW/m^2) was obtained and a current density of 0.99 mA/cm^2 using a pH 9 bicarbonate system. The power density was 38.6% higher when compared to the system using pH 7 phosphate buffer at the same concentration of 0.2 M (Fan et al. 2007).

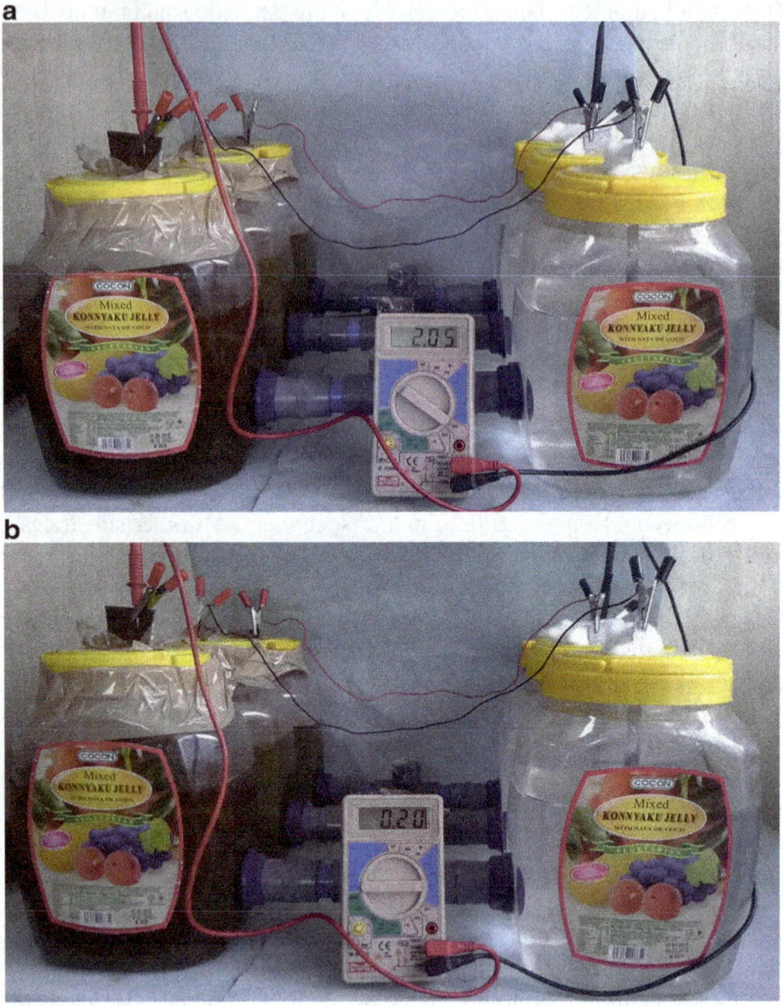

Fig. 10.1 (**a**) Voltage production of MFC connected in series. (**b**) Current production of MFC connected in series

10.1.1.1 Scale-Up of MFC with Rumen Fluid

Upon analyzing the individual parameters in small MFC, scale-up of rumen fluid MFC has been demonstrated with 3 L plastic bottles. The working volume was 2.5 L in each MFC. When three individual MFCs, namely, MFC 1, MFC II, and MFC III, were connected in series, it produced 2.05 V and 20 mA. When connected in parallel, they produced 0.73 V and 62 mA to the maximum. Figure 10.1 (a) gives the voltage and (b) current production of MFC connected in series. This denoted that to achieve a long-term voltage and current, parallel connection is favorable, and, for high voltage, series connection of MFCs is favorable. Similarly in a report, two individual MFCs were stacked together either in series or in parallel. The MFCs stacked in series produced a working voltage of 1.22 V (Gurung et al. 2012). Likewise, Aelterman et al. (2006), connected six individual MFCs in series and parallel which enabled an increase of the voltage by 2.02 V and current 255 mA while retaining high power output. The OCV of 0.67 and 4.16 V was obtained when they were connected in parallel and series, respectively (Aelterman et al. 2006).

The individual microbial fuel cell in the stacked series was observed for the potential and current readings separately, and the results were interpreted. Here among the three MFCs, MFC I gave a maximum production of 0.86 V. Though MFC III gave an initial peak in voltage of 0.85 V, it gradually decreased to 0.6 V in the course of time period. On the other side, MFC I also gave a stable current of 0.24 V. However, MFC II had the maximum production of 0.32 mA on the 10th day. The same observation was observed by Aelterman et al. (2006) where he has reported that during the connection of the individual MFCs together, the voltage diverged due to the microbial limitations at increasing currents. It is well known that a series connection could improve the voltage while maintaining the current (Aelterman et al. 2006). In a recent article, four membrane-electrode assembly MFCs were checked both individually and in series connection. Individually they showed 0.68 ± 0.05 V which sharply increased to 2.06 ± 0.03 V when connected in series (Kim et al. 2013). MFC stacked with bipolar plates made up of carbon blocks has been tested for their performance. Five single cells connected in series produced a maximum voltage of 2.5 V indicating that the individual cells generated 0.5 V (Shin et al. 2006). Figure 10.2 (a) gives the voltage and (b) current generation connected in series and parallel.

10.1.2 Comparative Analysis of Power Production of Pure, Co-culture, and Mixed Culture in Microbial Fuel Cell

10.1.2.1 Bacterial Strains

Bacterial strains which were isolated from the biofilm were streaked by quadrant plate method to obtain pure cultures. The isolated strains were named as Strain 1, Strain 2, Strain 3, Strain 4, and Strain 5. After the colony morphology observation, the strains were screened for various biochemical tests to infer the genus of the organism. Based on the gram staining, it was identified that Strains 1 and 3 are gram-positive rods, Strain 2 is a gram-negative coccobacillus, Strain 4 is a gram-negative rod, and Strain 5 is a gram-negative rod to ovoid. Based on the hanging drop technique, it was found that all the bacterial strains except the Strain 1 were motile confirming the presence of flagella or pili. This kind of projections is helpful for the electron transfer to the anode surface (Gorby et al. 2006). Based on the biochemical tests and 16s rRNA sequencing, the strains were identified as *Pseudomonas aeruginosa* DMR-3, *Bacillus tequilensis* DMR-5, *Bacillus thuringiensis* DRR-1, *Pseudomonas fragi* DRR-2, and *Paracoccus homiensis* DRR-3.

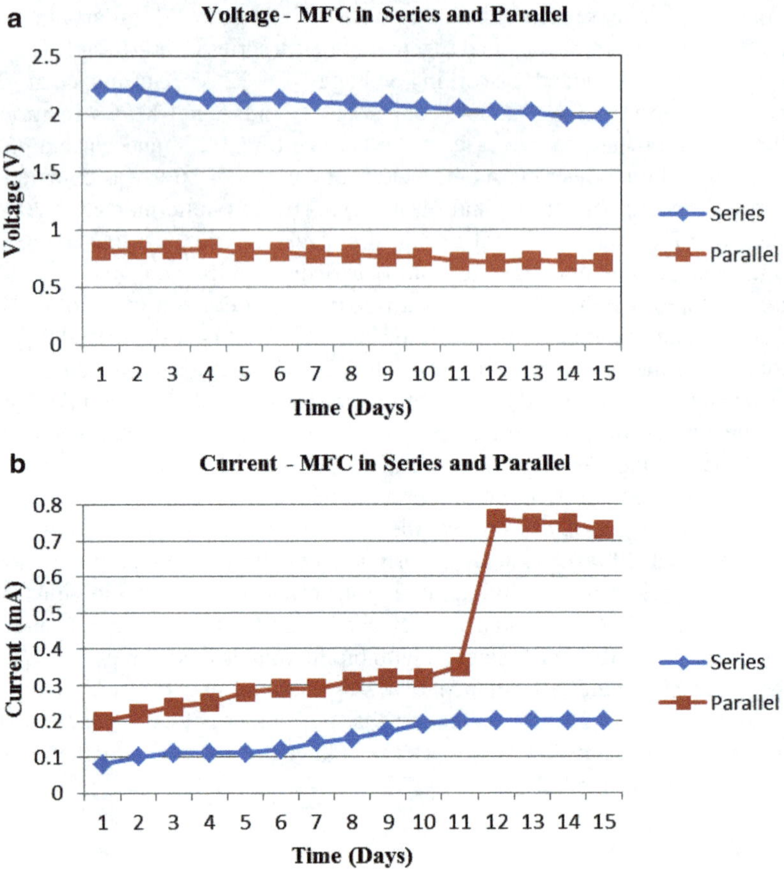

Fig. 10.2 (**a**) Voltage of MFC connected in series and parallel. (**b**) Current of MFC connected in series and parallel

10.1.2.2 Brief Pure Culture Study in Terms of Voltage Production and Cyclic Voltammogram

In this experiment, five cultures were inoculated as pure cultures in five separate MFCs. The readings were taken in multimeter for 12 days. Among the five cultures, *Paracoccus homiensis* and *Pseudomonas aeruginosa* produced the maximum voltage of 320 mV and 300 mV, respectively. *Bacillus thuringiensis* produced the least voltage of 150 mV. Likewise, *Paracoccus sp.* and *Pseudomonas sp.* gave the maximum current of 0.01 mA and 0.02 mA, respectively. Henceforth, *Paracoccus homiensis* was chosen for proton-exchange membrane study as a pure culture. Figure 10.3 (a) shows the potential and (b) current comparison between the five pure cultures.

Microbial fuel cell performance differs for each and every bacterium. *Saccharomyces cerevisae* and *Clostridium acetobutylicum* generated 10.89 mA and

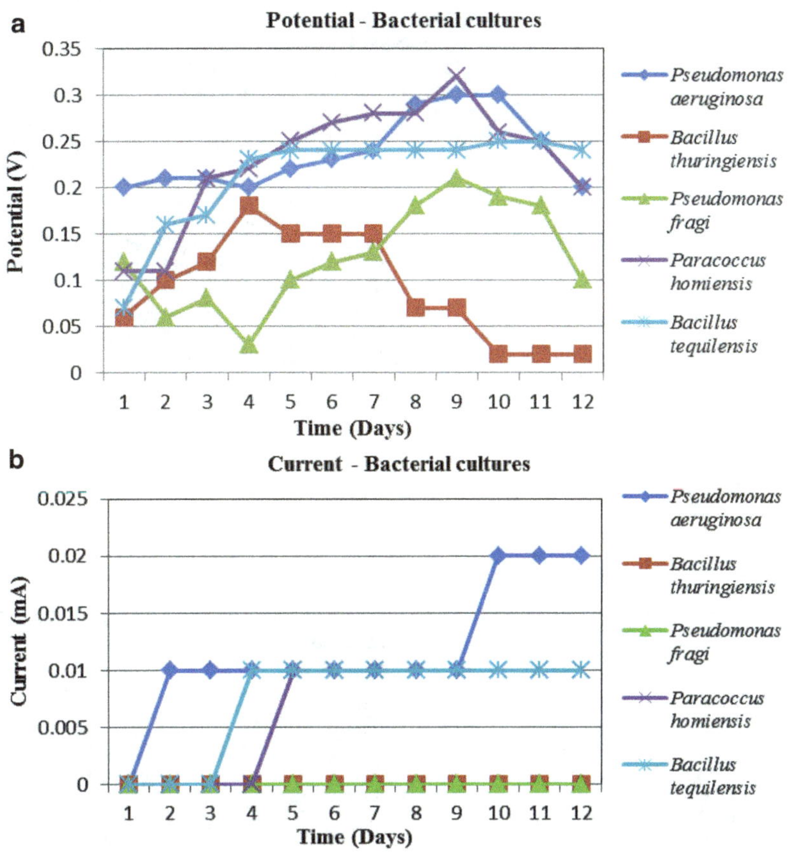

Fig. 10.3 (**a**) Potential of five pure bacterial strains. (**b**) Current of five pure bacterial strains

10.45 mA, respectively, after 10 days of operation (Mathuriya and Sharma 2009). On the other side, an aircathode MFC with *Enterobacter aerogenes* produced a maximum power density of 2.51 W/m^3 where no mediators were used (Zhuang et al. 2011). *Geobacter sulfurreducens* and *Geobacter metallireducens* exhibited lower current densities of 110 ± 7 A/m^3 (Call et al. 2009). *Shewanella oneidensis* DSP10 grown in medium with lactate exhibited 24 mW/m^2 for reticulated vitreous carbon, and once external mediators were used, the current and power increased by 30–100% (Ringeisen et al. 2006). *Hansenula anomala* yielded 2.34 W/m^3 with graphite felt as the anode material in a deaerated suspension of nutrient broth in anodic chamber (Prasad et al. 2007).

The cyclic voltammogram is a characteristic feature which confirms the electrochemical activity of the biofilm or individual bacteria. Hence, this technique has been widely used for the studies involving microbial fuel cell. The redox potential in the anode compartment and also information about the direct electron transfer

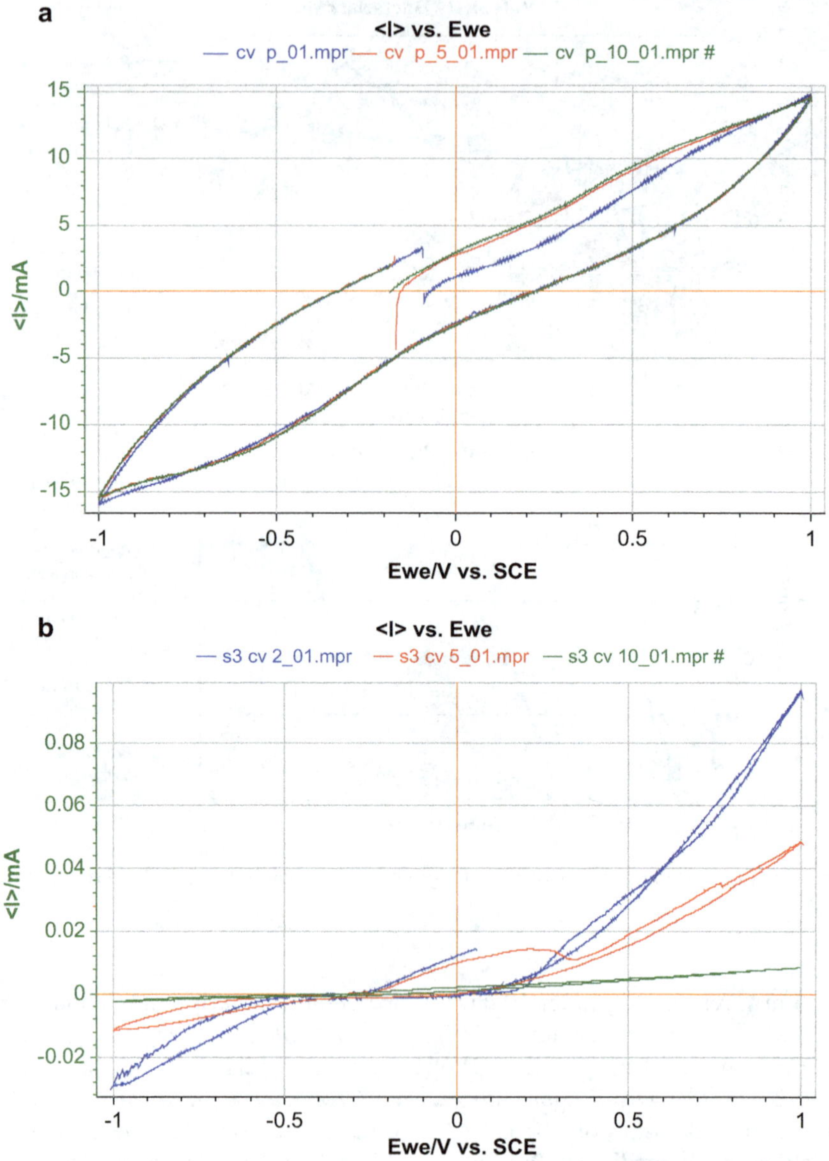

Fig. 10.4 (**a**) Cyclic voltammogram of *Pseudomonas aeruginosa* DMR-3. (**b**) Cyclic voltammogram of *Pseudomonas fragi* DRR-2. (**c**) Cyclic voltammogram of *Paracoccus homiensis* DRR-3

can be studied with the technique. For instance, the electrochemical activity of two enzymes has been demonstrated in a study where *Hansenula anomala* produced less peak currents when lactate has been added (Prasad et al. 2007). Figure 10.4a–c represents the cyclic voltammogram of *Pseudomonas aeruginosa*, *P. fragi*, and *Paracoccus homiensis*, respectively, showing prominent redox peaks which confirm

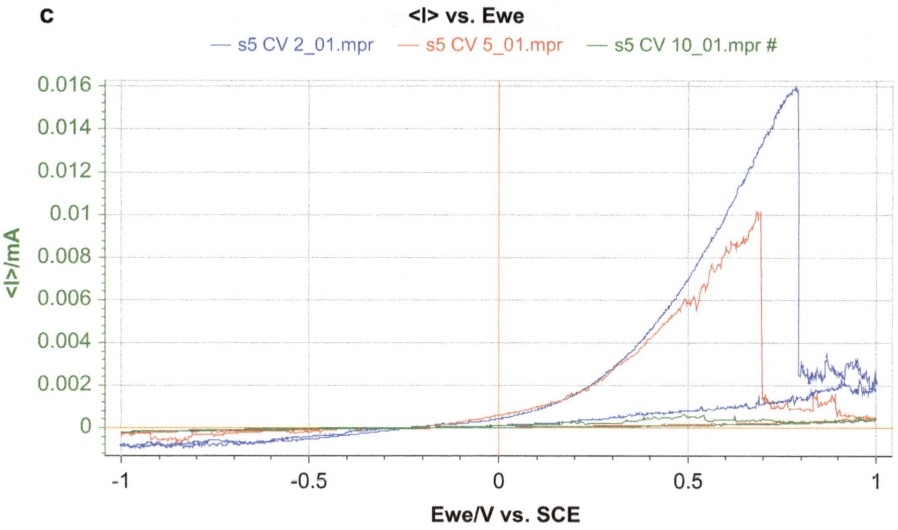

Fig. 4 (continued)

the electricity production in the voltage–current experiments ($V \times I$). *P. aeruginosa* showed an oxidation peak at −0.398 V and reduction peak at 0.587 V. *Pseudomonas fragi* showed a mild oxidation peak at −0.71 V and a reduction peak at 0.20 V. *Paracoccus homiensis* showed a reduction peak at high voltage of 0.77 V. *Bacillus thuringiensis* and *Bacillus tequilensis* did not show peaks in the voltammogram. These two bacteria produced less voltage in the previous experiment in MFC.

10.1.2.3 Co-culture and Mixed Culture Studies

The anodic chamber of MFC was inoculated with 110 × 10^5 CFU/mL of *Bacillus tequilensis*, 70 × 10^5 CFU/mL of *Pseudomonas aeruginosa*, and co-culture of *Bacillus tequilensis* and *Pseudomonas aeruginosa* (110 × 10^5 CFU/mL and 70 × 10^5 CFU/mL, respectively) in three separate MFCs on the same day. When inoculated as pure culture, *Pseudomonas aeruginosa* showed a maximum of 310 mV and 0.020 mA. *Bacillus tequilensis* produced a maximum of 250 mV and 0.010 mA. The co-culture of *Bacillus tequilensis* and *Pseudomonas aeruginosa* has shown a maximum of 450 mV and 0.040 mA. From the above results, it is evident that the co-culture produced high power density than the pure cultures.

In addition to microorganisms that can transfer electrons to the anode, the presence of other organisms appears to benefit MFC performance. It is reported that a mixed culture generates a current that was sixfold higher than a pure culture (Park and Zeikus (2002)). The anodic chamber was inoculated with 120 × 10^5 CFU/mL of *Paracoccus homiensis* and 100 × 10^5 CFU/mL of *Bacillus thuringiensis* and a co-

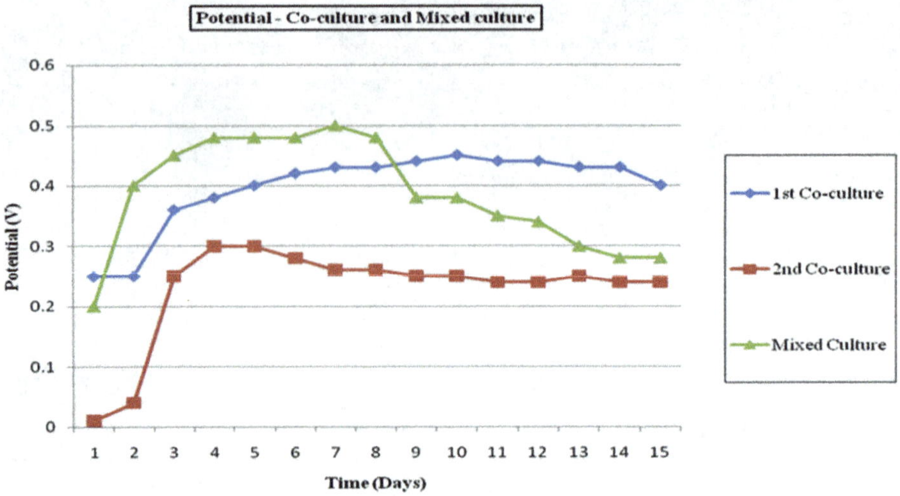

Fig. 10.5 Potential comparison of co-culture and mixed culture

culture of these bacteria in three separate MFCs. When *Bacillus thuringiensis* was tested as pure culture, it produced a maximum of 180 mV with no current, and *Paracoccus homiensis* produced a maximum of 300 mV and 0.010 mA. However, when the two bacteria were inoculated in the MFC, it produced 300 mV and 0.100 mA. Comparatively, the combination of the two cultures gave the maximum voltage and current. However, there is noticeable change in the current from 0.010 to 0.100 mA in co-culture. This shows that the pure cultures react on their own way, and when combined there might be some mechanism existing between the cultures which is the reason for the increase in current.

This experiment reveals the potential comparison between co-culture and mixed culture. The first co-culture is a combination of *Pseudomonas aeruginosa* and *Bacillus tequilensis*. The second co-culture is a combination of *Paracoccus homiensis* and *Bacillus thuringiensis*. Mixed culture is a combination of all the five bacterial strains used in this study. Figure 10.5 shows the comparison of potential between the cultures. Among the two different sets of co-culture, the first co-culture produced the maximum voltage of 450 mV. The second set of co-culture produced a maximum voltage of 300 mV. But compared to this, the mixed culture with five bacterial strains produced a maximum of 500 mV. Thus it is evident from the experiments that bacterial cultures in mixed form produce maximum power.

10.1.2.4 SEM Analysis

The anode subjected to scanning electron microscope analysis shows the biofilm formation attached on the surface of the electrode. Figure 10.6a shows the plain carbon sheet (control), and Fig. 10.6b shows bacteria (*Paracoccus homiensis*)

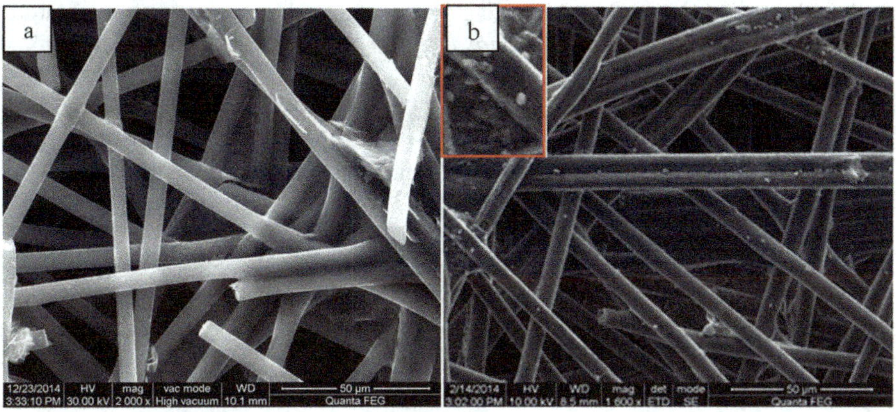

Fig. 10.6 (**a**) Scanning electron micrograph of the carbon sheet (control) (**b**) *Paracoccus homiensis* growth

adhering to the surface of the carbon sheet. From this image, it was found that biofilm was spreaded on the carbon sheet which facilitates bioelectricity production. A thick biofilm of *Aeromonas hydrophila* PA3 on the anode surface with uniform cells was observed through SEM. It has been reported that the biofilm has contributed to the maximum current (Pham et al. 2003).

10.1.2.5 Production of Bioelectricity in MFC by *Pseudomonas fragi* DRR-2 (Psychrophilic) Isolated from Goat Rumen Fluid

Over the period of time, MFC has been examined at ambient temperature with different microbes. There are many bacteria isolated from different places other than rumen such as *Rhodoferax ferrireducens* (Chaudhuri and Lovley 2003), *Shewanella putrefaciens* (Kim et al. 2002), *Geobacter sulfurreducens* (Bond and Lovley 2003), and *Desulfobulbus propionicus* (Holmes et al. 2004) which have been reported for power production in MFC. A recent study has focused on bioelectricity production from *Geopsychrobacter electrodiphilus* gen. nov., sp. nov., a psychrotolerant bacteria which can grow between 4 and 30 °C with an optimum temperature of 22 °C (Holmes et al. 2004). *Rhodoferax ferrireducens* is capable of transferring electrons to electrodes at 4 °C in a mediatorless microbial fuel cell (Chaudhuri and Lovley 2003). Previous studies show that mesophilic bacteria show higher growth rate, higher electron transfer, shorter lag phase, and lower respiration which are not found in low-temperature-adapted microbes (Hall et al. 2010). There are many cold-adapted microorganisms (psychrophilic) present in our environment which need to be explored for MFC research. The purpose of this study was to investigate bioelectricity generation by *Pseudomonas fragi*, psychrophilic bacterium growing in low temperature, so that they can be used in places of cold region. Based on the experimental results, it can be concluded that the bacteria showed higher growth rate, higher electron transfer, and shorter lag phase when subjected to low temperature. This is the first report on *Pseudomonas fragi* for the production of bioelectricity at low temperature.

10.1.2.6 Growth Curve and Protein Content of *Pseudomonas fragi* DRR-2 at Different Temperatures

The bacterial growth was measured every 24 h for a period of 15 days at different temperatures. The maximum growth was observed on the 6th day at 20 °C. The protein content was maximum on the 6th day at 10 °C, whereas for other temperature bacteria showed less content. This confirms that the optimum temperature for growth is 20 °C and the bacteria has the ability to grow in low temperatures (>4 °C). At all the temperatures, the total protein was observed to be highest between the 6th day and 10th day.

Under high-nutrient conditions, bacteria tend to alter their membrane lipid composition to adapt to the changing temperatures (van de Vossenberg et al. 1999), by the method known as homeoviscous adaptation (Sinensky 1974). According to Hall et al. (2010) report, at low temperatures, membranes can be highly firm and prevent the efficient function of transmembrane proteins, important for resource utilization. This is due to the membrane fluidity which plays a main role in the proton-motive force. However, in bacteria the cellular membrane is also used to create an electrochemical gradient, which makes the synthesis of ATP as protons move down the proton gradient into the cell. As membrane lipids play a main role in maintaining the membrane fluidity, it has been observed that the organisms dominated in cold environments are rich in MUFA or branched-chain fatty acids, while organisms in warmer environments have saturated fatty acids (SAFA) (Kaneda 1991). Mesophilic bacteria show higher growth rate, higher electron transfer, shorter lag phase, and lower respiration which are not found in low-temperature-adapted microbes (Hall et al. 2010). Our experimental results confirm that *Pseudomonas fragi* (psychrophilic) shows higher activity at low temperature (10 °C) where the protein concentration was found to be maximum.

Power Production of the Bacterium Under Different Temperatures Using Salt Bridge and Nafion 117

The bacterium produced a maximum voltage of 540 mV on the 10th day at 20 °C indicating that the favorable temperature for the growth gave the maximum electricity production. The maximum current was only 0.020 mA since salt bridge was used as the proton exchanger due to higher internal resistance. Compared to the room temperature, the bacteria produced more voltage in low temperatures.

A maximum voltage of 380 mV on the 10th day at 20 °C and a maximum current of 0.070 mA on the 7th day at 4 °C confirm that the bacteria are active at low temperatures between 4 and 20 °C. When compared to the salt bridge, Nafion 117 membrane gave a maximum current, indicating that the internal resistance of the fuel cell is decreased thereby improving the cell performance. *Geopsychrobacter electrodiphilus* produced a maximum current of 3.73 mA/cm^2 when acetate was provided as the electron donor (Holmes et al. 2004). Rumen microbes when they grow in low temperatures tend to produce less methane comparatively to the mesophilic conditions (Graham et al. 1959; Kennedy and Milligan 1978). Based on the above information, we prove that the isolated strain might have produced less methane and more hydrogen for the electron and proton transfer. This may be the reason for the increased bioelectricity production of *P. fragi* at low temperatures. Figure 10.7a and b represents the potential and current production of *P. fragi* at low temperatures with Nafion membrane.

Cyclic Voltammogram of the Strain in Low Temperatures

The cyclic voltammograms of the anodic biofilm clearly give an anodic potential and cathodic potential. This confirms that the bacteria grown in low temperatures exhibit a sigmoidal curve indicating that they are electrochemically active in nature. Figure 10.8 shows the voltammogram of the anodic biofilm at 20 °C. From the voltammograms, it has been observed that at 4 °C, a sharp oxidation peak at 0.04 V was found indicating the maximum substrate utilization of the microbe has taken place at low temperature. At the same time, a reduction peak at −0.2 V reveals that there the electron transfer has taken place. However, the electron transfer was found to be maximum at 20 °C, and the corresponding voltammogram confirms it with three reduction peaks in the reverse scan at −0.14, −0.8, and −0.6 V.

10.1.3 Performance of **Paracoccus homiensis** *DRR-3 in Microbial Fuel Cell with Membranes*

10.1.3.1 Power Production of *Paracoccus homiensis* DRR-3 with Nafion 117 in MFC

This research also focuses on to find an alternative membrane to the commercially available Nafion 117. Henceforth, Nafion 117 was tested for its efficiency in the 300 mL acrylic chamber which has a membrane holder. The other membranes which were tested are polyvinylidene difluoride (PVDF) and polycarbazole (PCZ) which are conductive in nature. This was the reason to choose them for the experiments.

Initially Nafion 117 was tested with three types of electrodes, namely, carbon cloth, carbon sheet, and graphite plate. The carbon paper produced the maximum potential and current with 0.8 V and 0.13 mA. The carbon cloth produced a maximum of 0.54 V and 0.7 mA, whereas graphite plate showed the least output of 0.24 V and 0.1 mA. This is due to the smooth surface of the graphite plate which did

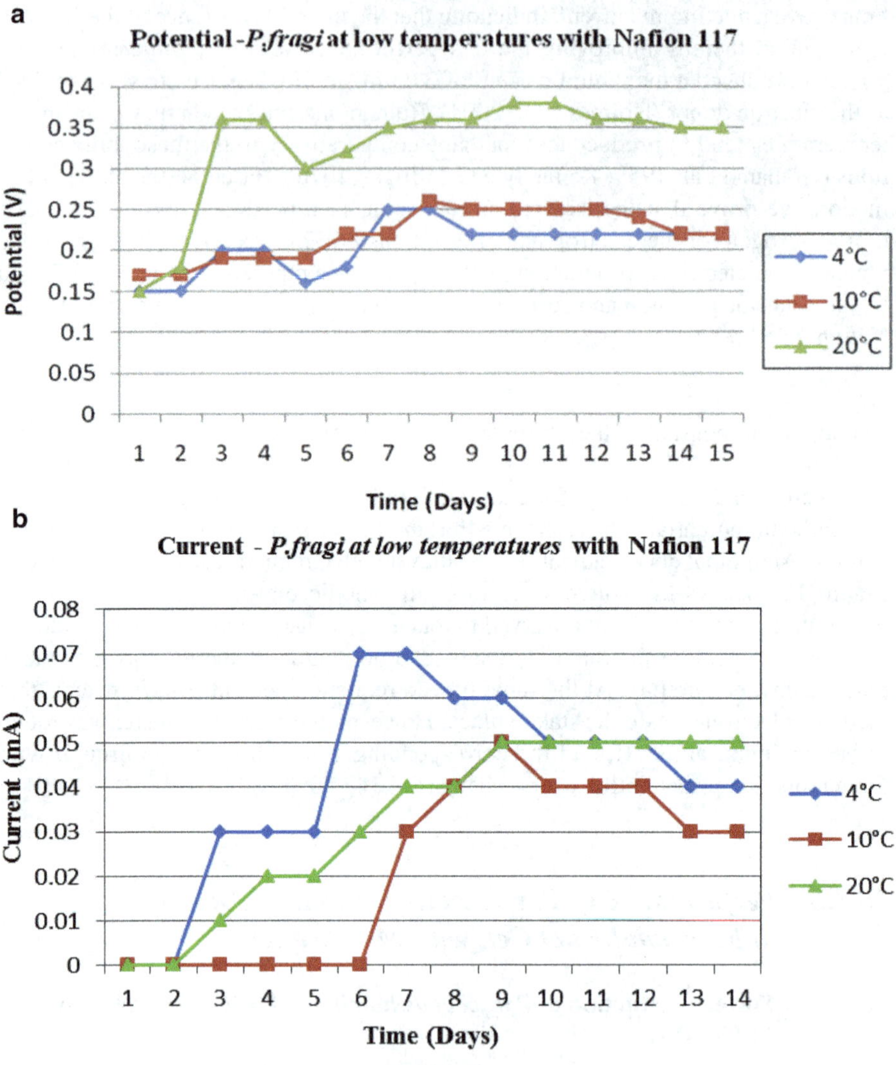

Fig. 10.7 (**a**) Potential curve of *P. fragi* at low temperatures using Nafion membrane. (**b**) Current curve of *P. fragi* at low temperatures using Nafion membrane

not help the bacterium to colonize the surface which is contradicting to the observation carried out by Junqiu Jiang where he observed the MFC yielding a maximum voltage of 0.687 V with a graphite fiber brush anode (Jiang et al. 2009). In a previous report, a modified CNT/PANI (carbon nanotube/polyaniline) increased the MFC performance with 1.18 V and current of 12.8 mA (Wang et al. 2013). In a MFC utilizing corn stover biomass, plain carbon paper was used as anode, and the cathode was made up of carbon cloth containing Pt catalyst. Reactors fed with the sample produced 437 mV (390 mW/m^2) (Wang et al. 2009). Carbon cloth of pro-

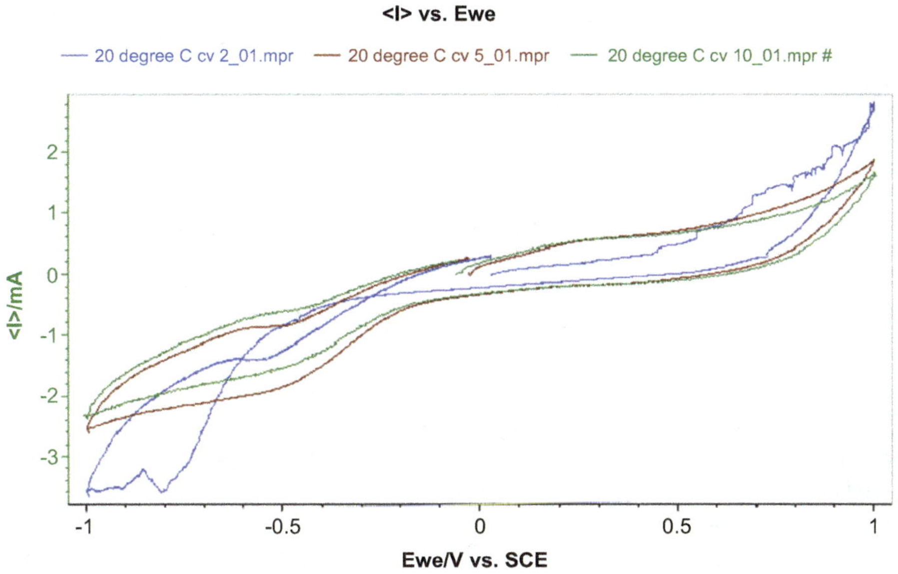

Fig. 10.8 Cyclic voltammogram Cyclic Voltammogram analyses of MFC inoculated with of *P. fragi* at 20°C with scan rates 1mV/s, 5mV/s & 10mV/s

jected surface area 7 cm^2 used in a MFC employing biodiesel waste as the organic matter produced a maximum of 450–500 mV (Yujie et al. 2011). These results confirm that carbon paper and modified carbon electrodes strongly play a main role in electron transfer when compared to other electrode materials.

10.1.3.2 Power Production of *Paracoccus homiensis* DRR-3 with PVDF and PCZ in MFC

Since carbon paper showed a maximum power production, it was used for the further experiments. *Paracoccus homiensis* produced a maximum of 0.64 V on the 8th day with the PVDF membrane as a proton exchanger. The voltage then gradually decreased to 0.37 V on the 17th day. Though PVDF could not achieve a high voltage as Nafion membrane (0.80 V), it produced a significant amount of power. However, PVDF produced a maximum of 0.16 A which is higher than that of Nafion membrane. Similarly, in a report polyether ether ketone was sulfonated and used as a proton-exchange membrane in a single-chamber MFC. *Escherichia coli* produced a maximum of 670 mW/cm^2 with SPEEK membrane, whereas Nafion 117 produced 300 ± 7 mW/cm^2 (Ayyaru and Dharmalingam 2011). This experiment has given us a hint that PVDF membrane might be a good alternative for Nafion in future in the field of microbial fuel cell.

Paracoccus homiensis gave a maximum voltage of 0.46 V on the 4th day which gradually decreased to 0.15 V on the 15th day. The maximum current production was 0.10 mA on the 9th day which gradually declined to zero. This membrane seems to produce less voltage when compared to the commercial Nafion and PVDF. However, it was taken into account for the further experiments to check the efficacy in terms of power production.

10.1.4 Membranes, Their Performance, Electrochemical Analysis in MFC

10.1.4.1 Cyclic Voltammogram of *P. homiensis* Using Membranes

Paracoccus homiensis in the presence of Nafion membrane has given the cyclic voltammogram with two oxidations peaks at −0.57 V and 0.37 V, respectively, and one reduction peak at 0.07 V. Similar kind of results were observed in *Shewanella oneidensis* MR-1 in the presence of buffer and lactate as anolyte showing a reduction peak at −500 mV (−0.5 V) with Nafion 424, DuPont membrane. An oxidation peak was observed at the potential of 200 mV (0.2 V) which is comparable to the present study (Manohar et al. 2008).

Paracoccus homiensis in the presence of PVDF membrane showed two reduction peaks at −0.59 V and 0.49 V, respectively, and an oxidation peak at 0.42 V which indicates the transfer of electrons at the anode chamber has taken place. Likewise, *Shewanella putrefaciens* used as an EAB (electrochemically active bacteria) showed a characteristic reduction peak at – 250 mV (i.e., −0.25 V), and in the anodic scan, it showed an oxidation peak at 0.09 V (Khilari et al. 2015).

The reduction peaks observed in the voltammogram in the presence of PCZ membrane signify the reducing activity of *Paracoccus homiensis.* Two oxidation peaks at 0.127 and 0.36 V were found in the anodic scan which indicates the oxidation of substrate by the bacterium. A reduction peak was observed at −0.37 V which confirms that the bacterium is electrochemically active and the membrane which has been used in the MFC is transferring electrons at a good rate. To summarize the membrane study, all the three membranes were working quite efficient in terms of electron transfer. However, to further elucidate a better performance, the internal resistance and conductivity should be taken into account.

10.1.4.2 Impedance Spectra of *P. homiensis* Using Membranes

EIS was used to measure the internal resistance in the MFC before and during the course of reaction. The results were plotted as Nyquist curves and further fitted with an equivalent circuit. EIS curves usually consist of well-defined semicircle followed by a straight line. The intercept of semicircle with the real impedance axis presents the total ohmic resistance (R_{ohm}) of the electrochemical cell including the solution resistance (R_s) and charge transfer resistance (R_{ct}) at the electrode-electrode interface (Dominguez-Benetton et al. 2012). The internal resistance had two major components, namely, the ohmic and non-ohmic resistance (Logan et al. 2006). The resistance produced by electrolyte and electrode material during electron transfer is known as ohmic resistance, and this is caused due to faradic reactions (He et al. 2006). The non-ohmic resistance due to the electrochemical reaction which happens on the surface of the electrode mainly because of the microbial metabolism is called as charge transfer resistance (Khan and Iqbal 2005). The present experimental study of the impedance spectra of *P. homiensis* with Nafion membrane MFC gave a solution resistance of 9.202 Ω. The polarization resistance is 70.34 Ω, and the charge transfer resistance is found to be 61.138 Ω.

The impedance spectrum obtained for PVDF membrane has given the possible circuit which depicts that a layer of biofilm has formed over the surface of the electrode. The solution resistance contributed by this MFC is 23.61 Ω, and the polarization resistance is 68.66 Ω. Figure 10.9 represents the Nyquist plot and the circuit of

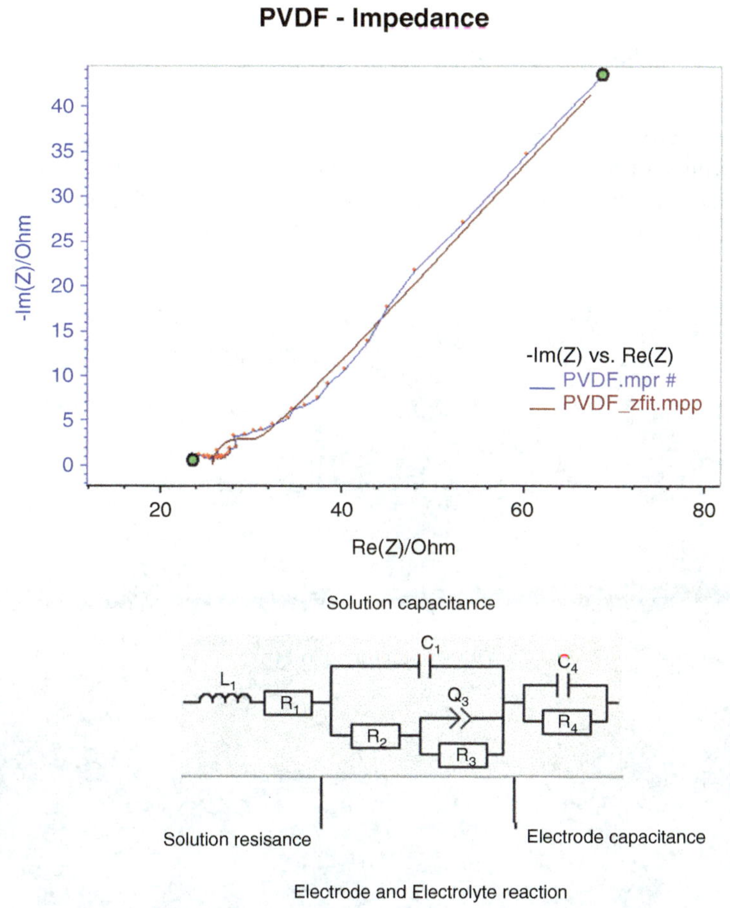

Fig. 10.9 Nyquist plot and equivalent circuit diagram of *Paracoccus homiensis* with PVDF membrane

MFC with *P. homiensis* employing PVDF membrane. Equivalent circuit modelling (ECM) was utilized to further explore the EIS results, specifically to determine the distribution of resistive and capacitive features in the operating MFC.

The solution resistance contributed in the MFC with the PCZ membrane is 23.45 Ω. The polarization resistance is 533.8 Ω. The charge transfer resistance is 510.35 Ω. The MFC performance using dairy waste with pure culture *E. coli* for 4 days operation was found to be maximum at low resistance (31.14 kΩ) with high conductivity as described by Patil et al. (2013). The measured ohmic resistance R_s for the SSFF-MFC, PANIche/SSFF-MFC, and PANIele/SSFF-MFC is 36.1 Ω, 36.5 Ω, and 32.7 Ω, respectively. The polarization resistance for the MFCs was 938.4 Ω, 279.1 Ω, and 215.6 Ω, respectively (Hou et al. 2015). From our EIS results, the PVDF membrane showed a better performance when compared to the others with a low resistance of 68.66 Ω.

10.1.5 Applications of Rumen Fluid MFC

Scale-up microbial fuel cell of four cells has been tested for various applications like glowing an 1.5 V LED, running a small fan, powering pocket calculator, powering digital wristwatch, and finally charging a mobile phone. The MFCs connected in series gave an output of 3.57 V and 60 mA. Figure 10.10 shows MFC powering

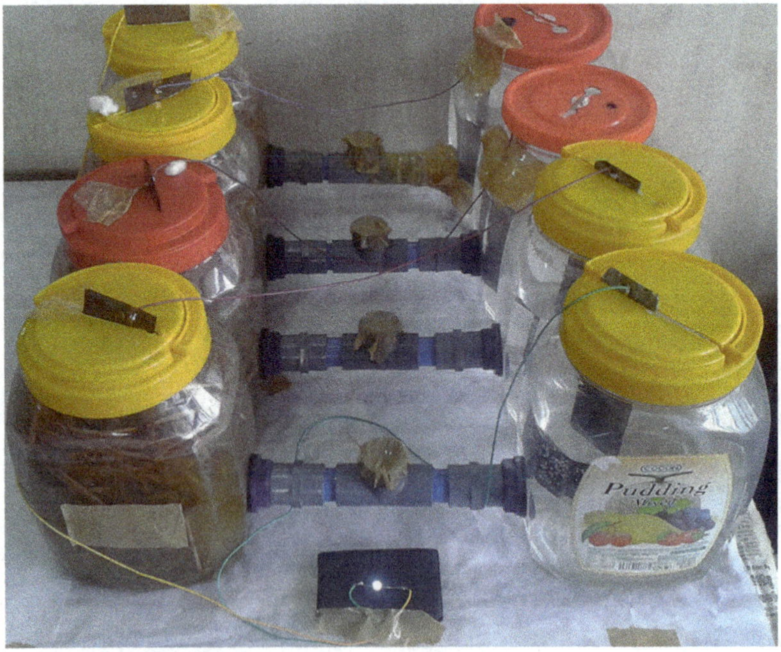

Fig. 10.10 Rumen fluid MFC glows a 1.5 V white LED

Fig. 10.11 MFC powering a pocket calculator. (**a**) Calculator soldered with the positive and negative ends of MFC. (**b**) Calculator getting powered by MFC

a 1.5 V white LED. Figure 10.11 shows MFC powering a calculator. In future, MFC can be used for various applications if they are worked in large scale.

10.2 Summary and Conclusion

MFC performance was primarly based on the reactor model, electrodes, organic matter, etc. Hence, various parameters such as electrodes, pH, substrates, catholytes, and buffers were tested to study the favorable conditions for the rumen MFC. The optimized parameters like carbon electrodes, pH 7.0, spinach, acetic acid, and acetate buffer used in a single MFC gave better efficiency. The cyclic voltammogram of the anodic biofilm confirmed the electrochemical activity of the biofilm. Scale-up of rumen MFC was done both in series and parallel connection where series connection gave 2.05 V and 20 mA. In parallel it gave 0.73 V and 62 mA. Totally five bacterial strains isolated from the biofilm were identified by biochemical tests and 16srRNA sequencing. The phylogenetic tree was constructed to study the family structure. Among the bacterial strains, *Pseudomonas aeruginosa*, *Pseudomonas fragi*, and *Paracoccus homiensis* produced consistent power and showed electrochemical activity. From the co-culture study, it was understood that a bacterium with high electricity production and a bacterium with low production when combined together give a much higher amount of bioelectricity, thus enabling a weaker bacterium to work better. The cyclic voltammograms support the I–V graphs. A special bacterium *Pseudomonas fragi* was also tested under different temperatures. Only at 20 °C, the bacteria produced higher bioelectricity production. A mixed culture of all the five bacterial strains was also carried out to check the efficiency. Though mixed cultures give a large amount of power, study of the individual bacterium might help us in carrying out this research to the next step such as genetic modification, identifying the functional gene, etc. Among the membranes tested, PVDF produced a significant power and less internal resistance in par with the commercial Nafion membrane: R_{in} of PVDF −68.66 Ω and R_{in} of Nafion −70.34 Ω.

Acknowledgments Dr. J. Deepika is very much thankful to the Head of the Department of Chemistry SRM University for helping her with the instrumentation facility for the cyclic voltammogram and electron impedance spectroscopy experiments. She wishes to convey special thanks to Nanotechnology Research Centre, SRM University, for helping in the scanning electron microscope images. Dr. L. Benedict Bruno thankfully acknowledges the Science and Engineering Research Board, National Postdoctoral fellowship scheme (Grant No. PDF/2017/001074), for financial support.

References

Aelterman P, Rabaey K, Pham HT, Boon N, Verstraete W (2006) Continuous electricity generation at high voltages and currents using stacked microbial fuel cells. Environ Sci Technol 40: 3388–3394

Ayyaru S, Dharmalingam S (2011) Development of MFC using sulphonated polyether ether ketone (SPEEK) membrane for electricity generation from waste water. Bioresour Technol 102:11167–11171

Bond DR, Lovley DR (2003) Electricity production by *Geobacter sulfurreducens* attached to electrodes. Appl Environ Microbiol 69:1548–1555

Call DF, Wagner RC, Logan BE (2009) Hydrogen production by *Geobacter species* and a mixed consortium in a microbial electrolysis cell. Appl Environ Microbiol 75:7579–7587

Chaudhuri SK, Lovley DR (2003) Electricity generation by direct oxidation of glucose in mediatorless microbial fuel cells. Nat Biotechnol 21:1229–1232

Chen ZS (2010) Evaluation and analysis of applying native fiber substrates I rumen microbial fuel cell. Masters thesis, Department of Mechanical and Electro-Mechanical Engineering, National Ilan University, Ilan

Chen WH, Guo JL, Huang WS, Wang GB (2008) The development of cellulosic ethanol technology. Agric Biotechnol 9:62–69

Cheng K-J, Costerton JW (1980) Adherent rumen bacteria-their role in the digestion of plant material, urea and epithelial cells. In: Ruckebusch Y, Thivend P (eds) Digestive physiology and metabolism in ruminants. MTP Press, Lancaster, pp 227–250

Cheng S, Xing D, Call DF, Logan BE (2009) Direct biological conversion of electrons into methane by electromethanogensis environ. Sci Technol 43:3953–3958

Dominguez-Benetton X, Sevda S, Vanbroekhoven K, Pant D (2012) The accurate use of impedance analysis for the study of microbial electrochemical systems. Chem Soc Rev 41:7228–7246

Fan Y, Hua H, Liu H (2007) Sustainable power generation of microbial fuel cells using bicarbonate buffer and proton transfer mechanisms. Environ Sci Technol 41(23):8154–8158

Feng Y, Wang X, Logan BE, Lee H (2008) Brewery wastewater treatment using air-cathode microbial fuel cells. Appl Microbiol Biotechnol 78:873–880

Feng Yujie, Qiao Yang, Xin Wang, Yankun Liu, He Lee, Nanqi Ren (2011) Treatment of Biodiesel Production Wastes with Simultaneous Electricity Generation using a Single-chamber Microbial Fuel Cell. Bioresour Technol 102: 411–415.

Gorby YA, Yanina S, McLean JS, Rosso KM, Moyles D, Dohnalkova A, Beveridge TJ, Chang IS, Kim BH, Kim KS, Culley DE, Reed SB, Romine MF, Saffarini DA, Hill EA, Shi L, Elias DA, Kennedy DW, Pinchuk G, Watanabe K, Ishii S, Logan B, Nealson KH, Fredrickson JK (2006) Electrically conductive bacterial nanowires produced by *Shewanella oneidensis* strain MR-1 and other microorganisms. Proc Natl Acad Sci U S A 103:11358–11363

Graham NM, Wainman FW, Blaxter KL, Armstrong DG (1959) Environmental temperature, energy metabolism and heat regulation in sheep. I. Energy metabolism in closely clipped sheep. J Agric Sci 52:13–17

Gurung et al (2012) The performance of serially and parallelly connected microbial fuel cells. Energy Sources A 34(17):1591–1598

Hall EK, Singer GA, Kainz MJ, Lennon JT (2010) Evidence for a temperature acclimation mechanism in bacteria: an empirical test of a membrane-mediated trade-off. Funct Ecol 24(4):898–908

He Z, Wagner N, Minteer SD (2006) An upflow microbial fuel cell with an interior cathode: assessment of the internal resistance by impedance spectroscopy. Environ Sci Technol 40(17): 5212–5217

Holmes DE, Nicoll JS, Bond DR, Lovley DR (2004) Potential role of a novel psychrotolerant member of the family *Geobacteraceae*, *Geopsychrobacter electrodiphilus* gen. nov., sp. nov., in electricity production by a marine sediment fuel cell. Appl Environ Microbiol 70(10):6023–6030

Hou J, Liu Z, Li Y (2015) Polyaniline modified stainless steel fiber felt for high-performance microbial fuel cell anodes. J Clean Energy Technol 3(3):165–169

Jiang J, Zhao Q, Zhang J, Zhang G, Lee D-J (2009) Electricity generation from bio-treatment of sewage sludge with microbial fuel cell. Bioresour Technol 100:5808–5812

Kaneda T (1991) Iso- and anteiso-fatty acids in bacteria – biosynthesis, function, and taxonomic significance. Microbiol Rev 55:288–302

Kassongo J, Togo CA (2011) Performance improvement of whey-driven microbial fuel cells by acclimation of indigenous anodophilic microbes. Afr J Biotechnol 10(40):7846–7852

Kennedy PM, Milligan LP (1978) Effects of cold exposure on digestion, microbial synthesis and nitrogen transformations in sheep. Br J Nutr 39:105–117

Khan MJ, Iqbal MT (2005) Modelling and analysis of electro-chemical, thermal, and reactant flow dynamics for a PEM fuel cell system. Fuel Cells 5(4):463–475

Khilari S, Pandit S, Varanasi JL, Das D, Pradhan D (2015) Bifunctional manganese ferrite/polyaniline hybrid as electrode material for enhanced energy recovery in microbial fuel cell. ACS Appl Mater Interfaces 7:20657–20666

Kim HJ, Park HS, Hyun MS, Chang IS, Kim M, Kim BH (2002) A mediator-less microbial fuel cell using a metal reducing bacterium, *Shewanella putrefaciens*. Enzym Microb Technol 30:145–152

Kim B, An J, Kim D, Kim T, Jang JK, Lee B-G, Chang IS (2013) Voltage increase of microbial fuel cells with multiple membrane electrode assemblies by in series connection. Electrochem Commun 28:131–134

Liu Z-D, Li H-R (2007) Effects of bio- and abio-factors on electricity production in a mediatorless microbial fuel cell. Biochem Eng J 36:209–214

Logan BE, Hamelers B, Rozendal R, Schrorder U, Keller J, Freguia S, Aelterman P, Verstraete W, Rabaey K (2006) Microbial fuel cells: methodology and technology. Environ Sci Technol 40(17):5181–5192

Manohar AK, Bretschger O, Nealson KH, Mansfeld F (2008) The polarization behavior of the anode in a microbial fuel cell. Electrochim Acta 53:3508–3513

Mathuriya AS, Sharma VN (2009) Bioelectricity production from various wastewaters through microbial fuel cell technology. J Biochem Technol 2(1):133–137

Momoh OLY, Neayor B (2010) Generation of electricity from Abattoir waste water with the aid of a relatively cheap source of catholytes. J Appl Sci Environ Manag 14(2):21–27

Park DH, Zeikus JG (2002) Impact of electrode composition on electricity generation in a single-compartment fuel cell using Shewanella putrefaciens. Appl Microbiol Biotechnol 59:58–61

Patil VD, Patil DB, Pawar SS, Otari SV, Deshmukh MB, Pawar SH (2013) Studies on electrochemical performance of microbial fuel cell based on dairy waste for energy conversion. Int J Chem Sci Appl 4(2):111–115

Pham CA, Jung SJ, Phung NT, Lee J, Chang S (2003) A novel electrochemically active and Fe (III)-reducing bacterium phylogenetically related to *Aeromonas hydrophila*, isolated from a microbial fuel cell. FEMS Microbiol Lett 223:129–134

Prasad D, Murugesan AS, Padmanaban M, Satyanarayanan S, Sheela RS (2007) Direct electron transfer with yeast cells and construction of a mediatorless microbial fuel cell. Biosens Bioelectron 22:2604–2610

Rabaey K, Clauwaert P, Aelterman P (2005) Tubular microbial fuel cells for efficient electricity generation. Environ Sci Technol 39:8077–8082

Reimers CE, Stecher HA III, Westall JC, Alleau Y, Howell KA, Soule L, White HK, Girguis PR (2007) Substrate degradation kinetics, microbial diversity, and current efficiency of microbial fuel cells supplied with marine plankton. Appl Environ Microbiol 73:7029–7040

Ringeisen BR, Henderson E, Wu PK, Pietron J (2006) High power density from a miniature microbial fuel cell using *Shewanella oneidensis* DSP10. Environ Sci Technol 40:2629–2634

Rismani-Yazdi H, Christy AD, Dehority BA, Morrison M, Yu Z, Tuovinen OH (2007) Electricity generation from cellulose by rumen microorganisms in microbial fuel cells. Biotechnol Bioeng 97:1398–1407

Shin S-H, Choi Y, Na S-H, Jung S, Kim S (2006) Development of bipolar plate stack type microbial fuel cells. Bull Kor Chem Soc 27(2):281–285

Sinensky M (1974) Homeoviscous adaptation – a homeostatic process that regulates the viscosity of membrane lipids in *Escherichia coli*. Proc Natl Acad Sci U S A 71:522–525

van de Vossenberg JLCM, Driessen AJM, da Costa MS, Konings WN (1999) Homeostasis of the membrane proton permeability in *Bacillus subtilis* grown at different temperatures. Biochim Biophys Acta 1419:97–104

Venkatamohan S, Mohanakrishna G, Purushotham Reddy B, Saravanan R, Sarma PN (2008) Bioelectricity generation from chemical wastewater treatment in mediatorless (anode) microbial fuel cell (MFC) using selectively enriched hydrogen producing mixed culture under acidophilic microenvironment. Biochem Eng J 39:121–130

Wang X, Feng Y, Wang H, Qu Y, Yu Y, Nanli N, Wang E, Lee H, Logan BE (2009) Bioaugmentation for electricity generation from corn stover biomass using microbial fuel cells. Environ Sci Technol 43:6088–6093

Wang C-T, Huang R-Y, Lee Y-C, Zhang C-D (2013) Electrode material of carbon nanotube/polyaniline carbon paper applied in microbial fuel cells. J Clean Energy Technol 1(3):206–210. https://doi.org/10.7763/JOCET.2013.V1.47

You S, Zhao Q, Zhang J (2006) A microbial fuel cell using permanganate as a cathodic electron acceptor. J Power Sources 162:1409–1415

Yujie F, Yang Q, Wang X, Liu Y, He L, Ren N (2011) Treatment of biodiesel production wastes with simultaneous electricity generation using a single-chamber microbial fuel cell. Bioresour Technol 102:411–415

Zhang X, Cheng S, Liang P, Huang X, Logan BE (2011) Scalable air cathode microbial fuel cells using glass fiber separators, plastic mesh supporters, and graphite fiber brush anodes. Bioresour Technol 102:372–375

Zhuang L, Zhou S, Yuan Y, Liu T, Wu Z (2011) Development of *Enterobacter aerogenes* fuel cells: from in situ biohydrogen oxidization to direct electroactive biofilm. Bioresour Technol 102:284–289

Chapter 11
Advances in Concurrent Bioelectricity Generation and Bioremediation Through Microbial Fuel Cells

Bikash Kumar, Komal Agrawal, Nisha Bhardwaj, Venkatesh Chaturvedi, and Pradeep Verma

11.1 Introduction

Industrialization and economic development of different countries can be measured in term of available energy source. For the past 200 years, fossil fuels were pillar of growth supporting the demand for energy, but each coin has two phases; therefore, along with advantages comes disadvantages. It damaged the environment leading to pollution, over-exploitation of natural resources and damaging the flora and fauna. These factors acted as promoter for different stakeholders to search for an affordable and environmental friendly alternative, such as biofuels and bioenergy. One such alternative was MFC; the concept came into existence in year 1910, when Michael C. Potter at University of Durham, UK, observed the ability of *Escherichia coli* to generate electricity (Potter 1911). It has emerged as a promising tool for bioenergy generation. MFC is a device that converts the chemical energy to electricity via biological pathways (Santoro et al. 2017; Zhang et al. 2016). The conventional chemical approaches are costly and require sophisticated infrastructure (Zaffar et al. 2016). MFC is a green tool for the treatment of different pollutants and simultaneous generation of electricity to meet the growing energy need of increasing human population. General schematic MFC found its application in wastewater treatment plants along with electricity and hydrogen generation (Wang et al. 2015), sediment bioremediation (Li and Yu 2015) and detoxification of polluted soil from toxic xenobiotic compounds (Rodrigo et al. 2014). The growing interests of scientific community in application of MFC and its further improvement can be easily observed

B. Kumar · K. Agrawal · N. Bhardwaj · P. Verma (✉)
Bioprocess and Bioenergy Laboratory, Department of Microbiology, Central University of Rajasthan, Ajmer, 305817, Rajasthan, India
e-mail: pradeepverma@curaj.ac.in; http://www.vermapradeep-lab.com/

V. Chaturvedi
SMW College, MG Kashi Vidyapeeth, Varanasi, 221002, Uttar Pradesh, India

V. Sivasankar et al. (eds.), *Microbial Fuel Cell Technology for Bioelectricity*,
https://doi.org/10.1007/978-3-319-92904-0_11

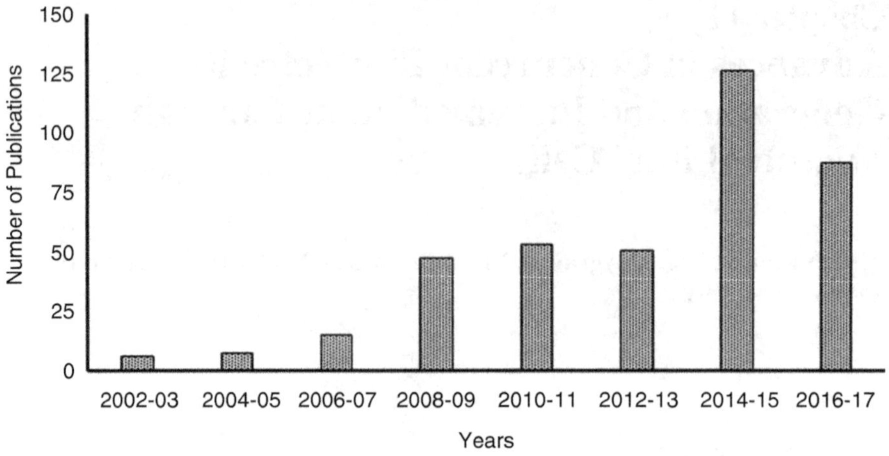

Fig. 11.1 Articles published on MFCs with major focus on bioremediation. The data is based on the article mentioning MFC and bioremediation in the SCOPUS citation database in November 2017

through growing number of publication in this area (Zhang et al. 2016) which has gained impetus in the past 10 years (Fig. 11.1). This chapter gives an insight into the recent development in MFC technology as pollutant treatment units besides generating electricity, its limitations and economic feasibility for commercialization.

11.2 Improvement in the Microbial Fuel Cell Technology for Bioremediation

The development of MFC technology has come a long way since the discovery "animal electricity" by Luigi Galvani in year 1780. In the early eighteenth century, Volta's experiment led to the invention of early battery, and in 1859 lead acid battery was invented by French physicist Gaston Plante. William Grove is considered as the father of fuel cell technology, and the concept acted as the theoretical base for the discovery of fuel cells in the future. Various vehicle manufacturers have shown keen interests in fuel cell technology which led to the discovery of different fuel cells such as soft oxide fuel cells (SOFCs), molten carbonate fuel cells (MCFCs) and proton exchange membrane fuel cells (PEMFCs). However these technologies had their own limitations such as high operating temperatures, slow start times, need of precious metals as catalysts, high temperature, high cost involved and highly corrosive media in some cases. The alternative to these technologies was MFC, an efficient alternative to the costly abiotic fuel cells, as it can be operated under ambient temperature and pressure. The MFCs technology also attracted the attention of NASA scientists in the year 1960s when they showed interests in turning of organic wastes into electricity during the space missions; however, that was a short-lived project. The MFC technology went on back seat until the work by Bennetto et al.

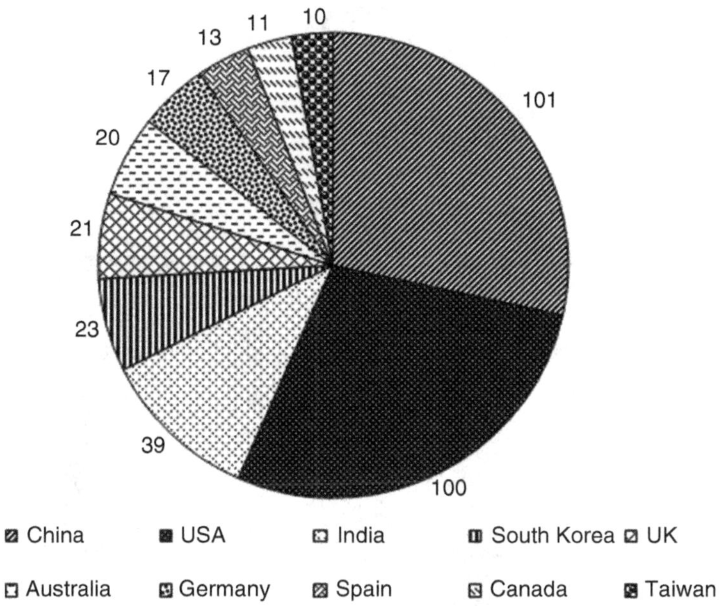

Fig. 11.2 Country-wise distribution in MFC research, number of articles on MFCs with focus on bioremediation. The data is based on the article mentioning MFC and bioremediation in the SCOPUS citation database in November 2017

(1983), where they reported the functioning of MFC with special focus on the use of mediators for electron transfer during electricity generation.

Since the advent of the twenty-first century, MFC attracted attentions of research groups around the world which is clearly visible from Fig. 11.1. Increase in number of publications of MFC with emphasis on bioremediation increased nearly sixty (60) times in year 2015–2016 as compared to year 2002–2003. It attracted attention of both developed and developing nation; China and the USA are in first place, and India is in third regarding MFC–bioremediation publications (Fig. 11.2).

In order to improve MFC, different strategies are used such as selection of electrodes, membranes materials, use of pollutants as substrates and designing new type of MFCs for specific applications which is explained later in the chapter.

11.3 Design of Microbial Fuel Cell

MFC can be divided into mainly types on the basis of design: single chambered and dual chambered. General schematic representation of MFC is described in Fig. 11.3 which contains separate cathodic and anodic chambers called as dual-chambered MFC whereas the single-chambered MFC contains both cathode and anode in a single chamber. Major components of MFC are cathode, anode and membrane.

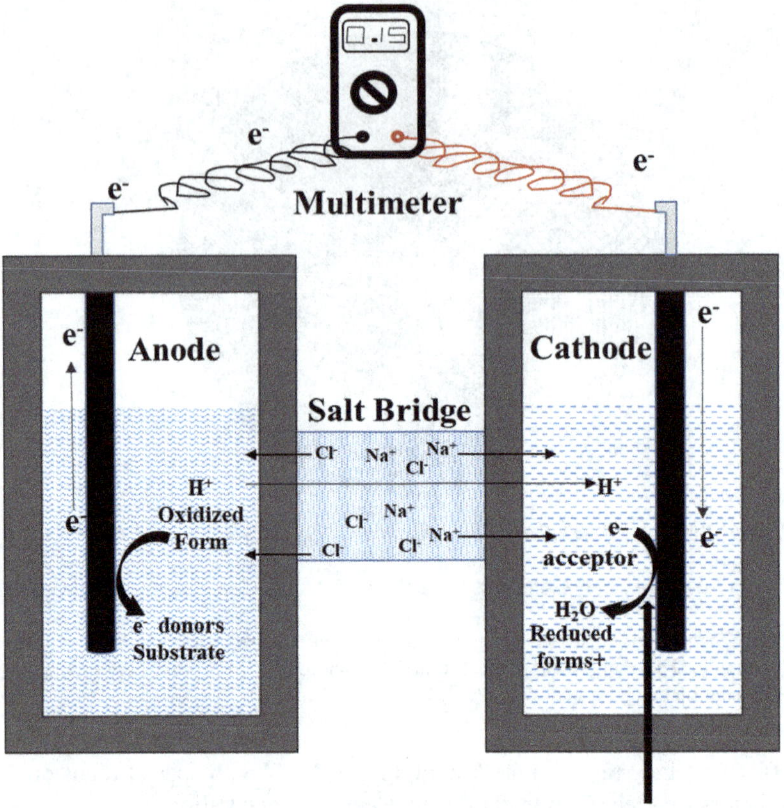

Fig. 11.3 Schematic representation of two-chambered microbial fuel cell and mechanism of electron and proton transfer in MFC (Based on Chaturvedi and Verma 2016; Zhang et al. 2016)

The choice of material for each component is an important factor in energy output and economics of MFC as technology. Each component and advancement in the selection of materials for the cathode anode and membrane is described in the next section.

11.4 Electrode Materials

The performance of MFC for electricity generation and bioremediation is based on selection of proper anode and cathode materials. The basis for electrode selection usually depends on bacterial adhesion, electron transfer and electrochemical efficiency (Mustakeem 2015). In general practice the cost of materials used for electrode must be low and power densities maximized (Mustakeem 2015). The criteria for selecting the anodic and cathodic material are as follow:

A. **Electrical Conductivity:** The electricity generation through MFC, an electron released from microbes, has to travel from the anode to the external circuit. The electrode material having less resistance will have higher conductivity for the electric current as it allow effortless flow of electron. The lower the resistance, the higher is the conductivity for the flow of electron and result generation of electricity. The low interfacial impedance also plays a key role in facilitating the electron transfer. In case of cathode, the higher ionic conductivity is required for facilitating the triple-phase boundary reaction (Mustakeem 2015; Natarajan and Van Nguyen 2004).

B. **Surface Area and Porosity:** The surface area of electrode affects the final output power of MFCs. The loss of current is directly proportional to the electrode resistance. The efficiency of the MFC can be enhanced by decreasing the resistance of the electrode material that can be done by increasing the effective surface area whilst keeping the volume same. The larger surface area will provide large area for reactions that will enhance electrode kinetics (Wang et al. 2011; Rismani-Yazdi et al. 2008). Higher porosity allows bacteria to access and colonize more that help in biofilm formation.

C. **Stability and Durability:** The material used for electrode must be capable of withstanding highly reducing and oxidizing environment in MFC that may lead to the swelling and decomposition of the materials. The electrode material must be highly durable which may be provided by increasing the surface roughness. However, excessive roughness may increase the probabilities of fouling that may decrease overall performance of the MFC in longer term. Therefore, a highly stable and durable electrode are required for overall enhanced performance of MFC (Santoro et al. 2017; Mustakeem 2015).

D. **Cost and Accessibility:** It is necessary to use such material which is easily accessible to develop at lab scale and will help in impletion at large scale. The cost of the electrode material will influence the capital involved in construction of the MFC to a large extent. Thus, for implementation of the MFC at commercial scale, the material should be cheap, sustainable and easily available. For example, platinum is widely used; however, it is expensive. Carbonaceous and non-precious metal materials such as composites might be a substitute to precious metals in electrodes in the future (Santoro et al. 2017; Mustakeem 2015).

E. **Biocompatibility:** It can be seen in term of compatibility of electrode surface for proper adhesion of the microbial biofilm. A highly biocompatible material will increase the bacterial adhesion and hence the life of the MFC and ease of current flow (Santoro et al. 2017; Mustakeem 2015).

11.4.1 Anode Materials

Anode is the electrode used in MFCs and is the site where electron donors undergo oxidization reactions (Zhang et al. 2016). The material used in anode preparation plays a significant role in the biofilm formation and the electron transfer between the microorganism and the electron acceptor. Various carbonaceous, metallic and composite materials are used for the construction of anode for increased power density and better energy output (Mustakeem 2015). List of different anode along with properties are tabulated in Table 11.1. The interaction between biofilm and

Table 11.1 List of anodic materials and their properties used in MFC

Anode materials	Properties	References
Carbon cloth	High surface area, porosity, electrical conductivity, flexibility and mechanical strength in forming complex 3D structure. But expensive	Guerrini et al. (2014); Santoro et al. (2013)
Carbon brush	Expensive material consists of titanium core with twisted carbon fibres, High surface area that provides an optimal area to volume ratio. The central titanium core guarantee the electrical conductivity	Liao et al. (2015)
Carbon rod	Most affordable anode used in MFC. Surface areas for carbon rod are low. Thus they are more preferred as current collectors rather being used as anode	Jiang and Li (2009); Liu et al. (2004)
Carbon mesh	Commercially available at low cost. Carbon mesh can be folded efficiently to prepare a 3D electrode, low electrical porosity conductivity, mechanical strength and poor durability	Wu et al. (2017); Wang et al. (2009)
Carbon veil	Very cheap carbonaceous material with high electrical conductivity and porosity. It can be easily folded to form a robust and porous 3D electrode. Carbon veil is fragile	Artyushkova et al. (2016); Boghani et al. (2014); Winfield et al. (2014)
Carbon paper	Planar and relatively porous carbonaceous material. Expensive and fragile with its application limited to lab-scale batch system.	Santoro et al. (2014); Srikanth et al. (2008)
Carbon felt	High electrical conductivity. Carbon felt has high porosity that allows the microbes to colonize the biofilm by penetrating through the large pores. High mechanical strength owing to thickness of the material and its cost is relatively low	Seviour et al. (2015); Roy et al. (2014); Calignano et al. (2015)
Granular activated carbon (GAC)	The granular activated carbon is used as packaging material rather than using GAC as stand-alone anode. Due to GAC's intrinsic property of very high surface area, it can help in the adsorption of organics pollutants or heavy metals. Low cost and electrical conductivity, high porosity	Yasri and Nakhla (2017); Zhao et al. (2016); Jiang et al. (2011)
Granular graphite	It has properties similar to GAC except the granular graphite has much lower surface area because of lack of activation. It is also used as packing material rather than using a stand-alone anode. High electrical conductivity	Feng et al. (2010)
Carbonized cardboard	3D materials consist of single wall corrugated cardboard from recycled paper. The material is very low cost and has high electrical conductivity and porosity	Kretzschmar et al. (2017); Chen et al. (2012b)
Graphite plate	It has relatively lower surface area, surface to volume ratio and electric output. Due to its high mechanical strength, it is often used as support for modified structures. Graphite electrode is simple anode with high electrical conductivity and relative low cost	Dewan et al. (2008); Heijne et al. (2008)

(continued)

Table 11.1 (continued)

Anode materials	Properties	References
Reticulated vitreous carbon	Reticulated vitreous carbon is highly conductive. The unique property of high porosity allows the biofilm to penetrate through the entire structure and colonize the entire electrode. The major disadvantage is the material is quite fragile and very expensive	Lepage et al. (2012)
Electrospun carbon fibres	Electrospun carbon fibres are prepared by layer-by-layer (LBL) electrospinning of polyacrylonitrile onto thin natural cellulose paper followed by carbonization. High-density layered biofilm propagation and high bioelectrocatalytic anodic current density	Chen et al. (2011); He et al. (2011)
Activated carbon nanofibres	Activated carbon nanofibres have an ultra-thin, porous interconnected structure with high bioaccessible surface area that promotes well-supported biofilm growth. Application in simultaneous power generation and removal of organics from wastewater	Delord et al. (2017); Karra et al. (2012)
Carbonized plant materials	Several plant materials (e.g. corn stem, different mushrooms) are carbonized and tested as anode. The carbonized plant materials have highly porous architecture support, high electron transfer rate and electroactive biofilm growth. Low-cost alternative to costly anodes. Very conductive, robust and cheap and easily accessible. Major drawbacks are low surface area, biocompatibility issues and corrosion	Karthikeyan et al. (2015); Chen et al. (2012a)
Metallic anode materials	Several metallic materials such as stainless steel (plate, mesh, foam or scrubber) have been used as anodic material. Copper, nickel, silver, gold and titanium have been also successfully tested as anodic material. However, copper and nickel ions can be poisonous for microbes forming biofilm due to its metal toxicity but high and stable performance	Baudler et al. (2017); Guo et al. (2016)
Composite material	Carbon nanotubes–polyaniline composite enhanced the electrocatalytic property and adhesion to the bacterial cell. Polypyrole-coated carbon nanotubes have enhanced power density by six times as compared to graphite electrode. PPy-coated CNT showed high electron transfer	Mustakeem (2015); Sharma et al. (2008)
Material treatments	Several surface treatments such as ammonia-treated electrode and graphene over stainless steel mesh. Acid treatment of electrode and electrochemical oxidation treatment helps in protonation of functional group and addition of functional group, respectively	Lim and Wilcox (2012); Cheng and Logan (2007); Lowy et al. (2006)

anode electrode is affected by surface morphology and chemistry. The bacterial attachment with the electrode can be regulated by controlling the surface chemistry properties such as surface charge (Santoro et al. 2015), hydrophilicity/hydrophobicity (Du et al. 2017), oxygen/nitrogen functional groups and immobilized mediators (Santoro et al. 2017; Li et al. 2014). Further attachments can be regulated by surface morphology and can be controlled at nano- and micro-scale level using chemical treatments, surface coatings and electrochemical and thermal treatments. Recently lot of surface modifications such as 3D electrode are preferred over 2D electrode, as theoretically it is suggested that surface area is directly proportional to electricity generation. However, several studies have suggested that there are limitations for 3D surface such as pH gradients and diffusion transport phenomenon associated with product and reactant (Blanchet et al. 2016). Whilst designing a MFC, anode electrodes must be selected such as to avoid clogging and dead zone, as it will help in long-term operation (Santoro et al. 2017).

11.4.1.1 Role of Anode in Bioremediation

Most of the organic pollutants present in waste and wastewaters are in reduced form, and anode acts as a site for oxidation of electron donors (Zhang et al. 2016). The chemical oxygen demand (COD) is adopted as mean to quantify the oxidation power of the pollutants. In the anodic chamber, COD is converted to carbon dioxide and water (Zhang et al. 2016; Kim et al. 2015a, b). The recalcitrant pollutants such as azo dyes (Thung et al. 2015), polycyclic aromatic hydrocarbons (PAHs) (Sherafatmand and Ng 2015), benzene derivatives (Zhang et al. 2015), inorganic wastewaters containing sulphide (Raschitor et al. 2015), industrial wastewaters (Abbasi et al. 2016) and several organic wastes such as chicken feather (Chaturvedi and Verma 2014), poultry droppings and human excreta (Kretzschmar et al. 2017; Ieropoulos et al. 2013) are used. It has also been used for the treatment of various waste treatments such dairy manure, polluted soil sample and landfill leachate. However, solid waste digestion efficiency of the anode is low, but several modifications have been carried out to improve its efficiency, e.g. polluted sediments/soil/groundwater can also be remediated by embedded anodes in the polluted matrix with an external cathode exposed to the air.

11.4.2 Cathode Materials

The oxygen reduction reaction (ORR) takes place at the surface of cathode under three phasic interfaces, i.e. electrode (solid), electrolyte (liquid), air (gas), to form water. The ORR is the limiting reaction of the MFCs, and typical MFC cathode can be divided into three layers, i.e. diffusion layer, conducting support layer and catalyst. The materials used for the cathode (Table 11.2) must bear certain properties to act as robust cathode such as (a) high mechanical strength, (b) catalytic property and (c) high electronic and ionic conductivity (Santoro et al. 2017; Mustakeem 2015). Most of the materials used for anode can be used as cathode; however, carbon-based materials have poor catalytic activity, and an additional catalyst material is required to boost the reduction process. Different types of material used as cathode are tabulated in Table 11.2:

(a) Cathode with Pt-based catalyst
(b) Cathode with non-Pt-based catalyst

Table 11.2 List of cathodic materials and their properties used in MFC

Cathode material	Cathode catalyst	Properties	References
Cathode with Pt-based catalyst/ platinum metal based	Carbon-Pt	Platinum is the most successful catalyst used for carbon cathode coating for oxygen reduction reaction because of its high surface area and low over potential for ORR. Limitations such as high cost, pH sensitivity, sulphide poisoning and non-sustainable	Watanabe (2008)
	Double-layer polydimethylsiloxane (PDMS)/carbon loaded with Pt	High-power density and hydrophobic materials like PDMS decrease the water diffusion into a single-chambered MFC	Zhang et al. (2016); Mustakeem (2015)
	Carbon nanotubes (CNTs) loaded with Pt	CNTs loaded with Pt by depositing it over CNTs surface. ORR catalytic activity was improved for CNT–Pt composite. The ORR was not affected considerably by decreasing Pt loading by 20-fold	Ghasemi et al. (2011); Cheng et al. (2006a)
Cathode with non-Pt-based catalyst/ platinum group metal-free	Iron phthalocyanine	Rapid electron transfer takes advantage of π–π interaction of metal and carbon	Yuan et al. (2011)
	Iron phthalocyanine-amino-functionalized multiwalled CNTs	High-power density	Yuan et al. (2011)
	Cobalt tetramethoxy-phenylporphyrin (CoTMPP)	Coulombic efficiency is comparable to Pt-based catalyst	Cheng et al. (2006b)
	Lead dioxide–carbon	Alternative to Pt as a catalyst, lead oxide has low cost and shows four times higher-power densities as compared to conventional Pt electrodes	Morris et al. (2007)
	Transition metal-based oxide	The cathode current with MnO_2-catalysed cathode was found to be much larger than graphite. β-MnO_2 deposited over glassy carbon using polyvinylidene fluorine (PVDF) as binder showed high catalytic efficiency and power densities as close to Pt based. CNTs covered with MnO_2 as cathode material showed high-power density increased by two order of magnitude as compared to plain stainless steel cathode	Amade et al. (2015); Cao et al. (2003)
	Palladium	Palladium, a Pt-like transition metal, has excellent catalytic. High stability and low cost as compared to Pt	Huang et al. (2011)
	Lead oxide	Four time higher-power density and eight time less cost compared to commercial Pt/C	Morris et al. (2007)
	Activated carbon	Large surface area, synthesized by electrospinning and pyrolysis of polyacrylonitrile precursor. Activated carbon fibre-treated electrodes was increased by more than 75%	Ghasemi et al. (2011)

(continued)

Table 11.2 (continued)

Cathode material	Cathode catalyst	Properties	References
Cathode with metal-free catalyst/ carbonaceous materials	Chemically modified carbon-based materials	Carbon nanotubes, graphene, and graphite foam can be used as catalyst for ORR	Su et al. (2010); Matter et al. (2006)
		Low cost, relatively high and stable performance	
	Nitrogen-doped graphene nanosheets	High catalytic activity, onset potential, higher electrocatalytic activity, lower over potential and long-term stability than commercially available Pt (10 or 20 wt%, E-TEX) for ORR	Lu et al. (2013); Ci et al. (2012); Gong et al. (2009)
Biocathode	Aerobic biocathode	In aerobic biocathode, oxygen acts as terminal electron acceptor, hydrogen peroxide acts as intermediate and transition metals such as iron and manganese act as electron mediators between the electrode and oxygen	Park and Zeikus (2003)
	Anaerobic type biocathode	In anaerobic type biocathode, oxygen is not present, whereas nitrates and sulphates act as terminal electron acceptors. Advantage by preventing the loss of electrons through oxygen also microorganisms at the cathode can act as biosensor for detection of biological oxygen demand (BOD) in water. Further microorganisms at biocathode can be used to produce as methane, ethanol and formic acid	Rabaey and Rozendal (2010); Kim et al. (2003)

(c) Cathode with metal-free catalyst
(d) Biocathode

List of cathodic materials used in MFC along with their properties are tabulated in Table 11.2

11.4.2.1 Role of Cathode in Bioremediation

In MFC the electron acceptors undergo reduction reactions at cathode. The oxidized substrates are reduced at the cathode if the available electric potential exceeds the threshold of oxidized substances (Zhang et al. 2016; Nancharaiah et al. 2015). The environment of cathodic chamber is highly reducing; therefore it is widely used in waste (landfill leachate) and wastewater treatment (liquid fraction of pig slurry, swine wastewater), organic substances (chlorobenzene and trichloroethylene) heavy metals like (Cr6+, V5+) and inorganic substances like ammonia (Sotres et al. 2015), xenobiotic compounds, etc.

11.4.3 Membrane Material

The architecture, choice of material and overall arrangement of membrane in MFC affects the performance, cost and multi-level applications. A large array of materials have been tested for its application as membrane such as natural rubber (Rajan et al. 2006), laboratory gloves rubber (Winfield and Chambers 2014), j-cloth, nylon fibres (Zhang et al. 2010), glass fibres, biodegradable shopping bags and ceramics (Winfield et al. 2013). The cation-exchange membranes (CEM), e.g. Naflon, is the most commonly used membrane system (Santoro et al. 2017). Dual-purpose ion-exchange bridge, monolithic 3D printed materials and porous materials (sufficient strength, chemical inertness and longevity) can be employed as the membrane materials (Santoro et al. 2017; Philamore et al. 2015). Several microporous filtration membranes (Zhuang et al. 2009), nylon infused membrane (Hernández-Fernández et al. 2015), photocopy paper (Winfield et al. 2015), ceramics and terracotta materials (Winfield et al. 2016) have also been tested as membrane.

11.5 Types of Waste Materials Used as Substrates in MFC

MFCs are considered as an efficient technology which effectively utilizes wastewater for energy generation (Winfield et al. 2016), Various waste materials used as substrates in MFC are as follows:

1. **Lignocellulosic Biomass:** Lignocellulosic materials are abundant and renewable natural resource. However it cannot be directly utilized as it has to be first converted to monosaccharides or low-molecular-weight compounds for the utili-

zation by the microorganism (Huang et al. 2011; Ren et al. 2007) for electricity generation. Cellulose and chitin are cheap, renewable and readily available biopolymeric materials which can be used for electricity generation (Rezaei et al. 2009). For direct conversion of cellulose, the microorganism(s) must be able to hydrolyse cellulose anaerobically, utilizing anode as an electron acceptor as well as oxidizing metabolites obtained after cellulose hydrolysis. Thus, using a solid substrate such as cellulose or chitin, the power production is limited due to a low rate of hydrolysis of the particulate material. However few studies have been carried out using particulate substrates in MFC.

2. **Synthetic Wastewater:** It has also been observed that synthetic or chemical wastewater having precise composition have also been used (Pant et al. 2010). However synthetic wastewater may contain redox mediators, such as cysteine and sulfur species (Aldrovandi et al. 2009). These redox mediators can act as abiotic electron donor and help in enhancing the production of electricity for a short while but it would not represent the true performance of the system (Aldrovandi et al. 2009) resulting in the ambiguity being generated in the results. Thus, this can be overcome by the use of minimal salt media with a single electron donor such as glucose or acetate.
3. **Brewery Wastewater:** Breweries wastewater has been used in MFCs, primarily because of its low strength, suitable for electricity generation due to food-derived nature of organic matter and lack of high concentrations of inhibitory substances, e.g. ammonia in animal wastewaters (Feng et al. 2008). It can be an ideal substrate for MFCs due to its nature of high carbohydrate content and low ammonium nitrogen concentration (Pant et al. 2010).
4. **Starch Processing Wastewater:** Starch processing wastewater (SPW) contains a relatively high content of carbohydrates sugars protein and starch which can be potentially converted to a wide variety of useful products (Jin et al. 1998). SPW was used as a fuel to enrich a microbial consortium generating electricity and current generation.
5. **Dye Wastewater:** Azo dyes constitute the largest class of synthetic dyes and are extensively present in the effluent of dye and textile industries. The removal of toxic substances present in the effluents before discharge is of paramount importance as they are effecting the environment adversely. In aquatic system intense colour of dyes has led to obstruction of light and transfer of oxygen into the water bodies which is having detrimental effect on the aquatic flora and fauna (Pant et al. 2007). Very recently, efforts were made to utilize these dyes as substrate in MFC leading to detoxification and generating electricity. The concentration of glucose and dyes plays an important role in the current generation; however, the microorganism are unable to decolorize high concentrations of dye (Sun et al. 2009). Another method which can be used includes simultaneous treatment of azo dye-containing wastewater and readily biodegradable organic matter-containing wastewater. The addition of later could help in improving the dye degradation efficiency due to microbial consortium associated with it along with the presence of nutrient material, thus saving both cost and energy. However, the system still requires considerable improvements in terms of finding appropri-

ate bacterial community that is capable of utilizing a mixture of dyes and other simple carbon sources in order to make MFCs a realistic solution for its extensive utility at large scale.

6. **Landfill Leachates:** The use of landfill effluent in a biological fuel cell for COD removal was first reported by Habermann and Pommer (1991). Landfill leachates are heavily polluted landfill effluents with a complex composition containing four major groups of pollutants: dissolved organic matter, inorganic macro-components, heavy metals and xenobiotic organic compounds (Kjeldsen et al. 2002). Recently, Greenman et al. (2009) demonstrated that it is possible to generate electricity and simultaneously treat landfill leachate in MFC columns. Gálvez et al. (2009) operated three MFCs fluidically connected in series for simultaneous leachate treatment and electricity generation.
7. **Inorganic and Other Substrates:** There are various substrates which have been used in the MFC as substrates such as paper-recycling plant wastewater (Huang and Logan 2008), unamended wastewater, phenol (Luo et al. 2009), carbon monoxide (Kim and Chang 2009), 1,2-dichloroethane (Pham et al. 2009), sulphate and thiosulphate (Zhao et al. 2009). MFC was used to remove the fermentation inhibitors in cellulosic biorefineries which included furfural, 5-hydroxymethylfurfural, vanillic acid, 4-hydroxybenzaldehyde and 4-hydroxyacetophenone with simultaneously generating electricity (Borole et al. 2009).
8. **Heavy Metals**

Hexavalent Chromium: Chromium has various industrial applications such as leather tanning, metallurgy, electroplating and wood preservatives (Chaturvedi and Verma 2016). Cr(VI) is more hazardous due to its mutagenic and carcinogenic properties (Humphries et al. 2004); therefore, there is a need for detoxification of hexavalent chromium [Cr(VI)]. Chemical or electrochemical reduction into nontoxic chromium is the most preferred method; however, other approaches such as ion-exchange resins, filtration and direct chemical reduction have also been employed (Kurniawan et al. 2006). These technologies require high-energy inputs and produces by-products which itself are pollutants. Therefore, reduction of chromium coupled with electricity generation using MFC and can be applied for Cr(VI) treatment (Tandukar et al. 2009) at the cathode. Various microorganism used in the process include *Trichococcus pasteurii* and *Pseudomonas aeruginosa*.

Selenite: Selenium and its derivatives such as selenite (SeO_3^{-2}) and selenate (SeO_4^{2}) are widely used in industries, e.g. glass manufacturing and electronic industries. Selenium can enter in the environment through sewage sludge, fly ash from coal-fired power plants, oil refineries and mining of metal ores (Tandukar et al. 2009). Selenite is more toxic than selenate to aquatic invertebrates and fishes (Hamilton 2004; Lemly 1997), and it accumulates in aquatic plants thus causing bioaccumulation in higher organisms which can cause both acute and chronic toxicity in aquatic organisms (Rovira et al. 2008). The application of MFC technology in reduction of selenium and production of electricity by using selenium-containing waste was investigated by Catal et al. (2009).

9. **Nitrate:** The use of nitrate-based fertilizers and animal waste has increased the presence of nitrate in water. Nitrate (nontoxic) can be transformed to nitrite (NO^{2-}) after entering human body, which can cause "blue baby syndrome" normally observed in infants, or it can be converted to N-nitroso compounds which are carcinogenic into humans. Some of the treatment methods employed include electrochemical treatment, ion exchange (IE), reverse osmosis (RO), electrodialysis (ED) and heterogeneous catalysis (HC) (Park and Yoo 2009). As the methods employed are expensive, the use of MFC for removal of nitrate has gained importance due to the feasibility of the process. Few workers have employed a metal catalyst (Polatides and Kyriacou 2005) or microorganisms as catalysts on cathode electrode (He and Angenent 2006). In few studies, electrochemical denitrification process to remove nitrate ions was employed at cathode chamber of bio-electrochemical denitrification system (Kondaveeti and Min 2013).
10. **Marine Sediments Rich in Acetate:** Marine sediments rich in acetate have been used as a substrate in MFC. Acetate presence helps in providing inertness towards alternative microbial conversions, fermentations and methanogenesis at room temperature (Sun et al. 2009; Aelterman 2009). Acetate can act as a carbon source to induce electroactive bacteria, and it is an end product in several metabolic pathways (including the Entner–Doudoroff pathway for glucose metabolism) for higher-order carbon sources (Biffinger et al. 2008; Bond and Lovley 2005). When Chae et al. 2009 compared the performance of four different substrates in terms of CE and power output, acetate-fed MFC showed the highest CE, followed by butyrate, propionate and glucose. According to Liu et al. (2009), acetate-based MFC achieved more electric power and external load resistance compared to those based on consortia induced by a protein-rich wastewater.
11. **Animal Wastes:** Animal wastes such as poultry dropping, cow dung, human faeces and urine are reportedly used as substrate for the bioelectricity generation. Barbosa et al. (2017) and Jimenez et al. (2016) suggested that animal waste can be considered as potential substrate for generation of electricity due its high daily production, high COD, high nutrient (N and P) concentrations and high solution conductivity. Based on these observation, Ieropoulos et al. (2016) suggested that the treatment of urine can be performed in MFCs whilst generating electricity for low-power devices, such as light-emitting diode (LED) lights or sensors. Ieropoulos et al. (2016) ran field trials by connecting a stack of MFC to public urinals and used it for lighting a room. This was an attempt for taking the technology beyond laboratory, and it helped understand how fast the technology evolves and is being able to address the issue of public hygiene, sanitation, direct treatment of human waste and energy generation. Melhuish et al. (2006) designed a robot "EcoBot-II", which was powered by on-board MFC with oxygen cathodes, it simultaneously utilized unrefined insect biomass and converted into useful energy. Kretzschmar et al. (2017) provided an experimental proof of concept using human faeces as substrate in MFC for power generation.
12. **Petroleum Hydrocarbons:** MFC can be a cost-effective and eco-friendly approach for bioremediation of petroleum hydrocarbons as these hydrocarbons

act as substrate for microbes. The MFC technique was used for detoxification of soil decontaminated with petroleum at an increased rate and is energy sufficient. A U-tube MFC designed by where they observed enhanced degradation rate of petroleum hydrocarbons by 120% high charge output. Morris et al. (2009) used MFC for enhancing biodegradation of diesel. They demonstrated that in MFC diesel removal rate was increased by four times using MFC because the electrode served as an electron acceptor. MFC utilizing diesel (v/v 1%) as sole carbon source also resulted in high-power density and current density (Cheng et al. 2017). Therefore, MFC technology can be effectively used for detoxification and enhancing biodegradation of polluted soil/wastewater containing petroleum contaminants in anoxic environments, thus, eradicating the need to adjust terminal electron acceptors such as oxygen.

11.6 Types of Microbial Fuel Cell for Bioremediation of Pollutants

11.6.1 Anaerobic Microbial Fuel Cell (ANMFC)

ANMFC require oxygen that may cause loss of electrons and lead to increase in energy demand to carry out the process (Abbasi et al. 2016). Therefore, anaerobic MFC can be an environmental friendly and cost-effective alternative for simultaneous electricity generation and bioremediation. Wastewaters generated from brewery industries (Akman et al. 2013; Pant et al. 2010); distillery and domestic wastewater (Jiang et al. 2011); pharmaceutical industry (Velvizhi and Venkata Mohan 2012), petrochemical, vegetable oil, food industry and animal carcass wastewater (Li et al. 2013); textile (Solanki et al. 2013) swine wastewater (Zhuang et al. 2012); and municipal wastewater (Zhang et al. 2013) could be treated using anaerobic MFC. Abbasi et al. (2016) designed an anaerobic MFC where they utilized wastewater samples from vegetable oil industries, metal works, glass and marble industries as substrate. This process has significant effect on wastewater treatment efficiency for COD in range of 85–90% at 96 h of hydraulic retention time (HRT). The coulombic efficiency of 5184.7 C with maximum voltage of 890 mV was generated when vegetable oil industries discharge was treated in MFC. A positive significant co-relation was observed between COD concentration and voltage generated.

11.6.2 Sediment Microbial Fuel Cell (SMFC)

SMFC comprises of an anode buried in sediment and a cathode in an oxygen-rich water (Rezaei et al. 2007; Xia et al. 2015). The SMFC have been successfully tested in removal of persistent organic compounds such as polycyclic aromatic hydrocarbons (PAHs), xenobiotic compounds and pesticides, etc. along with electricity generation (Sherafatmand and Ng 2015). Similarly Yu et al. (2017) reported maximum power density together with removal of anthracene, phenanthrene and pyrene by a closed SMFC utilizing PAH-polluted soils. They further studied the influence of electrode interval and role of microbial community in electricity generation and removal of PAHs. The decrease in electrode interval resulted in efficient electricity generation and removal of PAHs. The SMFC was dominated by the genus of *Geobacter* and enriched in electrogenic bacteria at the anode surface. The growth of certain microbes (except electrogenic bacteria in the soil) was improved by electrical stimulation. Similarly Xu et al. (2017) found that genus *Geobacter* are predominately found in SMFC and more electrogenic bacteria are found attached in biofilm of anode. They also demonstrated that addition of Fe (III) oxide in SMFC enhances the removal efficiencies for organic pollutants. Xia et al. (2015) reported enhanced biodegradation of organic chemicals of high polarity by operating a SMFC in heavily contaminated sediment and analysing its global organic chemical degradation profile. The study showed that SMFC prefers to stimulate the degradation of organic chemicals with higher polarity. Cao et al. (2015) constructed a SMFC in the top soil contaminated with hexachlorobenzene (HCB), a toxic refractory organic pesticide. Under anaerobic condition in the soil MFC, HCB was degraded via the reductive dechlorination pathway and an existence of the anode promoted electrogenic bacteria provided more electrons that subsequently improved electricity generation.

11.6.3 Benthic Microbial Fuel Cells (BMFC)

BMFC consists of an anode present in anoxic benthic sediment and a cathode in oxic overlying water which is further connected using an external electric circuit. BMFC are different from SMFCs as the latter does not require in situ deployed in real water bodies (Li and Yu 2015; Holmes et al. 2004). BMFC has emerged as sustainable and efficient technology for cleaning up of contaminated sediments and simultaneous energy generation. The bioremediation of the contaminated sediments is performed by utilizing the natural metabolic activities of microbes in detoxification, decomposition or immobilization of environmental contaminants present in the sediments. BMFC is at initial stage but have shown many potential benefits such as accelerated decontamination, relatively easy deployment, self-sustained operation and control and environmental benignity. The relatively lower efficiency, limitations in respect of system design, electrode selection, microbial control and selection of deployment environment severely limit its application (Li and Yu 2015).

11.6.4 Enzyme-Based Microbial Fuel Cells (EBC)

The fuel cells are of two types first employing living cells known as microbial biofuel cells and the other utilizing enzymes and referred as enzymatic biofuel cells (EBC). The microbial biofuel cells have long lifetimes ~5 years (Moon et al. 2006;

Kim et al. 2003) and can completely oxidize simple sugars to carbon dioxide (Bond et al. 2002). However they have low-power densities in W/cm^2 per unit electrode surface area which is due to slow transport across cellular membranes (Palmore and Whitesides 1994). By contrast, enzymatic biofuel cells typically possess higher-power densities (although still lower than conventional fuel cells) and can partially oxidize the fuel and have limited lifetimes (typically 7–10 days) owing to the sensitive nature of the enzyme (Kim et al. 2006; Barton et al. 2004) and to eliminate the need for a membrane separator. One of the most significant advances in enzymatic biofuel cells is the development of biocathodes and bioanodes that employ direct electron transfer (DET) instead of mediated electron transfer (MET). The DET is more preferred over the MET (Moore et al. 2004). The second method is the immobilization of the enzyme which has helped enzyme increase in the active lifetime of the enzymes (Topcagic et al. 2004). Recently, the active lifetimes have been extended beyond 1 year by encapsulation in micellar polymers (Akers et al. 2005; Moore et al. 2004; Topcagic et al. 2004). The EBC can be used extensively in biodegradation of toxic organic pollutant such as azo dyes, polyaromatic hydrocarbons, etc. The EBC have many disadvantages over traditional fuel cells and primary batteries; they remain limited by short lifetimes, catalytic inefficiencies, low fuel utilization and low-power densities (Minteer et al. 2007). However working solutions to short lifetimes and catalytic inefficiencies have been introduced, but advances in improved fuel utilization and power density are required.

11.6.5 Air-Breathing Cathode-Based Microbial Fuel Cells (ABC-MFC)

An air-breathing cathode consists of electrode substrate, catalyst layer and air diffusion layer (Wang et al. 2017). Single-chamber, air-breathing MFC with a flexible graphite sheet as the anode was designed by Sonawane et al. (2017), in which they used landfill leachate as substrate. They obtained open-circuit voltage (OCV) of 1.29 V, which is the highest reported OCV in the literature till date by utilizing landfill leachate as substrate. The reactor also resulted in generation of maximum cathode area-specific power density of 1513 mW m^{-2}. Jimenez et al. (2016) demonstrated the use of gas-diffusion air-breathing cathode-based MFC for generation of electricity utilizing urine as substrate. As discussed earlier cathode is site of ORR, and it is one of the limiting factor for MFC performance; therefore Kodali et al. (2017) used Mn-, Fe-, Co- and Ni-containing platinum group metal-free catalysts along with aminoantipyrine, AAPyr precursor for enhanced electricity generation. With increase in solution conductivity, it was observed that Fe-AAPyr was found to be the most suitable catalyst–precursor combination in air-breathing cathode-based MFC for enhanced electricity with high-power density.

11.6.6 Constructed Wetland Microbial Fuel Cells (CW-MFC)

Constructed wetlands (CWs) and MFCs are two different technologies, nevertheless very compatible technologies (Liu et al. 2013). Both the techniques are dependent on the actions of bacteria to remove pollutants from wastewater, but MFC has added advantage being energy generator. Therefore, the two techniques are combined in such a way that anode is buried in the anaerobic condition of constructed wetland and the cathode exposed to oxygen in the plant rhizosphere and collectively called as CW-MFC. The low oxygen availability at anode and higher redox gradient are an essential feature for generating electric current in CW-MFCs. The upflow regime of CW-MFC is such that it reduces the availability of dissolved oxygen (DO) at the anode whilst ensuring its maximum availability in the cathode region and also provide sufficient redox profile for MFC integration (Corbella et al. 2014; Fang et al. 2013). This natural redox gradient comes at cost of ohmic resistance of 120–500 (Doherty et al. 2015b; Villaseñor et al. 2013). A multi-electrode MFC with a separator electrode assembly was suggested by Ahn and Logan (2012) that help in reducing the ohmic resistance to 33 Ω. Glass wool separators can be used in construction of CW-MFC in order to minimize electrode spacing. This technology is excessively used in wastewater treatment with the aim of improving the wastewater treatment capacity of wetlands whilst simultaneously producing electrical power (Doherty et al. 2015a).

11.6.7 Thermophilic Microbial Fuel Cells (TMFC)

MFC used for the generation of electricity usually operated at ambient or mesophilic temperatures. Carver et al. (2011) suggested that thermophilic systems have potential for increased microbial activity rates on the anode that can subsequently enhance electricity generation. Air-cathode single chambers and two-chamber designs are usually used as for electricity generation, but the real role is played by thermophilic microbes associated with anode under anaerobic condition. Carver et al. (2011) described a thermophilic MFC design maintained at 57 °C with an anaerobic and thermophilic consortium. This suggested design minimized evaporation and associated microbes respired with glucose to generate a power density of 375 mW m^{-2} after 590 h. Voltage data and polarization showed that the design can work in both batch and continuous mode. Dai et al. (2017) constructed a two-chamber TMFC and utilized ethanol as an electron donor. They obtained an open-circuit potential of approximately 650 mV, maximum voltage of 550 mV and maximum power density of 437 mW m^{-2}, and the coulombic efficiency was 20.5 ± 6.0. They also analysed the microbial dynamics by high-throughput sequencing and 16S rRNA clone library sequencing; *Firmicutes* bacteria accounted for 90.9% of all bacteria associated with TMFC biofilm. The development of TMFC-involved biotechnologies will be beneficial for the wastewater treatment, production of valuable chemicals and generation of energy at the same time (Dai et al. 2017).

11.7 Commercial Application of MFC and Economic Feasibility

MFC technology has come a long way, and over the last decade, significant scientific and technological development had brought the MFC to the point of becoming commercialized technology. During commercialization of a technology, various market forces decide its success such as cheap, cost-effective, environmental friendly, larger consumer base and above all profitability. For assessing the profitability of the MFC and checking whether the technology is ready to enter market for commercial energy generation and waste treatment, Trapero et al. (2017) used classical evaluation criteria for investment decisions such as the net present value (NPV) and the internal rate of return (IRR). They have presented an economic assessment of a MFC in a juice processing plant where maximum power density of the cell using two different MFC cases, i.e. cathodes with and without Pt, was studied. The performance was then compared to the conventional activated sludge process. Three different scenarios, optimistic, pessimistic and most likely scenarios, were analysed. They also performed study to find the important factors and design influencing MFC performance. By a sensitivity analysis of the electrode area, and the annual growth rate of the electricity pricing, it was revealed that the electrode area parameter is the most influential factor. The results of the study clearly showed that the implementation of MFC is a favourable substitute to the use of classical aerated activated sludge, and it has potential economic benefits. Whilst designing for a current state-of-the-art MFC keeping in view the future challenges, research directions may be focused on selection of microbial consortium, along with operational consistency and stability of the developed system.

11.8 Future Prospects and Directions

MFC is a promising technique with great potential as it is an environment-friendly tool which does not involve the use of renewable natural resource and emission of pollution. In recent times substantial technological advancement and improvement are observed for waste/wastewater treatments in MFCs. MFCs have multiple applications which include energy generation, waste treatment and production of chemicals, sensors for pollution level in water/soil, etc. It can generate energy out of waste without any external/additional energy. This property will enable MFC technology to be used in remote areas for energy generation and waste treatment e.g., maintaining sanitation and hygiene, catering the need of poor people for affordable electricity which is of particular interest for countries and regions of the developing world. However, several aspects need to be addressed in order to improve the MFCs performance such as power output and efficiency, material costs for electrodes and separators, microbial consortium, suitable MFC design for treatment of specific waste and the optimization of the bio-electrochemical reactions. Keeping in mind the funding/development perspective, MFCs can be considered as nascent technology, but recent involvement of government laboratories, NGOs, philanthropist and business houses has provided helps and support for continuing research into MFC technology with a hope to find solutions to global environmental problems.

Acknowledgement The authors acknowledge the financial support provided by Central University of Rajasthan, Ajmer, India.

References

Abbasi U, Jin W, Pervez A, Ahmad Z, Tariq M, Shaheen S, Iqbal A, Mahmood Q (2016) Anaerobic microbial fuel cell treating combined industrial wastewater: correlation of electricity generation with pollutants. Bioresour Technol 200:1–7. https://doi.org/10.1016/j.biortech.2015.09.088

Aelterman P (2009) Microbial fuel cells for the treatment of waste streams with energy recovery. Dissertation, Ghent University, Faculty of Bioscience Engineering

Ahn Y, Logan BE (2012) A multi-electrode continuous flow microbial fuel cell with separator electrode assembly design. Appl Microbiol Biotechnol 93:2241–2248. https://doi.org/10.1007/s00253-012-3916-4

Akers NL, Moore CM, Minteer SD (2005) Development of alcohol/O_2 biofuel cells using salt-extracted tetrabutylammonium bromide/Nafion membranes to immobilize dehydrogenase enzymes. Electrochim Acta 50:2521–2525. https://doi.org/10.1016/j.electacta.2004.10.080

Akman D, Cirik K, Ozdemir S, Ozkaya B, Cinar O (2013) Bioelectricity generation in continuously – fed microbial fuel cell: effects of anode electrode material and hydraulic retention time. Bioresour Technol 149:459–464. https://doi.org/10.1016/j.biortech.2013.09.102

Aldrovandi A, Marsili E, Stante L, Paganin P, Tabacchioni S, Giordano A (2009) Sustainable power production in a membrane-less and mediator-less synthetic wastewater microbial fuel cell. Bioresour Technol 100:3252–3260. https://doi.org/10.1016/j.biortech.2009.01.041

Amade R, Vila-Costa M, Hussain S, Casamayor EO, Bertran E (2015) Vertically aligned carbon nanotubes coated with manganese dioxide as cathode material for microbial fuel cells. J Mater Sci 50:1214–1220. https://doi.org/10.1007/s10853-014-8677-2

Artyushkova K, Roizman D, Santoro C, Doyle LE, Fatima Mohidin A, Atanassov P, Marsili E (2016) Anodic biofilms as the interphase for electroactive bacterial growth on carbon veil. Biointerphases 11:31013. https://doi.org/10.1116/1.4962264

Barbosa SG, Peixoto L, Ter Heijne A, Kuntke P, Alves MM, Pereira MA (2017) Investigating bacterial community changes and organic substrate degradation in microbial fuel cells operating on real human urine. Environ Sci Water Res Technol 3:897–904. https://doi.org/10.1039/C7EW00087A

Barton SC, Gallaway J, Atanassov P (2004) Enzymatic biofuel cells for implantable and microscale devices. Chem Rev 104:4867–4886

Baudler A, Langner M, Rohr C, Greiner A, Schröder U (2017) Metal–polymer hybrid architectures as novel anode platform for microbial electrochemical technologies. ChemSusChem 10:253–257. https://doi.org/10.1002/cssc.201600814

Bennetto HP, Stirling JL, Tanaka K, Vega CA (1983) Anodic reactions in microbial fuel cells. Biotechnol Bioeng 25:559–568. https://doi.org/10.1002/bit.260250219

Biffinger JC, Byrd JN, Dudley BL, Ringeisen BR (2008) Oxygen exposure promotes fuel diversity for *Shewanella oneidensis* microbial fuel cells. Biosens Bioelectron 23:820–826. https://doi.org/10.1016/j.bios.2007.08.021

Blanchet E, Erable B, De Solan ML, Bergel A (2016) Two-dimensional carbon cloth and three-dimensional carbon felt perform similarly to form bioanode fed with food waste. Electrochem Commun 66:38–41. https://doi.org/10.1016/j.elecom.2016.02.017

Boghani HC, Papaharalabos G, Michie I, Fradler KR, Dinsdale RM, Guwy AJ, Ieropoulos I, Greenman J, Premier GC (2014) Controlling for peak power extraction from microbial fuel cells can increase stack voltage and avoid cell reversal. J Power Sources 269:363–369. https://doi.org/10.1016/j.jpowsour.2014.06.059

Bond DR, Lovley DR (2005) Evidence for involvement of an electron shuttle in electricity generation by *Geothrix fermentans*. Appl Environ Microbiol 71:2186–2189. https://doi.org/10.1128/AEM.71.4.2186

Bond DR, Holmes DE, Tender LM, Lovley DR (2002) Electrode-reducing microorganisms that harvest energy from marine sediments. Science 295:483–485

Borole AP, Mielenz JR, Vishnivetskaya TA, Hamilton CY (2009) Controlling accumulation of fermentation inhibitors in biorefinery recycle water using microbial fuel cells. Biotechnol Biofuels 2:7. https://doi.org/10.1186/1754-6834-2-7

Calignano F, Tommasi T, Manfredi D, Chiolerio A (2015) Additive manufacturing of a microbial fuel cell—a detailed study. Sci Rep 5:17373. https://doi.org/10.1038/srep17373

Cao YL, Yang HX, Ai XP, Xiao LF (2003) The mechanism of oxygen reduction on MnO_2-catalyzed air cathode in alkaline solution. J Electroanal Chem 557:127–134. https://doi.org/10.1016/S0022-0728(03)00355-3

Cao X, Song H, Yu C, Li X (2015) Simultaneous degradation of toxic refractory organic pesticide and bioelectricity generation using a soil microbial fuel cell. Bioresour Technol 189:87–93. https://doi.org/10.1016/j.biortech.2015.03.148

Carver SM, Vuoriranta P, Tuovinen OH (2011) A thermophilic microbial fuel cell design. J Power Sources 196:3757–3760. https://doi.org/10.1016/j.jpowsour.2010.12.088

Catal T, Bermek H, Liu H (2009) Removal of selenite from wastewater using microbial fuel cells. Biotechnol Lett 31(8):1211–1216

Chae KJ, Choi MJ, Lee JW, Kim KY, Kim IS (2009) Effect of different substrates on the performance, bacterial diversity, and bacterial viability in microbial fuel cells. Bioresour Technol 100:3518–3525. https://doi.org/10.1016/j.biortech.2009.02.065

Chaturvedi V, Verma P (2014) Metabolism of chicken feathers and concomitant electricity generation by *Pseudomonas aeruginosa* by employing microbial fuel cell (MFC). J Waste Manag 2014:9. https://doi.org/10.1155/2014/928618

Chaturvedi V, Verma P (2016) Microbial fuel cell: a green approach for the utilization of waste for the generation of bioelectricity. Bioresour Bioprocess 3:38. https://doi.org/10.1186/s40643-016-0116-6

Chen S, He G, Carmona-Martinez AA, Agarwal S, Greiner A, Hou H, Schröder U (2011) Electrospun carbon fiber mat with layered architecture for anode in microbial fuel cells. Electrochem Commun 13:1026–1029. https://doi.org/10.1016/j.elecom.2011.06.009

Chen S, He G, Liu Q, Harnisch F, Zhou Y, Chen Y, Hanif M, Wang S, Peng X, Hou H, Schroder U (2012a) Layered corrugated electrode macrostructures boost microbial bioelectrocatalysis. Energy Environ Sci 5:9769–9772. https://doi.org/10.1039/C2EE23344D

Chen S, He G, Hu X, Xie M, Wang S, Zeng D, Hou H, Schröder U (2012b) A three-dimensionally ordered macroporous carbon derived from a natural resource as anode for microbial bioelectrochemical systems. ChemSusChem 5:1059–1063. https://doi.org/10.1002/cssc.201100783

Cheng S, Logan BE (2007) Ammonia treatment of carbon cloth anodes to enhance power generation of microbial fuel cells. Electrochem Commun 9:492–496 https://doi.org/10.1016/j.elecom.2006.10.023

Cheng S, Liu H, Logan BE (2006a) Increased power generation in a continuous flow MFC with advective flow through the porous anode and reduced electrode spacing. Environ Sci Technol 40:2426–2432. https://doi.org/10.1021/es051652w

Cheng S, Liu H, Logan BE (2006b) Power densities using different cathode catalysts (Pt and CoTMPP) and polymer binders (Nafion and PTFE) in single chamber microbial fuel cells. Environ Sci Technol 40:364–369. https://doi.org/10.1021/es0512071

Cheng Y, Wang L, Faustorilla V, Megharaj M (2017) Chemosphere integrated electrochemical treatment systems for facilitating the bioremediation of oil spill contaminated soil. Chemosphere 175:294–299. https://doi.org/10.1016/j.chemosphere.2017.02.079

Ci S, Wu Y, Zou J, Tang L, Luo S, Li J, Wen Z (2012) Nitrogen-doped graphene nanosheets as high efficient catalysts for oxygen reduction reaction. Chinese Sci Bull 57:3065–3070. https://doi.org/10.1007/s11434-012-5253-5

Corbella C, Garfí M, Puigagut J (2014) Vertical redox profiles in treatment wetlands as function of hydraulic regime and macrophytes presence: surveying the optimal scenario for microbial fuel cell implementation. Sci Total Environ 470–471:754–758. https://doi.org/10.1016/j.scitotenv.2013.09.068

Dai K, Wen J-L, Zhang F, Ma X-W, Cui X-Y, Zhang Q, Zhao T-J, Zeng RJ (2017) Electricity production and microbial characterization of thermophilic microbial fuel cells. Bioresour Technol 243:512–519. https://doi.org/10.1016/j.biortech.2017.06.167

Delord B, Neri W, Bertaux K, Derre A, Ly I, Mano N, Poulin P (2017) Carbon nanotube fiber mats for microbial fuel cell electrodes. Bioresour Technol 243:1227–1231. https://doi.org/10.1016/j.biortech.2017.06.170

Dewan A, Beyenal H, Lewandowski Z (2008) Scaling up microbial fuel cells. Environ Sci Technol 42:7643–7648. https://doi.org/10.1021/es800775d

Doherty L, Zhao Y, Zhao X, Hu Y, Hao X (2015a) A review of a recently emerged technology: constructed wetland microbial fuel cells. Water Res 85:38–45. https://doi.org/10.1016/j.watres.2015.08.016

Doherty L, Zhao Y, Zhao X, Wang W (2015b) Nutrient and organics removal from swine slurry with simultaneous electricity generation in an alum sludge-based constructed wetland incorporating microbial fuel cell technology. Chem Eng J 266:74–81. https://doi.org/10.1016/j.cej.2014.12.063

Du Q, An J, Li J, Zhou L, Li N, Wang X (2017) Polydopamine as a new modification material to accelerate startup and promote anode performance in microbial fuel cells. J Power Sources 343:477–482. https://doi.org/10.1016/j.jpowsour.2017.01.093

Fang Z, Song HL, Cang N, Li XN (2013) Performance of microbial fuel cell coupled constructed wetland system for decolorization of azo dye and bioelectricity generation. Bioresour Technol 144:165–171. https://doi.org/10.1016/j.biortech.2013.06.073

Feng Y, Wang X, Logan BE, Lee H (2008) Brewery wastewater treatment using air-cathode microbial fuel cells. Appl Microbiol Biotechnol 78:873–880. https://doi.org/10.1007/s00253-008-1360-2

Feng Y, Lee H, Wang X, Liu Y, He W (2010) Continuous electricity generation by a graphite granule baffled air-cathode microbial fuel cell. Bioresour Technol 101:632–638. https://doi.org/10.1016/j.biortech.2009.08.046

Gálvez A, Greenman J, Ieropoulos I (2009) Landfill leachate treatment with microbial fuel cells; scale-up through plurality. Bioresour Technol 100:5085–5091. https://doi.org/10.1016/j.biortech.2009.05.061

Ghasemi M, Shahgaldi S, Ismail M, Kim BH, Yaakob Z, Wan Daud WR (2011) Activated carbon nanofibers as an alternative cathode catalyst to platinum in a two-chamber microbial fuel cell. Int J Hydrogen Energy 36:13746–13752 https://doi.org/10.1016/j.ijhydene.2011.07.118

Gong K, Du F, Xia Z, Durstock M, Dai L (2009) Nitrogen-doped carbon nanotube arrays with high electrocatalytic activity for oxygen reduction. Science 323:760–764

Greenman J, Gálvez A, Giusti L, Ieropoulos I (2009) Electricity from landfill leachate using microbial fuel cells: comparison with a biological aerated filter. Enzyme Microb Technol 44:112–119. https://doi.org/10.1016/j.enzmictec.2008.09.012

Guerrini E, Grattieri M, Trasatti SP, Bestetti M, Cristiani P (2014) Performance explorations of single chamber microbial fuel cells by using various microelectrodes applied to biocathodes. Int J Hydrogen Energy 39:21837–21846. https://doi.org/10.1016/j.ijhydene.2014.06.132

Guo K, Hidalgo D, Tommasi T, Rabaey K (2016) Pyrolytic carbon-coated stainless steel felt as a high-performance anode for bioelectrochemical systems. Bioresour Technol 211:664–668. https://doi.org/10.1016/j.biortech.2016.03.161

Habermann W, Pommer EH (1991) Biological fuel cells with sulphide storage capacity. Appl Microbiol Biotechnol 35:128–133. https://doi.org/10.1007/BF00180650

Hamilton SJ (2004) Review of selenium toxicity in the aquatic food chain. Sci Total Environ 326:1–31. https://doi.org/10.1016/j.scitotenv.2004.01.019

He Z, Angenent LT (2006) Application of bacterial biocathodes in microbial fuel cells. Electroanalysis 18:2009–2015. https://doi.org/10.1002/elan.200603628

He G, Gu Y, He S, Schröder U, Chen S, Hou H (2011) Effect of fiber diameter on the behavior of biofilm and anodic performance of fiber electrodes in microbial fuel cells. Bioresour Technol 102:10763–10766. https://doi.org/10.1016/j.biortech.2011.09.006

Heijne A, Hamelers HVM, Saakes M, Buisman CJN (2008) Performance of non-porous graphite and titanium-based anodes in microbial fuel cells. Electrochim Acta 53:5697–5703. https://doi.org/10.1016/j.electacta.2008.03.032

Hernández-Fernández FJ, Pérez de los Ríos A, Mateo-Ramírez F, Godínez C, Lozano-Blanco LJ, Moreno JI, Tomás-Alonso F (2015) New application of supported ionic liquids membranes as proton exchange membranes in microbial fuel cell for waste water treatment. Chem Eng J 279:115–119. https://doi.org/10.1016/j.cej.2015.04.036

Holmes DE, Bond DR, O'Neil RA, Reimers CE, Tender LR, Lovley DR (2004) Microbial communities associated with electrodes harvesting electricity from a variety of aquatic sediments. Microb Ecol 48:178–190. https://doi.org/10.1007/s00248-003-0004-4

Huang L, Logan BE (2008) Electricity generation and treatment of paper recycling wastewater using a microbial fuel cell. Appl Microbiol Biotechnol 80:349–355. https://doi.org/10.1007/s00253-008-1546-7

Huang YX, Liu XW, Sun XF, Sheng GP, Zhang YY, Yan GM, Wang SG, Xu AW, Yu HQ (2011) A new cathodic electrode deposit with palladium nanoparticles for cost-effective hydrogen production in a microbial electrolysis cell. Int J Hydrogen Energy 36:2773–2776. https://doi.org/10.1016/j.ijhydene.2010.11.114

Humphries AC, Nott KP, Hall LD, Macaskie LE (2004) Continuous removal of Cr (VI) from aqueous solution catalysed by palladised biomass of *Desulfovibrio vulgaris*. Biotechnol Lett 26:1529–1532

Ieropoulos IA, Ledezma P, Stinchcombe A, Papaharalabos G, Melhuish C, Greenman J (2013) Waste to real energy: the first MFC powered mobile phone. Phys Chem Chem Phys 15:15312. https://doi.org/10.1039/c3cp52889h

Ieropoulos IA, Stinchcombe A, Gajda I, Forbes S, Merino-Jimenez I, Pasternak G, Sanchez-Herranz D, Greenman J (2016) Pee power urinal – microbial fuel cell technology field trials in the context of sanitation. Environ Sci Water Res Technol 2:336–343. https://doi.org/10.1039/C5EW00270B

Jiang D, Li B (2009) Granular activated carbon single-chamber microbial fuel cells (GAC-SCMFCs): a design suitable for large-scale wastewater treatment processes. Biochem Eng J 47:31–37. https://doi.org/10.1016/j.bej.2009.06.013

Jiang D, Curtis M, Troop E, Scheible K, McGrath J, Hu B, Suib S, Raymond D, Li B (2011) A pilot-scale study on utilizing multi-anode/cathode microbial fuel cells (MAC MFCs) to enhance the power production in wastewater treatment. Int J Hydrogen Energy 36:876–884. https://doi.org/10.1016/j.ijhydene.2010.08.074

Jimenez IM, Santoro C, Carbonell SR, Greenman J, Ieropoulos I, Atanassov P (2016) Carbon-based air-breathing cathodes for microbial fuel cells. Catalyst 6:1–13. https://doi.org/10.3390/catal6090127

Jin B, van Leeuwen HJ, Patel B, Yu Q (1998) Utilisation of starch processing wastewater from production of microbial biomass protein and fungal α-amylase by *Aspergillus oryzae*. Bioresour Technol 66:201–206

Karra U, Manickam SS, Mccutcheon JR, Patel N, Li B (2012) Power generation and organics removal from wastewater using activated carbon nanofiber (ACNF) microbial fuel cells (MFCs). Int J Hydrogen Energy 38:1588–1597. https://doi.org/10.1016/j.ijhydene.2012.11.005

Karthikeyan R, Wang B, Xuan J, Wong JWC, Lee PKH, Leung MKH (2015) Interfacial electron transfer and bioelectrocatalysis of carbonized plant material as effective anode of microbial fuel cell. Electrochim Acta 157:314–323. https://doi.org/10.1016/j.electacta.2015.01.029

Kim D, Chang IS (2009) Electricity generation from synthesis gas by microbial processes: CO fermentation and microbial fuel cell technology. Bioresour Technol 100:4527–4530. https://doi.org/10.1016/j.biortech.2009.04.017

Kim BH, Chang IS, Gil GC, Park HS, Kim HJ (2003) Novel BOD (biological oxygen demand) sensor using mediator-less microbial fuel cell. Biotechnol Lett 25:541–545

Kim J, Jia H, Wang P (2006) Challenges in biocatalysis for enzyme-based biofuel cells. Biotechnol Adv 24:296–308. https://doi.org/10.1016/j.biotechadv.2005.11.006

Kim B, An J, Fapyane D, Chang IS (2015a) Bioelectronic platforms for optimal bio-anode of bio-electrochemical systems: from nano-to macro scopes. Bioresour technol 195:2–13

Kim KY, Yang W, Logan BE (2015b) Impact of electrode configurations on retention time and domestic wastewater treatment efficiency using microbial fuel cells. Water Res 80:41–46

Kjeldsen P, Barlaz MA, Rooker AP, Baun A, Ledin A, Christensen TH (2002) Present and long-term composition of MSW landfill leachate: a review. Crit Rev Environ Sci Technol 32:297–336. https://doi.org/10.1080/10643380290813462

Kodali M, Santoro C, Serov A, Kabir S, Artyushkova K, Matanovic I, Atanassov P (2017) Air breathing cathodes for microbial fuel cell using Mn-, Fe-, Co- and Ni-containing platinum group metal-free catalysts. Electrochim Acta 231:115–124. https://doi.org/10.1016/j.electacta.2017.02.033

Kondaveeti S, Min B (2013) Nitrate reduction with biotic and abiotic cathodes at various cell voltages in bioelectrochemical denitrification system. Bioprocess Biosyst Eng 36:231–238. https://doi.org/10.1007/s00449-012-0779-0

Kretzschmar J, Riedl S, Brown RK, Schröder U, Harnisch F (2017) eLatrine: lessons learned from the development of a low-tech MFC based on cardboard electrodes for the treatment of human feces. J Electrochem Soc 164:H3065–H3072. https://doi.org/10.1149/2.0121703jes

Kurniawan TA, Chan GYS, Lo W-H, Babel S (2006) Comparisons of low-cost adsorbents for treating wastewaters laden with heavy metals. Sci Total Environ 366:409–426. https://doi.org/10.1016/j.scitotenv.2005.10.001

Lemly AD (1997) Ecosystem recovery following selenium contamination in a freshwater reservoir. Ecotoxicol Environ Saf 36:275–281. https://doi.org/10.1006/eesa.1996.1515

Lepage G, Albernaz FO, Perrier G, Merlin G (2012) Characterization of a microbial fuel cell with reticulated carbon foam electrodes. Bioresour Technol 124:199–207. https://doi.org/10.1016/j.biortech.2012.07.067

Li W, Yu H (2015) Stimulating sediment bioremediation with benthic microbial fuel cells. Biotechnol Adv 33:1–12. https://doi.org/10.1016/j.biotechadv.2014.12.011

Li X, Zhu N, Wang Y, Li P, Wu P, Wu J (2013) Animal carcass wastewater treatment and bioelectricity generation in up-flow tubular microbial fuel cells: effects of HRT and non-precious metallic catalyst. Bioresour Technol 128:454–460. https://doi.org/10.1016/j.biortech.2012.10.053

Li B, Zhou J, Zhou X, Wang X, Li B, Santoro C, Grattieri M, Babanova S, Artyushkova K, Atanassov P, Schuler AJ (2014) Surface modification of microbial fuel cells anodes: approaches to practical design. Electrochim Acta 134:116–126. https://doi.org/10.1016/j.electacta.2014.04.136

Liao Q, Zhang J, Li J, Ye D, Zhu X, Zhang B (2015) Increased performance of a tubular microbial fuel cell with a rotating carbon-brush anode. Biosens Bioelectron 63:558–561. https://doi.org/10.1016/j.bios.2014.08.014

Lim DH, Wilcox J (2012) Mechanisms of the oxygen reduction reaction on defective graphene-supported Pt nanoparticles from first-principles. J Phys Chem C 116:3653–3660. https://doi.org/10.1021/jp210796e

Liu H, Ramnarayanan R, Logan BE (2004) Production of electricity during wastewater treatment using a single chamber microbial fuel cell. Environ Sci Technol 38:2281–2285. https://doi.org/10.1021/es034923g

Liu Z, Liu J, Zhang S, Su Z (2009) Study of operational performance and electrical response on mediator-less microbial fuel cells fed with carbon- and protein-rich substrates. Biochem Eng J 45:185–191. https://doi.org/10.1016/j.bej.2009.03.011

Liu S, Song H, Li X, Yang F (2013) Power generation enhancement by utilizing plant photosynthate in microbial fuel cell coupled constructed wetland system. Int J Photoenergy 2013:10. https://doi.org/10.1155/2013/172010

Lowy DA, Tender LM, Zeikus JG, Park DH, Lovley DR (2006) Harvesting energy from the marine sediment–water interface II: Kinetic activity of anode materials. Biosens Bioelectron 21:2058–2063 https://doi.org/10.1016/j.bios.2006.01.033

Lu Z-J, Bao S-J, Gou Y-T, Cai C-J, Ji C-C, Xu M-W, Song J, Wang R (2013) Nitrogen-doped reduced-graphene oxide as an efficient metal-free electrocatalyst for oxygen reduction in fuel cells. RSC Adv 3:3990. https://doi.org/10.1039/c3ra22161j

Luo H, Liu G, Zhang R, Jin S (2009) Phenol degradation in microbial fuel cells. Chem Eng J 147:259–264. https://doi.org/10.1016/j.cej.2008.07.011

Matter PH, Zhang L, Ozkan US (2006) The role of nanostructure in nitrogen-containing carbon catalysts for the oxygen reduction reaction. J Catal 239:83–96 https://doi.org/10.1016/j.jcat.2006.01.022

Melhuish C, Ieropoulos I, Greenman J, Horsfield I (2006) Energetically autonomous robots: food for thought. Auton Robots 21:187–198. https://doi.org/10.1007/s10514-006-6574-5

Minteer SD, Liaw BY, Cooney MJ (2007) Enzyme-based biofuel cells. Curr Opin Biotechnol 18(3):228–234

Moon H, Chang IS, Kim BH (2006) Continuous electricity production from artificial wastewater using a mediator-less microbial fuel cell. Bioresour Technol 97:621–627. https://doi.org/10.1016/j.biortech.2005.03.027

Moore CM, Akers NL, Hill AD, Johnson ZC, Minteer SD (2004) Improving the environment for immobilized dehydrogenase enzymes by modifying Nafion with tetraalkylammonium bromides. Biomacromolecules 5(4):1241–1247

Morris JM, Jin S, Wang J, Zhu C, Urynowicz MA (2007) Lead dioxide as an alternative catalyst to platinum in microbial fuel cells. Electrochem Commun 9:1730–1734. https://doi.org/10.1016/j.elecom.2007.03.028

Morris JM, Jin S, Crimi B, Pruden A (2009) Microbial fuel cell in enhancing anaerobic biodegradation of diesel. Chem Eng J 146:161–167. https://doi.org/10.1016/j.cej.2008.05.028

Mustakeem (2015) Electrode materials for microbial fuel cells: nanomaterial approach. Mater Renew Sustain Energy 4:1–11. https://doi.org/10.1007/s40243-015-0063-8

Nancharaiah YV, Mohan SV, Lens PNL (2015) Metals removal and recovery in bioelectrochemical systems: a review. Bioresour Technol 195:102–114

Natarajan D, Van Nguyen T (2004) Effect of electrode configuration and electronic conductivity on current density distribution measurements in PEM fuel cells. J Power Sources 135:95–109. https://doi.org/10.1016/j.jpowsour.2004.03.063

Palmore GTR, Whitesides GM (1994) Microbial and enzymatic biofuel cells. In: Enzymatic conversion of biomass for fuels production. ACS Publications, Washington, DC, pp 271–290

Pant D, Singh A, Satyawali Y, Gupta RK (2007) Effect of carbon and nitrogen source amendment on synthetic dyes decolourizing efficiency of white-rot fungus, *Phanerochaete chrysosporium*. J Environ Biol 29:79

Pant D, Van Bogaert G, Diels L, Vanbroekhoven K (2010) A review of the substrates used in microbial fuel cells (MFCs) for sustainable energy production. Bioresour Technol 101:1533–1543. https://doi.org/10.1016/j.biortech.2009.10.017

Park JY, Yoo YJ (2009) Biological nitrate removal in industrial wastewater treatment: which electron donor we can choose. Appl Microbiol Biotechnol 82:415–429. https://doi.org/10.1007/s00253-008-1799-1

Park DH, Zeikus JG (2003) Improved fuel cell and electrode designs for producing electricity from microbial degradation. Biotechnol Bioeng 81:348–355

Pham H, Boon N, Marzorati M, Verstraete W (2009) Enhanced removal of 1,2-dichloroethane by anodophilic microbial consortia. Water Res 43:2936–2946. https://doi.org/10.1016/j.watres.2009.04.004

Philamore H, Rossiter J, Walters P, Winfield J, Ieropoulos I (2015) Cast and 3D printed ion exchange membranes for monolithic microbial fuel cell fabrication. J Power Sources 289:91–99. https://doi.org/10.1016/j.jpowsour.2015.04.113

Polatides C, Kyriacou G (2005) Electrochemical reduction of nitrate ion on various cathodes – reaction kinetics on bronze cathode. J Appl Electrochem 35:421–427. https://doi.org/10.1007/s10800-004-8349-z

Potter MC (1911) Electrical effects accompanying the decomposition of organic compounds. Proc R Soc B Biol Sci 84:260–276. https://doi.org/10.1098/rspb.1911.0073

Rabaey K, Rozendal RA (2010) Microbial electrosynthesis revisiting the electrical route for microbial production. Nat Rev Microbiol 8:706–716

Rajan VV, Dierkes WK, Joseph R, Noordermeer JWM (2006) Science and technology of rubber reclamation with special attention to NR-based waste latex products. Prog Polym Sci 31:811–834. https://doi.org/10.1016/j.progpolymsci.2006.08.003

Raschitor A, Soreanu G, Fernandez-Marchante CM, Lobato J, Cañizares P, Cretescu I, Rodrigo MA (2015) Bioelectro-claus processes using MFC technology: influence of co-substrate. Bioresour Technol 189:94–98. https://doi.org/10.1016/j.biortech.2015.03.115

Ren Z, Ward TE, Regan JM (2007) Electricity production from cellulose in a microbial fuel cell using a defined binary culture. Environ Sci Technol 41:4781–4786. https://doi.org/10.1021/es070577h

Rezaei F, Richard TL, Brennan RA, Logan BE (2007) Substrate-enhanced microbial fuel cells for improved remote power generation from sediment-based systems. Environ Sci Technol 41:4053–4058. https://doi.org/10.1021/es070426e

Rezaei F, Richard TL, Logan BE (2009) Analysis of chitin particle size on maximum power generation, power longevity, and Coulombic efficiency in solid-substrate microbial fuel cells. J Power Sources 192:304–309. https://doi.org/10.1016/j.jpowsour.2009.03.023

Rismani-Yazdi H, Carver SM, Christy AD, Tuovinen OH (2008) Cathodic limitations in microbial fuel cells: an overview. J Power Sources 180:683–694. https://doi.org/10.1016/j.jpowsour.2008.02.074

Rodrigo J, Boltes K, Esteve-nuñez A (2014) Microbial-electrochemical bioremediation and detoxification of dibenzothiophene-polluted soil. Chemosphere 101:61–65. https://doi.org/10.1016/j.chemosphere.2013.11.060

Rovira M, Gim J, Mart X, de Pablo J, Mart V, Duro L (2008) Sorption of selenium (IV) and selenium (VI) onto natural iron oxides: goethite and hematite. J Hazard Mater 150:279–284. https://doi.org/10.1016/j.jhazmat.2007.04.098

Roy JN, Babanova S, Garcia KE, Cornejo J, Ista LK, Atanassov P (2014) Catalytic biofilm formation by *Shewanella oneidensis* MR-1 and anode characterization by expanded uncertainty. Electrochim Acta 126:3–10. https://doi.org/10.1016/j.electacta.2013.07.075

Santoro C, Li B, Cristiani P, Squadrito G (2013) Power generation of microbial fuel cells (MFCs) with low cathodic platinum loading. Int J Hydrogen Energy 38:692–700. https://doi.org/10.1016/j.ijhydene.2012.05.104

Santoro C, Guilizzoni M, Correa Baena JP, Pasaogullari U, Casalegno A, Li B, Babanova S, Artyushkova K, Atanassov P (2014) The effects of carbon electrode surface properties on bacteria attachment and start up time of microbial fuel cells. Carbon N Y 67:128–139 https://doi.org/10.1016/j.carbon.2013.09.071

Santoro C, Artyushkova K, Cornejo JA, Ista L, Bretschger O, Marsili E, Atanassov P, Schuler AJ (2015) Influence of anode surface chemistry on microbial fuel cell operation. Bioelectrochemistry 106:141–149

Santoro C, Arbizzani C, Erable B, Ieropoulos I (2017) Microbial fuel cells: from fundamentals to applications. A review. J Power Sources 356:225–244. https://doi.org/10.1016/j.jpowsour.2017.03.109

Seviour T, Doyle LE, Lauw SJL, Hinks J, Rice SA, Nesatyy VJ, Webster RD, Kjelleberg S, Marsili E (2015) Voltammetric profiling of redox-active metabolites expressed by *Pseudomonas aeruginosa* for diagnostic purposes. Chem Commun 51:3789–3792. https://doi.org/10.1039/C4CC08590F

Sharma T, Mohana AL, Chandra TS, Ramaprabhu S (2008) Development of carbon nanotubes and nanofluids based microbial fuel cell. Int J Hydrogen Energy 33:6749–6754. https://doi.org/10.1016/j.ijhydene.2008.05.112

Sherafatmand M, Ng HY (2015) Using sediment microbial fuel cells (SMFCs) for bioremediation of polycyclic aromatic hydrocarbons (PAHs). Bioresour Technol 195:122–130. https://doi.org/10.1016/j.biortech.2015.06.002

Solanki K, Subramanian S, Basu S (2013) Microbial fuel cells for azo dye treatment with electricity generation: a review. Bioresour Technol 131:564–571. https://doi.org/10.1016/j.biortech.2012.12.063

Sonawane JM, Adeloju SB, Ghosh PC (2017) Landfill leachate: a promising substrate for microbial fuel cells. Int J Hydrogen Energy 42:23794–23798. https://doi.org/10.1016/j.ijhydene.2017.03.137

Sotres A, Cerrillo M, Vias M, Bonmat A (2015) Nitrogen recovery from pig slurry in a two-chambered bioelectrochemical system. Bioresour Technol 194:373–382. https://doi.org/10.1016/j.biortech.2015.07.036

Srikanth S, Marsili E, Flickinger MC, Bond DR (2008) Electrochemical characterization of *Geobacter sulfurreducens* cells immobilized on graphite paper electrodes. Biotechnol Bioeng 99:1065–1073. https://doi.org/10.1002/bit.21671

Su DS, Zhang J, Frank B, Thomas A, Wang X, Paraknowitsch J, Schlögl R (2010) Metal-free heterogeneous catalysis for sustainable chemistry. ChemSusChem 3:169–180

Sun J, Hu Y-Y, Bi Z, Cao Y-Q (2009) Simultaneous decolorization of azo dye and bioelectricity generation using a microfiltration membrane air-cathode single-chamber microbial fuel cell. Bioresour Technol 100:3185–3192. https://doi.org/10.1016/j.biortech.2009.02.002

Tandukar M, Huber SJ, Onodera T, Pavlostathis SG (2009) Biological chromium(VI) reduction in the cathode of a microbial fuel cell. Environ Sci Technol 43:8159–8165. https://doi.org/10.1021/es9014184

Thung WE, Ong SA, Ho LN, Wong YS, Ridwan F, Oon YL, Oon YS, Lehl HK (2015) A highly efficient single chambered up-flow membrane-less microbial fuel cell for treatment of azo dye Acid Orange 7-containing wastewater. Bioresour technol 197:284–288

Topcagic S, Treu BL, Minteer SD (2004) Characterization/optiminization of oxygen biocathodes for membraneless biofuel cells. In: 206th ECS meeting, Honolulu, HI, 3–8 October 2004

Trapero JR, Horcajada L, Linares JJ, Lobato J (2017) Is microbial fuel cell technology ready ? An economic answer towards industrial commercialization. Appl Energy 185:698–707. https://doi.org/10.1016/j.apenergy.2016.10.109

Velvizhi G, Venkata Mohan S (2012) Electrogenic activity and electron losses under increasing organic load of recalcitrant pharmaceutical wastewater. Int J Hydrogen Energy 37:5969–5978. https://doi.org/10.1016/j.ijhydene.2011.12.112

Villaseñor J, Capilla P, Rodrigo MA, Cañizares P, Fernández FJ (2013) Operation of a horizontal subsurface flow constructed wetland – microbial fuel cell treating wastewater under different organic loading rates. Water Res 47:6731–6738. https://doi.org/10.1016/j.watres.2013.09.005

Wang X, Cheng S, Feng Y, Merrill MD, Saito T, Logan BE (2009) Use of carbon mesh anodes and the effect of different pretreatment methods on power production in microbial fuel cells. Environ Sci Technol 43:6870–6874

Wang HY, Bernarda A, Huang CY, Lee DJ, Chang JS (2011) Micro-sized microbial fuel cell: a mini-review. Bioresour Technol 102:235–243. https://doi.org/10.1016/j.biortech.2010.07.007

Wang D, Li Y, Li G, Wang C, Wang P, Zhang W, Wang Q (2015) Dye-sensitized photoelectrochemical cell on plasmonic Ag/AgCl @ chiral TiO_2 nanofibers for treatment of urban wastewater effluents, with simultaneous production of hydrogen and electricity. Appl Catal B 169:25–32. https://doi.org/10.1016/j.apcatb.2014.11.012

Wang Z, Dummi G, Wu Y, Zhao F (2017) Progress of air-breathing cathode in microbial fuel cells. J Power Sources 356:245–255. https://doi.org/10.1016/j.jpowsour.2017.02.004

Watanabe K (2008) Recent developments in microbial fuel cell technologies for sustainable bioenergy. J Biosci Bioeng 106:528–536 https://doi.org/10.1263/jbb.106.528

Winfield J, Chambers L (2014) Towards disposable microbial fuel cells: natural rubber glove membranes. Int J Hydrogen Energy 39:21803–21810

Winfield J, Chambers LD, Rossiter J, Ieropoulos I (2013) Comparing the short and long term stability of biodegradable, ceramic and cation exchange membranes in microbial fuel cells. Bioresour Technol 148:480–486. https://doi.org/10.1016/j.biortech.2013.08.163

Winfield J, Chambers LD, Stinchcombe A, Rossiter J, Ieropoulos I (2014) The power of glove: soft microbial fuel cell for low-power electronics. J Power Sources 249:327–332 https://doi.org/10.1016/j.jpowsour.2013.10.096

Winfield J, Chambers LD, Rossiter J, Greenman J, Ieropoulos I (2015) Urine-activated origami microbial fuel cells to signal proof of life. J Mater Chem A 3:7058–7065. https://doi.org/10.1039/C5TA00687B

Winfield J, Gajda I, Greenman J, Ieropoulos I (2016) A review into the use of ceramics in microbial fuel cells. Bioresour Technol 215:296–303. https://doi.org/10.1016/j.biortech.2016.03.135

Wu S, He W, Yang W, Ye Y, Huang X, Logan BE (2017) Combined carbon mesh and small graphite fiber brush anodes to enhance and stabilize power generation in microbial fuel cells treating domestic wastewater. J Power Sources 356:348–355. https://doi.org/10.1016/j.jpowsour.2017.01.041

Xia C, Xu M, Liu J, Guo J, Yang Y (2015) Sediment microbial fuel cell prefers to degrade organic chemicals with higher polarity. Bioresour Technol 190:420–423. https://doi.org/10.1016/j.biortech.2015.04.072

Xu X, Zhao Q, Wu M, Ding J, Zhang W (2017) Biodegradation of organic matter and anodic microbial communities analysis in sediment microbial fuel cells with/without Fe (III) oxide addition. Bioresour Technol 225:402–408. https://doi.org/10.1016/j.biortech.2016.11.126

Yasri NG, Nakhla G (2017) The performance of 3-D graphite doped anodes in microbial electrolysis cells. J Power Sources 342:579–588. https://doi.org/10.1016/j.jpowsour.2016.12.081

Yu B, Tian J, Feng L (2017) Remediation of PAH polluted soils using a soil microbial fuel cell: Influence of electrode interval and role of microbial community. J Hazard Mater 336:110–118

Yuan Y, Zhao B, Jeon Y, Zhong S, Zhou S, Kim S (2011) Iron phthalocyanine supported on amino-functionalized multi-walled carbon nanotube as an alternative cathodic oxygen catalyst in microbial fuel cells. Bioresour Technol 102:5849–5854. https://doi.org/10.1016/j.biortech.2011.02.115

Zaffar H, Irshad U, Pervez A, Naqvi TA (2016) Mode of action, toxicity and biodegradation of organochlorinated pesticides: a mini review. J Appl Environ Biol Sci 6:1–7

Zhang X, Cheng S, Huang X, Logan BE (2010) The use of nylon and glass fiber filter separators with different pore sizes in air-cathode single-chamber microbial fuel cells. Energy Environ Sci 3:659–664. https://doi.org/10.1039/B927151A

Zhang F, Ge Z, Grimaud J, Hurst J, He Z (2013) Long-term performance of liter-scale microbial fuel cells treating primary effluent installed in a municipal wastewater treatment facility. Environ Sci Technol 47:4941–4948. https://doi.org/10.1021/es400631r

Zhang B, Wang Z, Zhou X, Shi C, Guo H, Feng C (2015) Electrochemical decolorization of methyl orange powered by bioelectricity from single-chamber microbial fuel cells. Bioresour Technol 181:360–362. https://doi.org/10.1016/j.biortech.2015.01.076

Zhang Q, Hu J, Lee D (2016) Microbial fuel cells as pollutant treatment units: research updates. Bioresour Technol 217:121–128. https://doi.org/10.1016/j.biortech.2016.02.006

Zhao F, Rahunen N, Varcoe JR, Roberts AJ, Avignone-Rossa C, Thumser AE, Slade RCT (2009) Factors affecting the performance of microbial fuel cells for sulfur pollutants removal. Biosens Bioelectron 24:1931–1936. https://doi.org/10.1016/j.bios.2008.09.030

Zhao L, Li J, Battaglia F, He Z (2016) Computational investigation of the flow field contribution to improve electricity generation in granular activated carbon-assisted microbial fuel cells. J Power Sources 333:83–87. https://doi.org/10.1016/j.jpowsour.2016.09.113

Zhuang L, Zhou S, Wang Y, Liu C, Geng S (2009) Membrane-less cloth cathode assembly (CCA) for scalable microbial fuel cells. Biosens Bioelectron 24:3652–3656. https://doi.org/10.1016/j.bios.2009.05.032

Zhuang L, Zheng Y, Zhou S, Yuan Y, Yuan H, Chen Y (2012) Scalable microbial fuel cell (MFC) stack for continuous real wastewater treatment. Bioresour Technol 106:82–88. https://doi.org/10.1016/j.biortech.2011.11.019

Chapter 12
Microbial Desalination Cells: A Boon for Future Generations

Deepika Jothinathan

12.1 Introduction

As per the available information, the world's freshwater quantity has started decreasing tremendously due to the excess global warming, natural disasters, and also anthropogenic activities. Most of the water bodies are almost contaminated by our very own day-to-day events. Starting from our domestic waste to industrial waste such as effluents, pesticides, and radioactive waste have been continuously dumped into water bodies.

This has become a serious issue, as few of the countries have already started receiving alarming signals of water scarcity. This is the right time to implement these kinds of desalination technologies to receive a proper source of water.

The microbial fuel cell research has been extended in this field of desalination in the past 8 years. This is very much evident from the below graph depicting the number of articles published from 2010 to 2018 (Fig. 12.1). This contemporary change is due to many reasons like increased global warming, serious natural disasters, and immense water scarcity in parts of the world. These agitations have urged the researchers to find out alternative technology for desalinating the seawater. Comparative to the other technologies, microbe-driven desalination is a renewable one without any external power needed for the process. This can also provide a sufficient amount of bioelectricity. Although the conventional desalination serves the world with desalinated water, it consumes a huge amount of power.

MDCs have grabbed the attention of the imminent researchers due to its salient features such as non-expensive in terms of input energy, waste removal, desalinating water at a very economical way by combining MDCs and reverse osmosis (Yuan et al. 2015), and bio-restoring the heavy metal contaminated site (Ping et al. 2015).

D. Jothinathan (✉)
Department of Life Sciences, Central University of Tamil Nadu,
Thiruvarur, 610005, India

V. Sivasankar et al. (eds.), *Microbial Fuel Cell Technology for Bioelectricity*,
https://doi.org/10.1007/978-3-319-92904-0_12

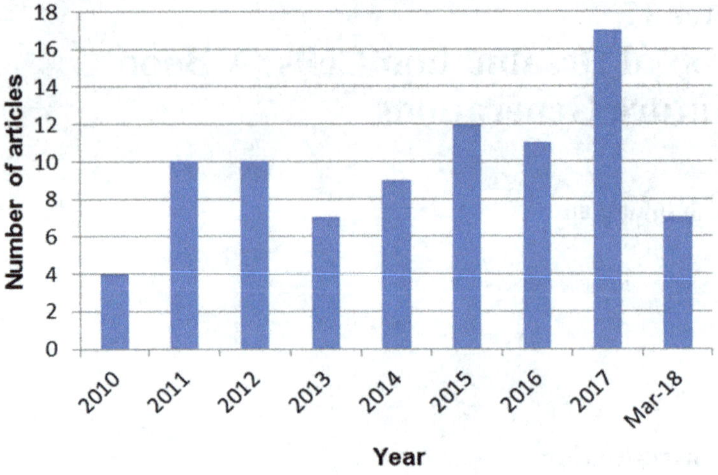

Fig. 12.1 Published articles in microbial desalination cell from 2010 to 2018 based on the keyword "microbial desalination cell" (Source: PubMed)

In order to improvise the technology in relation to desalination, stacked MDCs have been developed (Cao et al. 2009; Chen et al. 2011). Beyond these appreciable functions, they have their own constraints such as less possibility in concurrent organic removal and desalination in wastewater MDC (Zuo et al. 2013), desalination of only saltwater without any organics which is feasible (Lindstrand et al. 2000), and ineffective nitrogen removal from the anodic wastewater (Mehanna et al. 2010; Chen et al. 2012b).

12.1.1 Microbial Desalination Cell

Microbial desalination cell (MDC) is a technology where the microorganisms are fed with organic matter to produce bioenergy which in turn aids in desalinating the seawater. MDC is equipped with three chambers, anode, cathode, and a middle chamber filled with salt water. The anodic chamber is attached with anionic exchange membrane (AEM), and cathode is attached with cationic exchange membrane (CEM). During the oxidation of organic matter provided in the anode, protons are released into the anolyte, and positively charged species are prohibited from leaving the anode by the AEM. Meanwhile, the negatively charged particles from saltwater in the middle chamber are attracted toward the positive species in anode. The protons are disbursed in the cathodic chamber resulting in transfer of positively charged ions from the middle chamber to cathodic compartment. By this way, the ionic salts are eliminated from the middle chamber resulting in desalinated water. This process is free of physical or chemical methods (Cath et al. 2006). It is similar to electrodialysis process, but usage of energy is not engaged in MDC. The schematic representation of MDC has been given in Fig. 12.2.

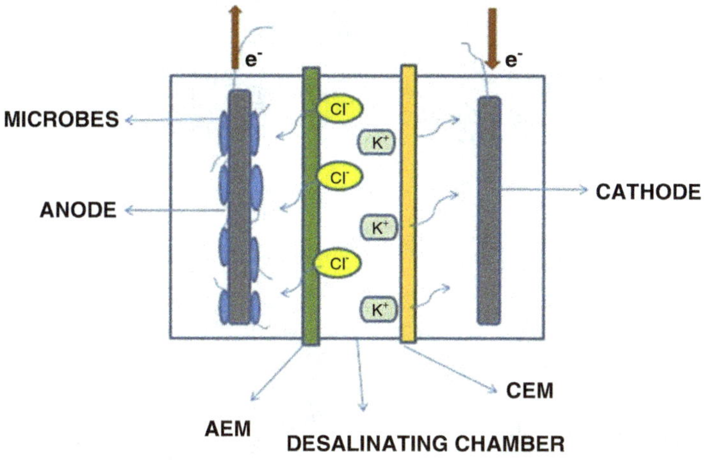

Fig. 12.2 Schematic representation of MDC

12.1.1.1 Materials: Electrodes, Anolyte, Separating Membrane

The electrodes in majority of the MDCs are of carbon nature such as carbon felt, carbon cloth, graphite felt, etc. Since carbon material has been proven to be the best material in microbial fuel cell with high performance, it has been used in MDC for increasing the efficiency. The type of MDC, electrode materials, anolytes, and electron exchange membrane along with their desalination efficiency has been listed in Table 12.1.

The anionic and cationic exchange membranes used in most of the studies are AMI7001/Membranes International and CMI7000/Membranes International, respectively. Depending upon the wastewater used in the anode, the substrate and anolyte type will vary.

12.1.1.2 Substrate/Anolyte/Catholyte

Substrate plays a vital role in the performance of MDC. Most the studies pertaining to microbial desalination employed sodium acetate as the anolyte. It turned out to be an efficient substrate in MDC, and by increasing its concentration, the desalination efficiency was improved (Cao et al. 2009; Kim and Logan 2011). Recent reports suggests that phosphate-buffered sodium acetate enhanced the salt removal (Jacobson et al. 2011). In photosynthetic MDC, synthetic wastewater with aerobic sludge acted as anolyte and microalgae as biocatalyst in cathode (Kokabian and Gude 2013). In a recent report, a stacked MDC was fortified with xylose (1 g/L) in 50 mM PBS. It was isolated as a waste product during the corn stover hydrolysis in biodiesel production has been used as a fermentable substrate (Qu et al. 2013).

Engine oil was used as an organic substrate which showed a considerable energy production and desalination in a recent MDC technology (Sabina et al. 2014).

Table 12.1 MDC configurations with its features

S. no.	MDC type	Electrodes	AEM	CEM	Anode solution	Desalination efficiency (%)	References
1	Three-chamber MDC	Anode and cathode carbon felt	DFI120/Tianwei Membrane	Ultrex CMI7000, Membrane International	Sodium acetate	90	Cao et al. (2009)
2	Biocathode MDC	Anode and cathode carbon felt	AMI-7001S/Membranes International	CMI7000S/Membranes International	Sodium acetate	92	Wen et al. (2012)
3	rMDC	Anode carbon graphite fiber brushes, air cathode carbon cloth	DF120/Tianwei Membrane	Ultrex CMI7000/Membrane International	Xylose	–	Qu et al. (2012)
4	PMDC	Anode and cathode graphite paper	AMI7001/Membranes International	CMI7000/Membranes International	Synthetic wastewater with aerobic sludge	40	Kokabian and Gude (2013)
5	Stacked MDC	Anode carbon felt, cathode carbon cloth	DF120/Tianwei Membrane	Ultrex CMI7000/Membrane International	Sodium acetate	99.4	Chen et al. (2011)
6	UMDC	Anode graphite granules, cathode carbon cloth	AMI7001/Membranes International	CMI7000/Membranes International	Synthetic wastewater (phosphate buffered sodium acetate)	>99	Jacobson et al. (2011)
7	cMDC	Anode graphite brush, cathode carbon cloth	CEM/ACC/Ni or Cu (CMXSB, Astom Corporation, Japan/ activated carbon cloth	Ni or Cu/ACC/CEM (Ni or Cu mesh current/activated carbon cloth/CMXSB, Astom Corporation	Phosphate buffered sodium acetate	88	Forrestal et al. (2012)

A multistage MDC (M-MDC) with two alternating anodes and cathodes was operated in two operating modes utilizing with domestic wastewater. In anode-anode-cathode-cathode mode, the wastewater has produced comparatively more current and desalination efficiency when compared to anode-cathode-anode-cathode mode (Zuo et al. 2016). Domestic wastewater has been utilized as the anolyte and phosphate-buffered ferricyanide as a catholyte in a MDC (Luo et al. 2012).

12.1.2 MDC Designs

The MDC has been developed in various designs such as biocathode, photosynthetic, stacked, ion-exchange resin packed, capacitive, recirculation, osmotic, submerged microbial desalination-denitrification cell, etc. Although these models have achieved a much performance, the operability in large scale has not been tested.

12.1.2.1 Biocathode MDC

Biocathode has turned out to be a sustainable electrode in promoting the performance by utilizing microbes as a catalyst. By using microbial catalyst, the biocathode MDC has reduced the construction price and been effective in producing valuable chemicals (Zhang et al. 2012). An aerobic biocathode MDC is more continuous and is inexpensive to operate than an abiotic cathode MDC. In a study done by Zhang et al. (2016), with biocathode MDC, it was observed that the salt removal, power density, and columbic efficiency were 44%, 77%, and 27%, respectively. Biofouling was found to be a major cause for the reduced performance. The researchers had the option of changing membranes after every cycle of MDC operation (Zhang et al. 2016).

12.1.2.2 Photosynthetic MDC

Another interesting research work based on algae powered MDC has thrown light in the area of biocathodes. This study has much emphasized on the use of biocathode to increase the system performance combined with reduced usage of the expensive catalysts. As similar to the environment, algae in PMDCs release oxygen as a terminal electron acceptor. Compared to other biocathodes, algae cathode has potential benefits such as the valuable products from the biomass after the process completion (Kokabian and Gude 2015). This is far better when compared to other MDCs where sludge formation takes place. Before utilizing algae in MDC, few factors such as carbon dioxide concentration, light availability, medium components, and other conditions have to be studied. Constant light source in lab level is achievable, but in large scale, it will increase the cost. It is clearly understood that light plays a major factor in the photosynthesis and actual growth of algae (Markou and Georgakakis 2011).

12.1.2.3 Stacked MDC

This type of MDC is made by placing multiple ion-exchange membranes in between the anode and cathode in order to achieve the maximum desalination. This will aid in increased charge transfer efficiency and salt removal via the membrane pairs (Gude et al. 2013). When compared to other MDCs, stacked MDC is an economical method to extract more energy. Stacked MDC had its own limitations like high saltwater – desalination and operable only in small scale. However, an attempt was made by Zuo et al. (2014) in order to scale up this technology. He used a >10 L stacked MDC packed with mixed ion-exchange resins with 0.5 g/L NaCl concentration and operated the system in batch mode. A desalination efficiency of 95.8% was achieved with 0.02 g/L of final effluent concentration (Zuo et al. 2014).

12.1.2.4 Supercapacitive MDC

Capacitive deionization (CDI) is based on exploitation of high surface area carbon materials at two electrodes. The potential difference applied between the anode and cathode in this MDC accelerates the consecutive process of adsorption and desorption by which the ions are detached from the saltwater (Suss et al. 2015; Anderson et al. 2010). The electrodes are self-polarized by the reduction-oxidation reactions, and thus the cathode acts as a positive electrode and anode acts as a negative electrode and of the internal supercapacitor. In supercapacitive microbial desalination cell (SC-MDC), CDI and MDC were combined to increase the power production. The conductivity of the solution and pH of the solution was observed with increase in time. An addition of capacitive electrode on the cathode side assisted the system to overcome the ohmic losses. To reduce the cost of the system, platinum was not used during the fabrication (Santoro et al. 2017).

12.1.3 Pros and Cons of MDC

pH plays a significant role in the process of desalination. The reduced pH hinders the microbial activity and thus results in a low current production. This can be eliminated by adding catholytes of pH (Lakshminarayanaiah 1969), recycling of anode solution (Chen et al. 2012a) and cultivating bacteria that would grow in low pH in anode (Ping and He 2013). Biofouling is one among the major limitation in MDC where the AEM is highly affected because of the biofilm formation over the membrane surface. It can be prevented by covering the membrane surface with some material that would not favor the microbial growth (Logan 2008).

12.1.4 *Future of MDC*

MDC has become a popular technology in terms of energy production, water softening, and desalinating saltwater. It has improvised on its own way in terms of different modes of operation, electrode materials, conjugated membranes, different biochemical pathways, MDC construction, etc. However, there are certain challenges that should be overcome for an effective operation. The performance of a microbial desalination cell is strongly influenced by the oxidation of organics by microbes and internal resistance. MDC has also expected to be applied in various fields such as bioremediation of industrial effluents, heavy metals, xenobiotic removal, etc. in the near future. Researchers are working toward a sustainable future for achieving 100% desalination efficiency with the maximum power production.

12.2 Summary and Conclusion

Microbial desalination cell has its own potential in desalinating saltwater apart from producing bioelectricity. In the future, this technology will surely lead us to a good path. The study of microbe's contribution in MDC is much understood. The microorganism's nature of degrading organic matter must be well studied before attempting the same in large scale. Improvisation of ion-exchange membranes with cost-effective materials, reduced membrane fouling, and catalyst addition will surely add a feather to the cap of commercial MDC applications.

Acknowledgments I would like to acknowledge Dr. E.M. Shankar, Head of the Department, Department of Life Sciences, for encouraging me in writing this chapter. I would also like to render my sincere thanks to Dr. Sivasankar Venkataraman who has been a constant support.

References

Anderson MA, Cudero AL, Palma J (2010) Capacitive deionization as an electrochemical means of saving energy and delivering clean water. Comparison to present desalination practices: will it compete? Electrochim Acta 55:3845–3856

Cao X, Huang X, Liang P, Xiao K, Zhou Y, Zhang X, Logan BE (2009) A new method for water desalination using microbial desalination cells. Environ Sci Technol 43(18):7148–7152

Cath TY, Childress AE, Elimelech M (2006) Forward osmosis: principles, applications, and recent developments. J Membr Sci 281:70–87

Chen X, Xia X, Liang P, Cao X, Sun H, Huang X (2011) Stacked microbial desalination cells to enhance water desalination efficiency. Environ Sci Technol 45(6):2465–2470

Chen S, Liu G, Zhang R, Qin B, Luo Y, Hou Y (2012a) Improved performance of the microbial electrolysis desalination and chemical-production cell using the stack structure. Bioresour Technol 118:507–511

Chen X, Liang P, Wei Z, Zhang X, Huang X (2012b) Sustainable water desalination and electricity generation in a separator coupled stacked microbial desalination cell with buffer free electrolyte circulation. Bioresour Technol 119:88–93

Forrestal C, Xu P, Ren Z (2012) Sustainable desalination using a microbial capacitive desalination cell. Energy Environ Sci 5:7161–7167

Gude V, Kokabian B, Gadhamshetty V (2013) Beneficial bioelectrochemical systems for energy, water, and biomass production. J Microb Biochem Technol 6:2

Jacobson KS, Drew DM, He Z (2011) Efficient salt removal in a continuously operated upflow microbial desalination cell with an air cathode. Bioresour Technol 102:376–380

Kim Y, Logan BE (2011) Series assembly of microbial desalination cells containing stacked electrodialysis cells for partial or complete seawater desalination. Environ Sci Technol 45:5840–5845

Kokabian B, Gude VG (2013) Photosynthetic microbial desalination cells (PMDCs) for clean energy, water and biomass production. Environ Sci: Processes Impacts 15:2178–2185

Kokabian B, Gude VG (2015) Sustainable photosynthetic biocathode in microbial desalination cells. Chem Eng J 262:958–965

Lakshminarayanaiah N (1969) Electroosmosis in ion-exchange membranes. J Electrochem Soc 118:338–342

Lindstrand V, Sundström G, Jönsson AS (2000) Fouling of electrodialysis membranes by organic substances. Desalination 128(1):91–102

Logan BE (2008) MFCs, 1st edn. Wiley-Interscience, Hoboken

Luo H, Xu P, Roane TM, Jenkins PE, Ren Z (2012) Microbial desalination cells for improved performance in wastewater treatment, electricity production, and desalination. Bioresour Technol 105:60–66

Markou G, Georgakakis D (2011) Cultivation of filamentous cyanobacteria (blue-green algae) in agro-industrial wastes and wastewaters: a review. Appl Energy 88(10):3389–3401

Mehanna M, Saito T, Yan J, Hickner M, Cao X, Huang X, Logan BE (2010) Using microbial desalination cells to reduce water salinity prior to reverse osmosis. Energy Environ Sci 3(8):1114–1120

Ping Q, He Z (2013) Improving the flexibility of microbial desalination cells through spatially decoupling anode and cathode. Bioresour Technol 144:304–310

Ping Q, Abu-Reesh IM, He Z (2015) Boron removal from saline water by a microbial desalination cell integrated with donnan dialysis. Desalination 376:55–61

Qu Y, Feng Y, Wang X, Liu J, Lv J, He W, Logan BE (2012) Simultaneous water desalination and electricity generation in a microbial desalination cell with electrolyte recirculation for pH control. Bioresour Technol 106:89–94

Qu Y, Feng Y, Liu J, He W, Shi X, Yang Q, Lv J, Logan BE (2013) Salt removal using multiple microbial desalination cells under continuous flow conditions. Desalination 317:17–22

Sabina K, Fayidh MA, Archana G, Sivarajan M, Babuskin S, Babu PA, Radha KK, Sukumar M (2014) Microbial desalination cell for enhanced biodegradation of waste engine oil using a novel bacterial strain *Bacillus subtilis moh3*. Environ Technol 35(17):2194–2203

Santoro C, Abad FB, Serov A, Kodali M, Howe KJ, Soavi F, Atanassov P (2017) Supercapacitive microbial desalination cells: new class of power generating devices for reduction of salinity content. Appl Energy 208:25–36

Suss ME, Porada S, Sun X, Biesheuvel PM, Yoon J, Presser V (2015) Water desalination via capacitive deionization: what is it and what can we expect from it? Energy Environ Sci 8(8):2296–2319

Wen Q, Zhang H, Chen Z, Li Y, Nan J, Feng Y (2012) Using bacterial catalyst in the cathode of microbial desalination cell to improve wastewater treatment and desalination. Bioresour Technol 125:108–113

Yuan H, Abu-Reesh IM, He Z (2015) Enhancing desalination and wastewater treatment by coupling microbial desalination cells with forward osmosis. Chem Eng J 270:437–443

Zhang G, Zhao Q, Jiao Y, Wang K, Lee DJ, Ren N (2012) Biocathode microbial fuel cell for efficient electricity recovery from dairy manure. Biosens Bioelectron 31:537–543

Zhang H, Wen Q, An Z, Chen Z, Nan J (2016) Analysis of long-term performance and microbial community structure in bio-cathode microbial desalination cells. Environ Sci Pollut Res 23(6):5931–5940

Zuo K, Yuan L, Wei J, Liang P, Huang X (2013) Competitive migration behaviors of multiple ions and their impacts on ion-exchange resin packed microbial desalination cell. Bioresour Technol 146:637–642

Zuo K, Cai J, Liang S, Wu S, Zhang C, Liang P, Huang X (2014) A ten liter stacked microbial desalination cell packed with mixed ion-exchange resins for secondary effluent desalination. Environ Sci Technol 48(16):9917–9924

Zuo K, Liu F, Ren S, Zhang X, Liang P, Huang X (2016) A novel multi-stage microbial desalination cell for simultaneous desalination and enhanced organics and nitrogen removal from domestic wastewater. Environ Sci: Water Res Technol 2(5):832–837

Chapter 13
The Performance of Microbial Fuel Cells in Field Trials from a Global Perspective

Ponmudi Priya and Vajiravelu Sivamurugan

13.1 Microbial Fuel Cells (MFC): A Sustainable Solution for Energy Demand

The global energy demand is exponentially increasing everyday, and it has been currently managed by the consumption of fossil fuels and their products. It is a well-known fact that the persistent utilisation of the fossil fuels has made awful damage to our environment. In this context, from last few decades, the motivation towards the development of sustainable energy is strongly driven due to growing demand for global requirement of energy to meet the technological, economical and social welfare needs of the community. The rapid depletion of fossil fuels and emerging environmental awareness on global warming have given impetus to search for the alternative energy sources with sustainable supply, which has become one of the major research objective in many countries. Hence, the energy gleaned especially from renewable resources or cost-effective resources would pave a plausible way for the sustainable energy development (Carla Jones and Stephen 2016; Ravinder and Pradeep 2017).

The wastes generated from agricultural or plant-based byproducts are termed as 'biomass' which can be effectively converted into biofuels. Unlike the chemical fuel cells, the chief compositions of these biomass wastes are carbohydrates and proteins. Microorganisms consume these wastes and degrade them into harmless materials (Elizabeth et al. 2011; Ritu and Sanjeev 2017). The basis for the construction of microbial fuel cells (MFCs) relies on the conversion of chemical energy stored in

P. Priya
Post Graduate and Research Department of Zoology, Pachaiyappa's College, Chennai, 600030, India

V. Sivamurugan (✉)
Post Graduate and Research Department of Chemistry, Pachaiyappa's College, Chennai, 600030, India

V. Sivasankar et al. (eds.), *Microbial Fuel Cell Technology for Bioelectricity*,
https://doi.org/10.1007/978-3-319-92904-0_13

the form of chemical bonds suitably into electrical energy through redox reactions catalysed by microorganisms (Bruce Logan et al. 2006; Frank and Higson 2007; Rachnarin and Roshan 2017).

The microorganisms, viz. *Shewanella* sp. and *Geobacter* sp., have been used for the construction of MFCs that degrade the inorganic and organic compounds such as sugars, proteins, cellulosic materials and polyphenols through oxidation-reduction reactions (Padma and Dirk Hays 2012; Zhuwei et al. 2007). The MFCs are one among the cost-effective renewable energy sources, where the electricity is generated from domestic and industrial effluents. More than two decades, the development of MFCs not only has contributed to the energy sector but also augment in efficiently converting industrial and domestic wastes into electricity through microbes. Many reviews, monographs and book chapters pertaining to MFCs are published in the reputed journals by several scientific researchers (Carlo et al. 2017; Mostafa et al. 2015; Oliveira et al. 2013).

13.2 Why Microbial Fuel Cells (MFCs)?

The awareness on the protection of environment since the dawn of the twenty-first century has purported the search for alternative fuels around the world with attention focusing on the MFCs because of its greener and bioenergy production. The main objective of the MFC development is to treat the industrial and domestic effluents in a cost-effective manner by using microbes, which are capable of producing electricity from these wastes through the oxidation and reduction reactions catalysed by them (Kim et al. 2008; Yi-Chi et al. 2015). For instance, the complete oxidation of one glucose molecule to CO_2 in the presence of air/O_2 produces 24 electrons, which can be utilised for the generation of electrical energy. Thus, any energy source obtained from various biomasses, which are rich in carbohydrates, proteins, alcohols, hydrocarbons and organic acids, could be used as a fuel for the MFCs. In addition, the polymeric carbohydrates such as cellulose and starch can also be utilised for fuel. The organic matter present in theses wastes could be oxidised by these self-replicating bacteria also known as *exoelectrogens* through electron transfer reactions to produce the bioenergy.

The major advantage of MFCs is not only providing electrical energy but also to treat wide range of agricultural wastes such as cornhusks, rice husks, whey and also animal or human sewage. For instance, the wastewater sludges containing carbohydrates especially glucose, sucrose, glucoronic acid, starch and xylose generate current density in the range of 0.7 and 1.3 mA/cm^2 at the concentration of 6 to 7 mM/L (Pant et al. 2010).

Further, MFCs also succour to continuous monitoring of quality of wastes and minimal investment on the fuels. Accordingly, there has been tremendous efforts made across globe which can be clearly observed through the huge number of research articles published in the last few years (2010–2016) dealing with the various aspects of the MFCs (Carlo et al. 2017; Ravinder et al. 2017b). Figure 13.1

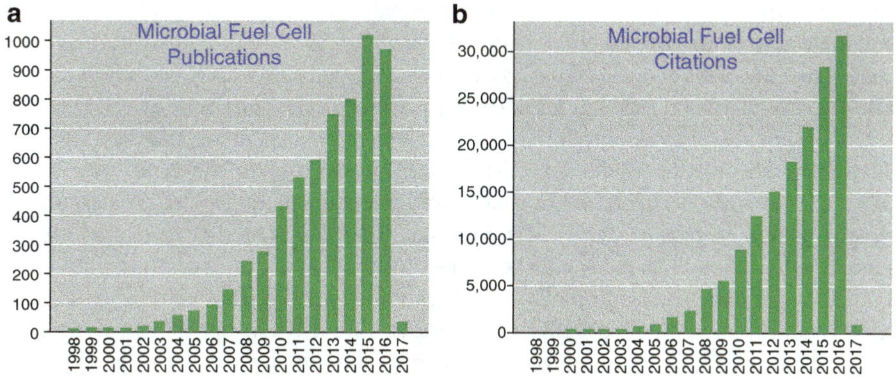

Fig. 13.1 (**a**) Number of publications appeared during 1998–2017. (**b**) Number of citations on MFCs during 1998–2017. (Credit: Carlo et al. 2017)

summarises the number of research articles published on MFCs since 1998 till 2017(Carlo et al. 2017). A major breakthrough in the year 1998 was the construction of mediator-less MFC that made a greater contribution in the advancement of the MFCs (Kim et al. 1999a, b). The development of MFCs has enormous opportunities in various domains for sustainable energy production.

13.3 From Laboratory to Pilot Scale: In Nutshell

Galvani was the pioneer who identified the relationship between electricity and bioactivity (bioelectromagnetics). He observed the muscle of dead frog's leg twitched when an electric spark was applied to it. In 1911, Chesse Potter noted that when platinum electrode was in contact with the *E. coli*, potential difference was generated. This observation led Cohen to construct the first microbial battery, which generated potential of 35 V (Cohen 1931). Lower efficiency and the higher construction costs were the major issues confronted with the development and manufacturing of MFCs at the commercial level in earlier 1990s. However, there have been tremendous efforts undertaken to solve the issues associated with efficiency and life cycle of the MFCs. From the dawn of the twenty-first century, the focus on improving the efficiency with low construction cost has gained the highest priority among the research community working in the various domains of sustainable energy-related fields. The progress of innovation in MFCs from laboratory level to pilot scale largely depends on the design of reactor, proton exchange membrane and low-cost electrode materials. On the other hand, the innovations made at laboratory level should be successfully adopted for pilot-scale MFC plants.

For the pilot scale, the maximum current density has to be generated with reduced cost of the electrode materials. Though the electrode materials such as platinum and graphite showed high performance, they are expensive, and hence they must be replaced with low-cost metals like iron or cobalt and carbon materials like carbon felts, carbon cloth and carbon brushes (Bruce Logan 2010). Thus, the electrode materials cumulatively contribute to the current density as well as commercial feasibility. Issues related to the generation of maximum current density can be resolved by adopting air-cathode model using efficient microorganism and suitable fuel materials. The electrode designs also equally contribute to the efficiency of the MFCs. They must be designed in such a way to cover the large surface area to volume ratio which can be achieved by tubular cathodes coated with conducting or catalytically active materials and should be closely packed. For successful commercialisation of MFCs, adoption of suitable reactor designs, viz. mediator-less design, air-cathode model, stacked MFCs, continuous flow MFCs, upflow tubular model and flat-pad MFCs in addition to single- and double-chamber designs, is desirable. One of the predominant techniques adopted for the treatment of wastewater relies on the commercialisation or construction of pilot plant of MFCs (Kelly and John 2009).

13.4 Qualities of MFCs

The MFCs are often referred to as bioelectrochemical cells and have seen many obstacles since its inception. Consequently, numerous innovations have been appearing periodically in the form of research articles, reviews and patents (Fig. 13.1) to improve the quality of MFCs. Primarily, MFCs operate through the biochemical redox reactions catalysed by the microorganisms that involve transfer of exogenous electrons from the organic matter chiefly from the domestic or industrial effluents and trapping of these electrons in an external circuit (Ravinder et al. 2016; Prashant et al. 2016). Both in laboratory and at pilot-scale level, MFCs offer to fulfil the energy demand by generating the electricity from wastes and also provide an excellent solution for treating the wastewater or waste materials (Rene Rozendal et al. 2008; Micheal and Thomas 2013; Minghua et al. 2013; Oliveira et al. 2013; Tonia and Giorgia L 2017).

MFCs are sustainable source of energy in terms of economical and environmental perspectives with superior performance than the ordinary electrochemical cells. The advantages of the MFCs are as follows:

(i) Electroactive bacteria or microorganism that acts as a catalyst for redox reactions.
(ii) These fuel cells work in the temperature range of 15–45 °C, which is in the level of ambient conditions.
(iii) MFCs are neither acidic nor basic; thus, they can efficiently work under neutral pH conditions.
(iv) Utilisation of large volume of organic biodegradable mass fuels for MFCs, which efficiently manages the solid wastes or effluents that are generated from the domestic and industrial activities. The bioelectrochemical cells have been

classified based on their applications and the products generated during the reactions (Ravinder et al. 2017a).

13.5 Source of Green Energy

The demand for energy is satisfied by renewable energy sources such as hydropower, wind, geothermal and solar which are commonly referred to as 'green energy sources'. However, in the current scenario, the energy demand may not be fully supplied by the above resources. Thus, the quest for the alternative energy sources has gained much momentum. The renewable energy sources derived from biomass such as wood waste, municipal solid waste, landfill biogas, ethanol and biogas are also capable of producing 'green energy'. Among the biomass that is utilised for the energy production, municipal solid wastes together with wastewater containing organic and inorganic components can be successively employed as a replacement for fossil fuels. Apart from the usual treatment of wastes with physical and chemical methods, the biological treatment of wastes proves to be more of an energy efficient process. The typical biological treatment involves the degradation of organic matter by the microorganism under the ambient conditions. However, most of the biological degradation processes proceed through oxidation-reduction reactions. Thus, the idea of MFCs emerged from the use of exogenous electrons released during the redox degradation of the organic matter (Heming and Zhiyong 2013).

More interestingly, *Enteromorpha prolifera*, a green alga that causes serious environmental issues in southern region yellow sea, has been utilised as an energy source for the MFCs (Min et al. 2013). The algae have the ability to grow rapidly both in freshwater and seawater and cause green tide. This rapid growth of alga created huge biomass that has found varied applications. The alga is composed of 50% carbohydrate, and the hydrolysed biomass of *E. prolifera* is used as a good source of energy for the construction of MFCs. In this study, the MFC has been constructed using carbon cloth as anode and air cathode and single cylindrical chamber as the reactor. The fuel cell showed power density value of 1027 mW m^{-2}, and the chemical oxygen demand (COD) has been significantly reduced to 71%. It is quite evident that the biomass generated by *E. prolifera* could be successfully used for the generation of the electricity using MFC technology.

13.6 Generating Power While Treating Wastes

The advancement of MFCs at commercial level generally depends on the effective degradation of various kinds of organic/inorganic wastes. The redox biochemical reactions of microorganisms are pivotal to construct the fuel cells to generate electricity using organic wastes derived from domestic and industrial sources (Rene Rozendal et al. 2008; Pant et al. 2010; Hai-Liang et al. 2015; Quanguo et al. 2016).

The microorganisms present in the wastes effectively catalysed the electrochemical oxidation of the organic impurities or pollutants. The typical working principle of the MFCs involves the oxidation of organic and inorganic components by the microorganisms and the electrons generated in the process transferred to the anode and protons (H^+) to the cathode thereby completing the electrical circuit. After the organic and inorganic wastes have been completely degraded or utilised, fresh wastes can be replenished or can be percolated to the MFCs for the continuous generation of electricity. Besides the power generation, the quality of wastewater from domestic or industrial activities could also be effectively monitored by the use of MFCs. The investment incurred on the construction of MFCs should be minimised for the pilot or commercial scale process (Bruce Logan et al. 2006).

The utilisation of MFC technology is of great benefit for the wastewater treatment and could be an inexpensive way of monitoring the quality of the wastewater or ground water regularly (Zhuwei et al. 2007; Abilasha Singh 2016; Valesquez-orta et al. 2017). Substantially, the wastewater generated from dairy industries (Xiaonan et al. 2011; Ana Faria et al. 2017), agro-food industries (Daniele et al. 2017), textile industries (Anam et al. 2014), microbreweries industries (Ellen et al. 2016), removal of heavy metal from wastewater (Syed Zaghum et al. 2017), winery industry (Cusick et al. 2011), beer brewery industry (Parawira et al. 2005), domestic wastewater (Shijia et al. 2016), municipal wastes, food wastes, animal wastewater (Jeffery et al. 2010) and urine (Couler et al. 2016) can be utilised as a source of fuel for the MFC technology (Fig. 13.2).

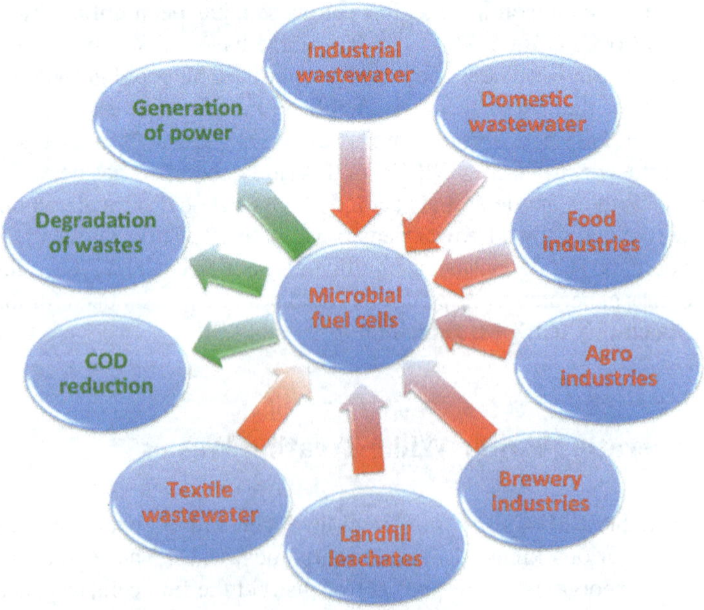

Fig. 13.2 MFC is a source of green energy

The use of landfill leachate is explored as a source material for MFC. An air-breathing cathode and carbon felt anode-based MFCs have been fabricated using landfill leachate as a fuel. The cell is constructed at laboratory-scale level and showed maximum open circuit volt of about 1.29 V over the study period of 17 days. The maximum power density was achieved up to 1513 mW m^{-2}, and it was the highest value reported so far for any MFCs using landfill leachate as an energy source (Jayesh Sonawane et al. 2017). *Mesophilic* bacteria were found to be the catalyst for the redox reaction of the organic wastes. The optimised cell parameters and use of cost-effective electrodes in the present study are prerequisite for the commercial or pilot-scale investigations. However, intensive research on the composition of wastes and exploration of the various species of microbes are necessary for better construction of the MFCs.

In addition to the wastewater or municipal solid wastes, the farm or agricultural wastes, such as humus, cattle manure (Haugen et al. 2015), peat moss, saw dust, rice husk and whey etc., could also be used as a biodegradable source for the MFC applications (Pant et al. 2010). Using sawdust, a single-chamber MFC has been constructed using carbon felt as anode and air-breathing cathode. It showed stable power generation as compared to the other solid wastes over the period of 40 days (Ademola et al. 2017). On the other hand, the MFC using humus as carbon source showed excellent power density but showed poor performance after 40 days. Interestingly, the combination of humus and sawdust increased the life span of the MFC up to 9 months. Similarly, rice bran, which is a waste product generated during the milling process, could also be used as carbon source for the MFC (Takahashi et al. 2016). The study showed that a single-chamber MFC was constructed with a capacity of 15 mL at laboratory-scale level using graphite felt as anode and polytetrafluoroethylene (PTFE) coated with platinum catalyst as cathode. The MFC showed maximum power density from 360 to 520 mWm^{-2} in the presence of pure water and mineralised water. These studies revealed that there are plenty of carbon sources available for the construction of the MFCs at commercial-scale level.

13.7 Reactor Design for Pilot-Scale Process

Many of the pilot-scale MFCs have been constructed for the wastewater treatment either from domestic or industrial sources (Micheal and Thomas 2013; Bruce Logan 2010; Nastro 2014). The design of MFCs would affect the efficiency of the generation of sufficient current density for the commercial implementation. Currently, there are various designs of MFCs, viz. single-chamber, double-chamber, upflow cylindrical–/tubular-type, flat bed-type (FPMFC), stacked-type (Oliveira et al. 2013; Prashant et al. 2016) and sediment microbial fuel cells (Atieh et al. 2015, Valeria et al. 2017), developed for the large-scale applications. With respect to the nature and type of land and the composition of the wastewater, the appropriate model of the reactor has to be chosen for the better performance of the MFCs (Carlo et al. 2017). The laboratory trials of MFCs showed considerable amount of current density at 10 A m^{-2} surface area of electrode, which is sufficient for the construction of MFCs for commercial-scale power production (Rene Rozental et al. 2008). Several studies have been reported on the MFCs for laboratory levels, but scaling up or commercial process would be successful if the MFCs with high power density are constructed for the real-time applications like wastewater treatment. In this section, various models of MFCs constructed for the pilot scale is discussed.

13.7.1 Single-Chamber MFCs

Although many designs of MFCs have been reported, one of the most widely used MFCs are the membraneless single-chamber (MLSC) MFCs. These MFCs owing to its simple reactor constructions and least expensive membranes are extensively employed for the large-scale applications. A single-chamber MFC (SCMFC) with carbon cloth as both anode and cathode, coated with platinum catalyst, utilises brewery wastewater as a fuel source (Yujie et al. 2008). The cell showed maximum power density up to 205 mW/m^2 at 30 °C with COD removal efficiency up to 37%. Further, addition of phosphate buffer at 50 mM and 200 mM concentration increased the power density to 438 and 528 mW/m^2, respectively.

Daqian and Baikun (2009) reported single-chamber MFC (SCMFC) made of granular activated carbon (GAC) chamber containing graphite rod as anode and platinum-coated carbon cloth as air cathode. The wastewater collected from the University of Connecticut having COD value of 200 mg/L and pH 7.2 served as fuel. Excess of sodium acetate was added to achieve the desired COD value of 1500 mg/L, and the MFC was operated at 30 °C. The SCMFC containing four sets of multiple anode and cathode was also constructed to achieve the maximum power density. The SCMFC generated maximum power density up to 7 W/m^2 at electrode distance of 2 cm, and COD level (100–200 mg/L) reduced up to 89%. The multiple anodes and cathodes containing GAC-SCMFC generated 3.25 mA current, while single-anode GAC-SCMFC showed output of 3 mA current. However, the overall performance of MFCs greatly depends on the work ability of the cathode. The power density of MFCs has been increased by the use of MFC containing multi-anode and cathode (Baikun et al. 2011). Daqian et al. (2011) reported the increment in the power density when multi-anode/cathode containing MFC were used for the treatment of wastewater. The MFCs were constructed and evaluated at a domestic wastewater treatment plant in the Gloversville Johnson Joint Wastewater Treatment Facility (GJJWWTF) (New York, USA) (Fig. 13.3). The study revealed that the power density increased from 300 to 380 mW m^{-2} by using the MFC containing 12 anodes and cathodes. In this study, an attempt has been made to replace the expensive platinum electrode by metal-doped MnO_2 cathode. With respect to platinum electrode, Cu- and Co-doped MnO_2 showed higher power density up to 465 and 500 mW/m^2, respectively. Though the cathode showed advantage, gradually the decrease in power density was observed due to the cathode fouling by the precipitation of calcium and sodium.

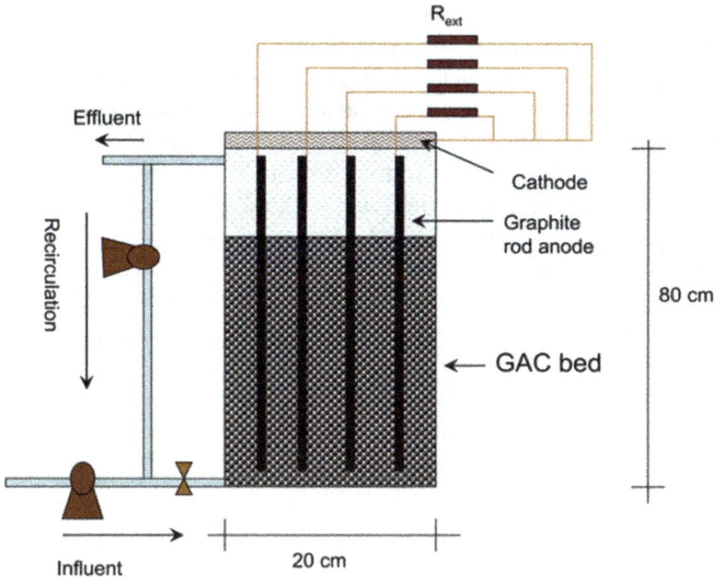

Fig. 13.3 Schematic view of MAC-MFC at pilot scale. (Credit: Daqian et al. 2011)

A single-chamber air-cathode MFC has been developed for the continuous wastewater treatment containing variable COD values. A synthetic wastewater containing 0.1 to 0.4 g/L of glucose was treated with novel submerged-air-cathode MFC (SE-AMFC) with working volume of 5.7 L (Yu et al. 2012). The study revealed that while increasing COD values from 100 to 400 mg/L, the power density of SE-MFC increased from 191 to 754 mW m^{-2}. However, the increasing COD values declined the COD removal efficiency.

Yongwon et al. (2013) developed a single-chamber hexagonal-shaped large-scale MFC for the treatment of activated sludge from the Jungnang Sewage Treatment Centre, Seoul, Korea. The performance of the cell was evaluated by the COD removal efficiency and current density, and the durability of the MFC was tested for more than 5 months. The cell constructed with a capacity of 1.29 L consists of 30 wt% wet-proofed carbon cloth as air cathode and 20% Pt/C-coated carbon cloth as anode. The cell was operated at neutral pH in 30, 35 and 40 °C. The cell showed 94% COD removal efficiency at 40 °C, and this study uncovered that the efficiency of COD removal could be increased from 88.5% to 94% by increasing the operating temperature from 30 to 40 °C. In addition, the columbic efficiency also increased from 3% to 21% by increasing the operating temperature. The study unearthed the fact that the power density greatly depends on the reactor size and its efficiency could be reduced with an increasing size which may be due to an increase in internal resistance in larger size reactor. Thus for the successful implementation of the MFCs for the larger-scale applications, the internal resistance has to be controlled.

A pilot-plant MFC with 45 L capacity has been constructed and combined with the effluent from wastewater treatment plant operated by Emschergenossenschaft, a non-profit organisation in Germany mainly working on the river basin management (Heinz et al. 2016). For the real-time application, a membraneless single-chamber MFC (SCMFC) was connected to the primary treatment plant. 45 L capacity was achieved by successively connecting four individual chambers with a capacity of 11.2 L. Each compartment was connected with the platinum-coated PTFE as air cathode and graphite fibre brush as anode. The performance of the MFC was evaluated by the removal efficiency of COD, total nitrogen (TN) and total suspended solids (TSS). The COD value of the wastewater was found to be 130 mg/L, which is very low due to the dilution by the river water, and conductivity between 3.0 and 4.2 S/cm was established.

13.7.2 Two-Chamber MFCs

The usual construction of MFCs contains two chambers, and the chambers are separated by proton exchange membrane (PEM). In one chamber, the oxidation of organic materials occurs on carbon cloth or platinum-based electrodes, which serve as anode. Electrons generated are transferred to the external circuit. The electrons are collected by another chamber where oxygen get reduced (cathode) to O^{2-} and combine with proton to form water as byproduct (Booki and Bruce Logan 2004) (Fig. 13.4).

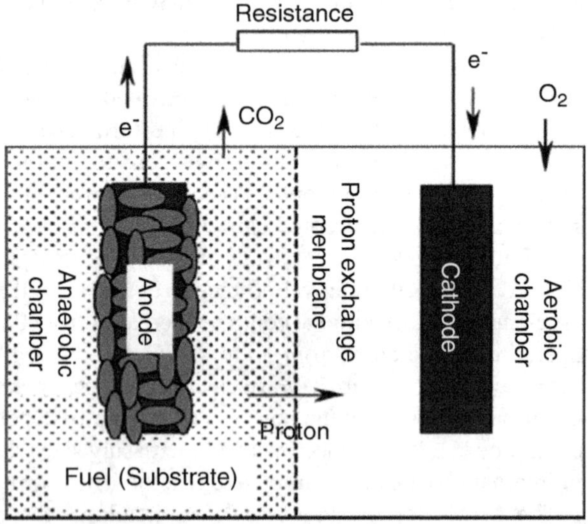

Fig. 13.4 Typical two-chamber microbial cells. (Credit: Zhuwei et al. 2007)

The dairy wastewater is treated by using dual chamber as reported by Daniele et al. (2017). The chambers are filled with granular graphite material and connected with electrodes made of graphitic rods. The reactor was operated for 2.5 months and supplied with raw dairy effluent. The COD value of the dairy industries was found to be in the range of 650–3000 mg L^{-1}. The DCMFC showed COD removal efficiency up to 80–90% with columbic efficiency achieved up to 60%. The study showed excellent organic matter removal efficiency with maximum recovery of energy in the form of power density up to 27 Wm^{-3}.

13.7.3 Vertical or Upflow Chamber MFCs

Upflow anaerobic sludge bed (UASB) reactors are mainly constructed for the biogas production or for the purpose of preparing composite from the organic waste material like cow and dairy wastes (Hina et al. 2015). In Lahore, Pakistan, the domestic wastewater obtained from the Garden Town wastewater-pumping agency was treated using UASB. The UASB is a cylindrical-type reactor built with a capacity of 4.6 L at the top, and in the bottom is a gas-liquid-solid separator built with the volume of 11.2 L. At 25 °C both cow dung manure and dairy wastewater showed 77% and 68% COD reduction, respectively, after 120 days. Since, the reactor design showed effective biogas and biomass production, it could be successfully adopted for making the MFCs. Another usage of UASB has been explored for the treatment of opaque beer brewery waste sludge collected from the largest brewery industry in Harare, Zimbabwe (Parawira et al. 2005). The reactor with a capacity of about 500 m^3 was built for treating the wastes having COD value of 6 kg/m^3 per day. Before treating the wastes, the digester chamber was seeded with active municipal sludge and acclimatised for 3 months and the brewery waste sludge fed at hydraulic retention time of 24 h. The performance of the reactor was assessed by examining the parameters such as pH, COD, total dissolved solids (TDS), TSS and settleable solids. The study showed that the anaerobic treatment of the wastes using the microbes efficiently reduced the COD in the average range of 30–70% (initial COD value from 16 to 4 g/L) over the study period of 2 years.

Thus based on the COD removal efficiency from the sludge, a series of cylindrical-type longitudinal MFCs with a total internal volume of 1 L have been constructed for the treatment of sucrose wastes as model substrate (Jung Rae et al. 2011). The MFCs consist of 0.5 mg m^{-2}-coated membrane as air cathode and carbon veil as anode, and it was operated for more than 7 months. The active microbes cultured from the sludge were obtained from Cog Moors Sewage Treatment Works, Cardiff, UK, and were acclimatised for more than 3 weeks in the digester compartment. Independently connected MFC modules showed maximum power density up to 6% and 36% at 0.8 and 0.08 g/L of sucrose containing wastes, respectively, as compared to parallel connected MFC modules. The study emphasised that the increasing number of tubular MFC modules would maximise the organic removal as well as power generation efficiency. The tubular design MFCs has higher possibility for the scaled-up process.

Haugen et al. (2015) has reported the use of computer modelling for the full-scale pilot plant optimal design of anaerobic digestion reactor (UASB) for the treatment of dairy farm wastes especially dairy manure produced from herd. The planned full-scale pilot plant has the volume of 250 L, and solid particles removed were 25% wet dairy wastes. The reactor has produced more than 70% of methane gas. The mathematical optimisation of the reactor design and optimisation of parameters revealed that at the reactor temperature of 36 °C, the maximum power surplus up to 49.8 MWh/y with the hydraulic retention time of 6.1 days was achieved.

In another instance, the use of upflow reactor design coupled with ionic liquid-type membrane for the continuous electricity production from the treatment of industrial wastewater has been investigated. In this study, the reactor with 1.7 L was built containing a single-chamber air-cathode electrode coated with 0.5 mg/cm^2 platinum, and the combination of graphite bar and carbon granules was used as anode (Salar-Garcia et al. 2016). Instead of the usual Nafion membrane, ionic liquid based on the triisobutyl(methyl)phosphonium tosylate and methyl trioctylammonium chloride was used for the preparation of polymer inclusion membranes.

13.7.4 Stacked MFCs

Reaching the maximum power density of the MFCs is a major task for the commercial or large-scale applications. Besides the single-chamber, dual-chamber, upflow tubular MFCs, the development of stackable MFCs is proved to be a successful design to achieve the maximum power density (Ravinder et al. 2017a, b). By definition, the stacked MFCs are constructed by joining multiple SCMFC, DCMFC, and UCMFC in series or in parallel to increase the output of the MFCs. Thus, in the stacked MFCs, the multiplication of the power output of single MFCs is possible with respect to the number of MFCs connected together. In addition to the power generation, the maximum reduction in COD values can be achieved because of the efficient degradation of the organic matters present in the wastewater.

For instance, six single continuous MFCs showed power output up to 258 W m^{-2} in the stacked configuration and when the MFCs connected in parallel or in series showed an increase in voltage up to 2.02 V at 228 W m^{-2} (Peter et al. 2006) (Fig. 13.5). Both anode and cathode materials consist of graphite granules and graphite rod, and the rod was operated for more than 7 months. By connecting six individual units of MFCs, the total volume of 360 mL was achieved. Based on the open circuit voltage (OCV) output, the individual MFCs showed 693 mV, and nearly sixfold increase in the OCV was achieved when the MFCs were connected in series, whereas the parallel connection showed 668 mV OCV.

In another example, 10 units of single MFCs were connected in parallel to generate OCV of up to 13.03 V using the activated sludge as the fuel and were operated in continuous mode (Pablo et al. 2013). The stack was designed to power the

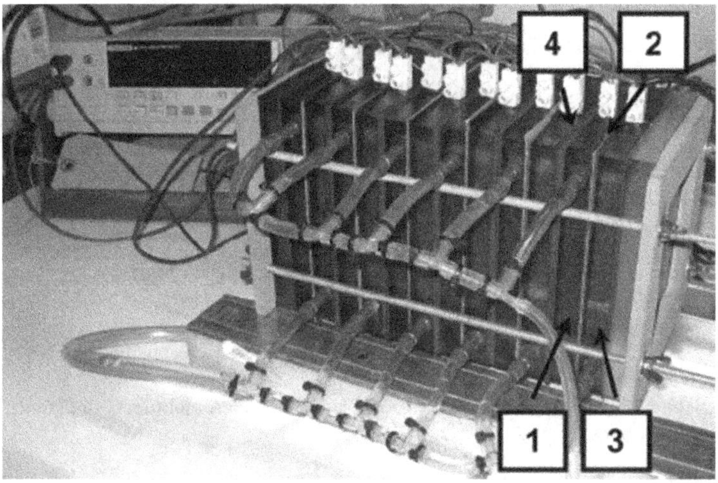

Fig. 13.5 Parallel connected stack microbial fuel cell (MFC) consisting of six individual microbial fuel cells (MFCs) with (1) a granular graphite anode, (2) an Ultrex cation exchange membrane and (3) a 50 mM hexacyanoferrate cathode separated by (4) a rubber sheet. (Credit: Peter et al. 2006)

microcontroller, and it is programmed for self-monitoring as a self-sustainable model. On the other hand, the stacks of the MFCs showed more than 96% reduction in the COD value using the continuous operation mode. Similarly, Yujie et al. (2014) have constructed a stacked MFC for the real-time application to treat the municipal wastewater collected from Harbin Institute of Technology, Harbin, China. A single horizontal MFC with a capacity of 250 L has been built and connected with 0.25 mg/cm^2 platinum-coated carbon mesh as air cathode and carbon brush combined with titanium wire as the anode. Four of the above individual modules were connected independently without affecting the other module performance or operation. The maximum power output up to 116 mW was generated by the above model, and in addition to that, a maximum COD reduction up to 79% was observed. However, the development of internal resistance is the limiting factor in the large-scale applications.

A 90 L pilot-scale stackable MFC has been constructed for the treatment of brewery wastewater (Yue Dong et al. 2015). In this study, five individual modules were slotted in the 100 L reactor, and each module is connected with activated carbon-coated PTFE as cathode and carbon brush woven with titanium wire as anode, and the system was operated at 25 °C (Fig. 13.6). Brewery wastewater collected from Harbin Brewery Co., Harbin, China, was used as the influent with hydraulic retention time of 3 days, and the performance of the MFC was evaluated for 6 months.

The above design effectively reduced the suspended solids (SS) up to 81.7 and 86.3% in the diluted sewage (stage 1) and raw sewage water (stage 2), respectively. Likewise, tremendous reduction in COD values were also observed up to 84.7 and

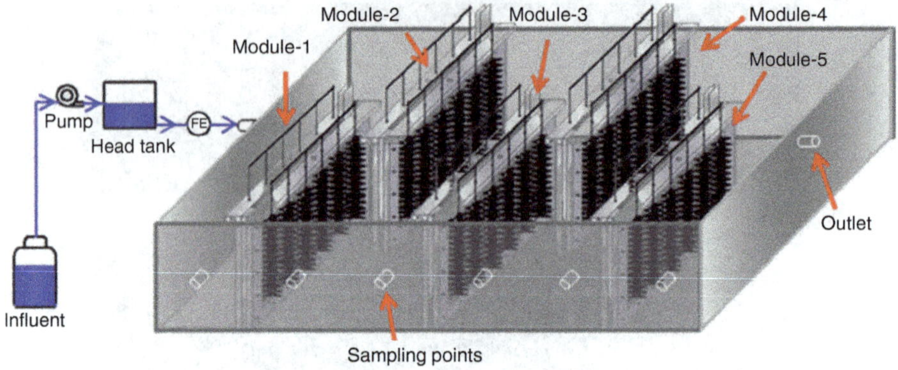

Fig. 13.6 Schematic drawing of the 90 L stackable baffled microbial fuel cell. (Credit: Yue Dong et al. 2015)

87.6% in the stage 1 and stage 2 processes. In addition, the system proved to be self-sustained since it generated 0.097 kWh m^{-3} that is capable of powering the pumping system (0.027 kWh m^{-3}). The results are encouraging, and the above design could be successfully adopted for the larger-scale or commercial applications.

By using stacked MFC design, Byung-Min et al. (2016) developed the stacked MFC for the treatment of ethanolamine containing synthetic wastewater at pilot-scale level. The working volume of the reactor was set in 1 L in which the reactor is equally divided into five parts using the baffles and connected to two sets of electrodes to improve the efficiency of the cell. The system utilised wettable carbon cloth as anode and 30% wet proof carbon cloth as air cathode. The air-facing side of the electrode was coated with activated carbon/PTFE and 10% platinum-coated side used for solution-facing side. The anodic chamber is filled with the synthetic wastewater containing 1000 mg/L of ethanolamine. The performance of the cell was evaluated based on the COD reduction and power-generating efficiency at 25 °C. The use of dual anode and cathode in the stacked MFC design has generated power up to 0.86 W m^{-2} with maximum carbon and ammonia removal efficiency up to 95.30 and 95.70%, respectively.

Since the reactor designs have the major role on the performance of the MFCs, the development of novel designs would be always beneficial for the improvement of the power-generating as well as waste-treating efficiency. Also, each design has its own flaws, which limits their application in the commercial-scale or pilot-scale level and in turn affects the practicability of the MFCs. For instance, the anode fouling affecting the biofilm on the anode is a major issue in SCMFC, and proton accumulation is a shortcoming of DCMFCs. So far, the stacked MFC design includes either SCMFC or DCMFC or baffled designs were investigated. But the hybrid of SCMFC and DCMFC in the stacked design has so far not been investigated. However, recently hybrid fuel cell stacks have been constructed at the laboratory level for the treatment of synthetic wastewater using single and dual chamber (Wei et al. 2016). The hybrid design showed improvement in the power generation of

0.8 V with operational stability up to 16 h, which is comparatively stable than the stacked SCMFC and DCMFC alone.

13.7.5 Flat-Plate Microbial Fuel Cells (FPMFCs)

The raise in internal resistance is a major issue in the large-scale application of MFCs. Thus, the FPMFC design is mainly developed to overcome the internal resistance to ease the pilot-scale operation of the MFCs with continuous operation. In the FPMFC model, the anode and cathode are kept very close and separated by a cation exchange membrane. For instance, wastewater that is consistently generated either from domestic or industrial process is required for the continuous treatment in order to maintain the equality of groundwater. Also, using MFCs for the treatment of wastewater can continuously generate the electricity. The typical FPMFC is constructed using single electrode consisting of platinum- or carbon-based electrodes separated by proton exchange membrane (Fig. 13.7) (Booki and Bruce Logan 2004). The domestic water or fresh organic materials can be employed as energy sources. The FPMFC, with total cell volume of 22 cm^3 is constructed with two chambers separated by Nafion membrane, in which carbon paper connected to one chamber act as anode and 10% platinum coated electrode connected to other chamber acts as cathode. For the real-time application, the domestic wastewater collected from Pennsylvania State University water treatment plant used as fuel and also carbon sources such as glucose, acetate, dextran, starch and butyrate

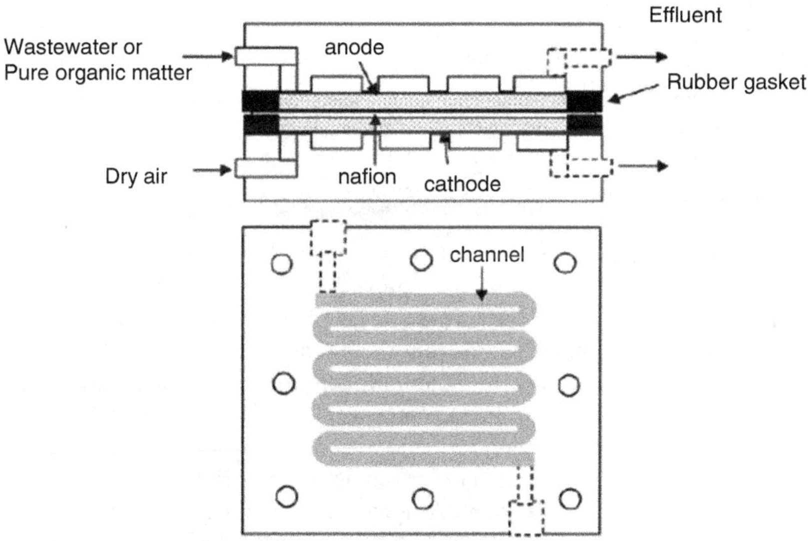

Fig. 13.7 Schematic view of FPMFC. (Credit: Booki and Bruce Logan 2004)

were added to make up the final COD value of 1000 mg/L. The FPMFC was operated at 30 °C.

Marjolein et al. (2012) reported a plant-based FPMFC using *Spartina anglica* as a plant source. The design of FPMFC uses graphite felt as both anode and cathode, and the electrodes are separated by cation exchange membrane. The working volume of the MFC is about 650 mL. The performance of the FPMFC was monitored for nearly 2 years. The study revealed that the use of FPMFC model reduced the internal resistance considerably as compared to the tubular model. The performance of membrane used for the separation of the electrodes is of great importance for the effective functioning of the FPMFC. The performance of FPMFC constructed using various commercially available membranes has been evaluated (Sona et al. 2016). In this study, the membrane having high coefficients of oxygen and ethanol mass transfer showed lower power density when the electrodes were kept in close proximity. However, the Nafion 117 showed high performance as compared to other separators used in the study. For instance, the cell constructed with Nafion 117 with electrodes spacing of 2 mm showed the OCV up to 0.75 V and 10.7 ± 0.5% columbic efficiency when compared to other MFCs. In addition, a marginal increment was observed in OCV and columbic efficiency when the electrode spacing was increased from 2 mm to 4 and 8 mm.

Unless the issues such as expensive electrodes, internal resistance and electrode fouling are resolved, the continuous operation at the larger scale would not be commercially viable. The presence of proton exchange membrane (PEM) in the FPMFC model would possibly eliminate the biomass as well as oxygen that would enhance the MFC efficiency to generate the maximum power during the continuous operation. Adoption of the FPMFC model made of graphite felt electrodes along with air-breathing cathode could solve the issues related to the electrodes. For instance, the FPMFC is constructed using graphite felt anode, platinum-coated carbon cloth as air cathode separated by PEM containing activated sludge collected from Howe Sound Pulp and Paper Mill, British Columbia, Canada, for the treatment of synthetic wastewater at batch process as well as a continuous process (Sona et al. 2015). In the batch mode process, the cell has produced maximum power density up to 40 mW m^{-3}, and the same has showed better performance by producing 95 mW m^{-3} during 250 h. The COD removal efficiency was achieved up to 60% during the continuous operation of the MFC.

The anaerobic fluidised bed microbial fuel cells (AFB-MFCs) have been developed for the continuous operation using air cathode. The synthetic wastewater circulated through the column of carbon particles, and the wastewater was continuously pumped through a peristaltic pump from the storage tank (Xuyun et al. 2015). In this model the cell was capable of generating 900 mV by gradual increment up to 80 h, and after that the power steeply increased to 900 mV and stabilised up to 120 h.

Thus, based on the application and the requirement, various MFCs model can be chosen for the further development. Among the several designs, cylindrical-type, FPMFC and combined stacked model MFCs have tremendous opportunities for the larger-scale applications. Several instances on the construction design of various types of MFCs and its effective operation at pilot scale have been discussed above. The majority of the large-scale applications of MFCs largely dealt with treatment of wastewater together with the generation of power successfully. It would be a worthy option to discuss the various types of wastewater treated using the MFC technology.

13.8 Field Trials of MFCs

The commercialisation of MFCs has much complexity since the efficiency of MFC is still under development when working with real industrial effluents (Padma and Hays 2012; Escapa et al. 2016). Firstly, effluents containing various kinds of organic wastes have been used as fuel for anode in MFC, but the mechanism of interaction between the electroactive microorganism and fuel could not be clearly established, and still it remains a challenging task for investigators (Ganesh et al. 2017; Hai-Liang et al. 2015). Secondly, MFCs generate relatively lower energy when compared with electrochemical cells though they offer many advantages over the latter. In addition, the knowledge on the nature and type of the effluents generated from industries and anthropogenic activities is necessary, and feasibility of treating them with electroactive microbes should also be checked for the construction of MFCs.

Thus this section focuses on the contribution on the treatment of the wastewater using MFCs which has been summarised in a nutshell. In addition, the problems associated with the commercialisation and solutions are also discussed at the end of the chapter as solutions at laboratory level. Further, the latest reports on economical cathode material development are also discussed.

13.8.1 Application of MFC for Wastewater Treatment

The uses of microbial fuel cell technology have immense potential for the wastewater treatment, and it has the history more than a decade (Dan et al. 2015; Dimou et al. 2014). Based on the origin and availability, one of the MFC designs may be adopted for the continuous or batch process applications. A number of reviews appeared frequently that deal with the advancements and challenges in the treatment of various kinds of wastewater (Zhuwei et al. 2007; Rene Rozendal et al. 2008; Bruce Logan 2010; Micheal and Thomas 2013; Minghua et al. 2013; Oliveira et al. 2013; Haishu et al. 2016; Prashant et al. 2016; Iwona et al. 2016; Gude 2016a, b; Carlo et al. 2017), and MFC technology is used to enhance the removal efficiency of organic matters to restore the sediments (Henan et al. 2017)

13.8.2 Constructed Wetlands

Constructed wetlands (CWs) are best known as artificial wetlands that are created mainly for the treatment of wastewater of domestic, industrial and agricultural sources generated from small communities (Wu et al. 2014). The wastewater generated from the anthropogenic activities may contain various kinds of organic wastes that could be a viable source for the generation of electricity if the microbial fuel cell technology is successfully implemented. The installation of MFC at pilot scale in the CW would treat the wastes as well as the system would become self-sustainable. In addition, it would help in monitoring the quality of wastewater influent and effluent. The construction of wetlands majorly uses the model of horizontal subsurface flow (HSSF-CW) at 0.3–0.6 m depth. The performance of the HSSF-CW is evaluated based on the design parameters. The efficiency of the treatment depends on the percentage removal of COD, nitrogen content and sulphate (Corbella and Puigagut 2015). An added advantage for the construction of the MFC in CW is to provide redox gradient of about 0.5 V between the upper and lower layers.

For instance, as shown in Fig. 13.8, a pilot-scale HSSF-CW combined with MFC model has been constructed in the Institute of Chemical and Environmental Technology, Castilla La Mancha University, Spain (Villasenor et al. 2013). The above CW combined MFC system to a 150 L wastewater tank and peristaltic pump to supply wastewater at variable flow rate. The graphite material is used as cathode and anode connected to the upper and lower layer, respectively. The CW-MFC system was operated with respect to variable COD and organic loading rate by continuous mode for about 6 months. The above system showed average power density up to 15.6 mW m^{-2} and maximum power density up to 43 mW m^{-2} with COD removal efficiency up to 90–95% during 110–130 days of the overall study period.

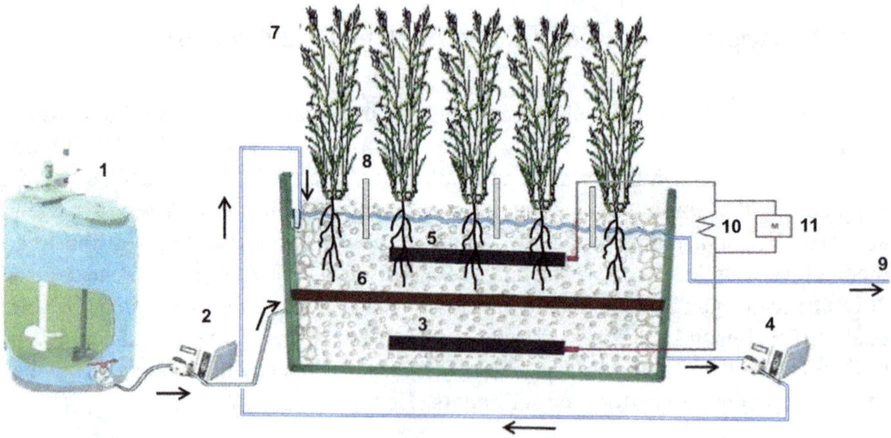

Fig. 13.8 Experimental installation. (1) Wastewater tank; (2) peristaltic pump; (3) anode; (4) peristaltic pump; (5) cathode; (6) bentonite layer; (7) reed plants; (8) sampling points; (9) treated effluent; (10) resistance; (11) multimeter. (Credit: Villasenor et al. 2013)

Doherty et al. (2014) has reported the CW based on alum sludge combined with MFC for the treatment of swine slurry. The performance of the system has been evaluated with respect to the electrode spacing, sludge flow rate and ammonia, COD and phosphate removal efficiency. The maximum power density achieved up to 0.268 W m^{-3} with 75% ammonia and 64% COD removal efficiency by upflow-downflow regime. On the other hand, continuous upflow operation delivered 80% boost in power density with 79% ammonia and 81% COD removal efficiency.

Two small-scale CW-MFC systems with a capacity of 3.7 L have been constructed for the treatment of swine wastewater dedicated for operation at batch and continuous mode (Zhao et al. 2013). The continuous upflow mode of operation showed improved performance by 76.5% average COD removal efficiency with maximum power density of 9.4 mW m^{-2}, whereas batch mode operation showed 71.5% removal efficiency with maximum power density of 12.83 μW m^{-2}.

Very limited investigations have been made in the evaluation of MFC combined with constructed wetlands. However, some interesting preliminary results showed MFC has potential application in CWs as treating the wastes as well as generating power.

13.8.3 Small Island

Kiran and Praneet (2015) reported the performance of the pilot-scale MFCs built in Trinidad and Tobago, which are small-island developing states. Generation of wastewater and shortage of groundwater quality seriously affect these small islands. Thus, the states require the state-of-art MFC technology, which can generate power by using wastewater stream as a fuel. The above system is specially designed for small island countries like Caribbean islands. The construction of MFC consists of two major chambers separated by proton exchange membrane (Fig. 13.9). The domestic wastewater collected from various sources is flown to the anodic chamber after thorough screening of wastewater. On the other hand, the seawater collected from the Atlantic Ocean and Gulf of Paria is taken in the cathodic chamber. The microbes present in the domestic wastewater are attached to anode. The study revealed that the constructed MFC showed decrease in biological oxygen demand (BOD) and increase in chemical oxygen demand (COD) from 30% to 75%. Both the systems delivered power density up to 84 and 96 mW/m^2.

The above system is advantageous since it generates power from the domestic wastewater while treating them with the microbes. Thus, the above MFC could be economically viable as well as offers solution for wastewater treatment.

Bruce Logan (2010) reported that the power density of MFCs could reach up to 1 KW/m^3 and 6.9 W/m^2 per surface area of anode at the laboratory level. The main obstacle is applying the laboratory parameters to the production at pilot scale, and it required further modification of the laboratory parameters. Novel and economically feasible electrodes viz., cathodes, membranes and separators are crucial to achieve better performance at the commercial scale. Currently, the fabrication of air-cathode

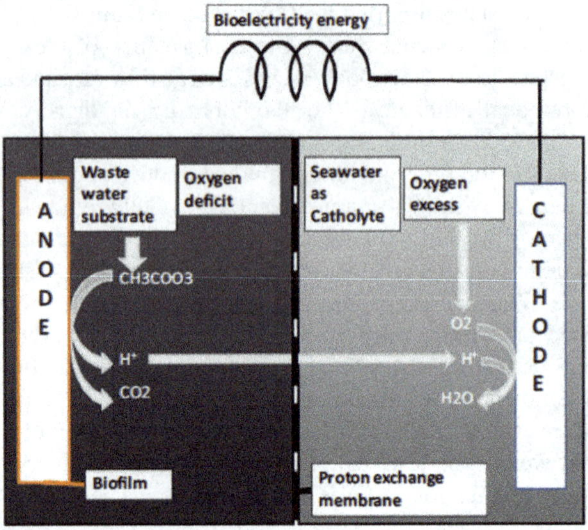

Fig. 13.9 Two-chamber MFCs constructed at Trinidad and Tobago. (Credit: Kiran and Praneet 2015)

design could be successfully implemented for large systems like wastewater treatment using the MFC technology.

Padma and Dirk (2012) have reported the possible types of MFC construction and development of electrode materials. The MFCs are mainly classified into four types. The first type is the two-chamber MFC, where anode and cathode are kept in different chamber and separated by proton exchange membrane. The second type is the single-chamber MFC, in which anode and cathode are kept in the same chamber and separated by proton exchange membrane. It's construction is simple when compared to other types and is therefore, less expensive. However, it has drawbacks like microbial contamination and short circuit. The third type is a vertical-type MFC, where anode is kept at the bottom and cathode at the top. Glass wool and glass beads were used to separate the electrodes. This system can be employed for large-scale wastewater treatment and power production. In the fourth type, a series of single-chamber MFC were connected to form stacked MFCs. By using the stacked-type MFC, high power density can be achieved. In addition to the usual designs, flat-plate microbial fuel cell design can also be used for the treatment of wastes and could also be successfully operated in the small islands. However, optimisation on the design parameters such as reactor size, membrane and electrode materials is necessary as well as thorough understanding of wastewater parameter would help to choose the design and construction of suitable MFCs.

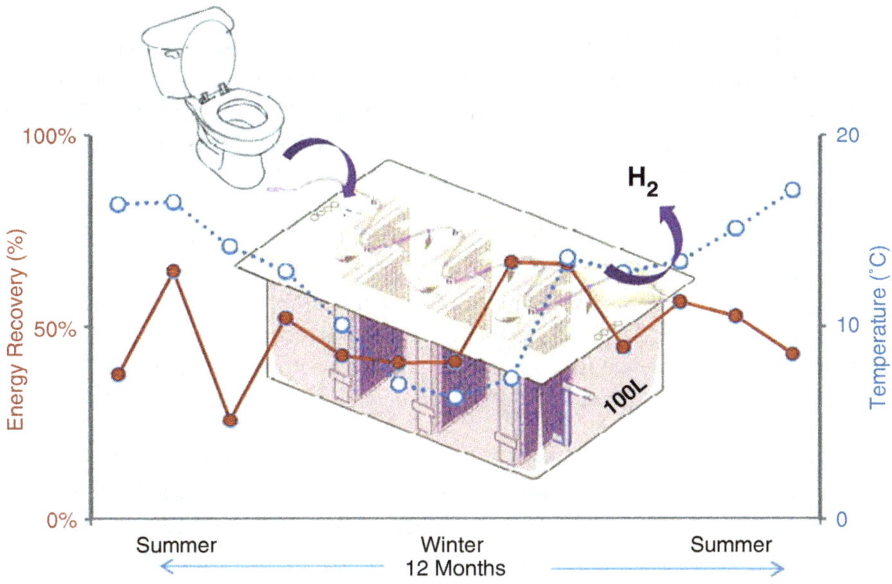

Fig. 13.10 Construction of microbial electrochemical cells (MEC) using domestic wastewater. Also, the plot of energy recovery and temperature is included in the image. (Credit: Elizabeth et al. 2014)

13.8.4 Domestic Wastewater

The organic and inorganic wastes generated in water due to day to day human activities has to be regularly monitored for their qualities (Valesquez-Orta et al. 2017). Waste water generated from different sources has to properly treated prior its entry into the river or main water stream. By using MFC technology, domestic wastewater can be treated while generating power, and the system may become self-sustainable, and in addition, this technology also avoids expensive and non-energy recoverable treatment methods (Booki and Bruce Logan 2004; Jeffrey et al. 2010; Castro 2014; Cotteril et al. 2017; Elizabeth et al. 2013)

Elizabeth et al. (2014) reported on the pilot-scale microbial electrolysis cell (MEC), which is performed on domestic wastewater as a fuel. The wastes generated from anthropogenic activities have been used to generate electricity by constructing the MEC with a capacity of 100 L (Fig. 13.10). The performance of the MEC was continuously monitored for 12 months in temperature range of 1–22 °C. The cell can be capable of producing 0.6 L of hydrogen per day, and the hydrogen production was found to be decreasing with respect to increase in time. However, nearly 48% on average electrical energy was recovered with an efficiency of 41.20%. On the other hand,

Castro (2014) developed a large-scale MFC system for the treatment of human wastes (The Green Latrine) in Ghana. The construction of MFC for the human faeces and urine is expected to deliver treated effluent, compost as well as energy recovery. In this investigation, hydraulically separated three-chamber MFC has been built. The performance of the MFC was evaluated by nitrogen and organic matter removal efficiency. By using the above model, the COD removal efficiency achieved was 90% and nitrogen removal efficiency achieved was 76.8 ± 7.1% and delivered maximum power production was 3.4 ± 0.01 nW m^{-2}.

A 200 L pilot-scale MFC system is constructed for the treatment of municipal wastewater collected from primary clarifier, Pepper's Ferry wastewater treatment plant, Radford, VA, USA (Zheng et al. 2015). The system consists of 2 L tubular reactor connected in series by various arrangements to make 100 L effective working volume as well as to achieve the maximum power recovery efficiency. The MFC system consists of cathode made of nitrogen-doped activated carbon with an HRT of 18 h. The system showed maximum conversion efficiency up to 80% with 11.4 mW power output.

Cotteril et al. (2017) developed a large-scale MEC for the treatment of domestic wastewater in collaboration with Northumbrian Water Ltd., Northwest England, UK. The large-scale MEC module with an area of 1 m^2 has been built to treat wastewater having an average COD value of 340 mg L^{-1} without the addition of any acetate or phosphate buffers. The MEC system consists of 316 stainless steel mesh as cathode (total surface area 0.8 m^2) and graphite felt as anode (anodic area 1 m^2 × 3 modules) with a tank volume of 175 L, and the MEC is operated for the span of 217 days. The large MEC has generated hydrogen of about 0.8 L per day. The COD removal efficiency was found to have an average value of 63.5%. However, the rise of pilot-scale MEC to commercial scale needs process optimisation in terms of reactor design, electrode materials and thorough understanding of the wastewater.

13.8.5 Brewery and Winery Industries

The wastewater generated from winery and brewery industries is rich in various kinds of sugars, proteins and other organic matters, which led to the rise in COD value of wastewater and wastes being non-toxic in nature (Prashant et al. 2016). Because of the high COD value, the winery and brewery wastewater could be an excellent recoverable energy source if they are properly treated using suitable MFC design.

Parawira et al. (2005) reported on the large-scale anaerobic treatment using upflow anaerobic sludge blanket (USAB) reactor for the treatment of organic matter collected from the opaque beer brewery wastewater. The system was installed close to the beer brewery effluent plant, and the study continued for about 2 years. The study is intended for the degradation of organic matter and focused on the COD removal efficiency and other wastewater parameters. The use of USAB effectively reduced the COD value up to 57%. However, the study is not extended to energy recovery using MFC technology.

A prototype upflow single-chamber MFC has been constructed for the treatment of diluted brewery wastewater (1:20 by volume) collected from Hexham Municipal Sewage Treatment Plant, Northumberland, UK (Krishna and Scott 2010). The COD value of the wastewater is around 430 mg L^{-1}, and the MFC was operated in batch mode for 455 h. The cell showed stable maximum density up to 330 mW m^{-2}.

Cusick et al. (2011) evaluated the pilot-scale MFC constructed for the treatment of winery wastewater using continuous mode of operation. The wastewater generated from winery industry is rich in various types of carbohydrates and is the major source of energy. A single-chamber MEC design is used for the field study with reactor working volume of 1000 L and graphite felt and SS304 as anode and cathode, respectively. The electrolysis cell delivered an average COD removal efficiency up to 62 ± 20% with hydraulic retention time of 1 day. The current generation reached up to 7.4 A m^{-3} at the end of 100 days of study.

As discussed earlier in this chapter, Yue dong et al. (2015) constructed a 90 L single-chamber MFC containing five modules of electrodes fixed in the stacked design for the treatment of brewery wastewater. The study revealed that this pilot-scale MFC produced power while degrading the organic wastes present in the effluent by efficiently reducing the COD up to 88% without any energy input. The use of MEC for the treatment of wastewater collected from the craft brewery, Ontario, Canada, has been explored by Ellen et al. (2016). The wastewaters generated from the microbreweries are rich in organic matters and have the highest COD value up to about 2250 mg L^{-1}. A two-chamber MFC has been developed for the treatment of 84 L per day of wastewater. The system has achieved 91.9% COD removal efficiency with the generation of 26.4 mWh electricity.

Based on the studies above, the carbohydrate-rich wastewater collected from brewery, winery and also sugar industries, which are naturally having higher COD value, would be a viable source for the operation of MFCs.

13.8.6 Agro-Food and Dairy Industries

The use of MFC technology has been explored for the treatment of the wastewater generated from food-processing industries such as rice mills, cassava mills, palm-oil mills and mustard tuber mills as well as from the dairy industry wastes such as cheese whey, milk wastes and yoghurt wastes. Among the food industry wastes, dairy wastewater contains the highest COD values because of the presence of sugar as a major constituent. The performance of the MFC has been evaluated based on the reduction of COD. In addition, various designs of the MFC have been explored for the treatment of agro-food industry wastewater (Prashant et al. 2016; Wen-wei et al. 2013).

Daniele et al. (2017) discussed on the energy recovery from the dairy wastewater using dual chamber MFC (DCMFC). The DCMFC reduced the COD value up to 90% with generation of 27 W m^{-3} power density. XiaoNan et al. (2011) explored on the possibility of using continuous flow MLMFC for the treatment of cow manure, milk wastes, cow slurry wastes and feed wastes. The use of MLMFC effectively reduced the COD values up to 98% under continuous mode of operation and recovered more than 80% of energy as hydrogen from the wastes.

A laboratory-level single-chamber MFC using the rice bran wastes as carbon has been reported by Takahashi et al. (2016). The capacity of MFC is 15 mL and equipped with graphite felt as anode and PTFE coated with platinum catalyst as air cathode. The mineral solution containing paddy soil and rice bran was used as anolyte, and the cell was operated at 30 °C. The power output increased from 0.1 V to 0.5 V after 30 days of acclimatisation. The study revealed that *Trichococcus* and *Geobacter* are specifically responsible for the oxidative degradation of the organic matter such as rice bran. Instead of rice bran, other agricultural wastes like whey could also be used as a carbon source.

13.9 Problems Associated with Pilot-Scale Studies

The major challenges which remain in the construction of commercial- or pilot-scale MFCs are expensive electrode material, screening the complexity of fuels, type of microbes, operating temperature, power density and longevity. The upscaling of MFCs using the parameters as developed in the laboratory mainly depends on the operating conditions such as temperature, pH, wastewater type and operation time. In addition to the above, the properties of the electrode materials, surface area of the electrode, electrode-microorganism interaction and the reactor size has to be considered for the successful development of MFCs. The cost of the electrode materials should be minimised substantially, and more economically viable electrode without losing the efficiency of the power generation would be commercialised effortlessly (Bruce Logan et al. 2006). On the other hand, the clear understanding of the nature and composition of the wastewater is essential for establishing the mechanism of electron transfer. In addition, the thorough investigation on the microorganisms present in the wastewater should be made (Vinay and Kundu 2010; Venkata Mohan et al. 2014).

13.10 Solutions at Laboratory Level

The efficient anode is necessary for the upscaling of MFCs. The characteristics of the anode should possess good electrical conductivity, stronger adhesion towards microorganisms, larger surface area, excellent stability towards severe weathering condition/continuous operation and low-cost materials. The development of low-cost anodic materials would tremendously reduce the construction investment of the MFCs. Subsequently, the carbon cloth or nitrogen doped gra phene aerogel that are low-cost carbon based materials can be used.

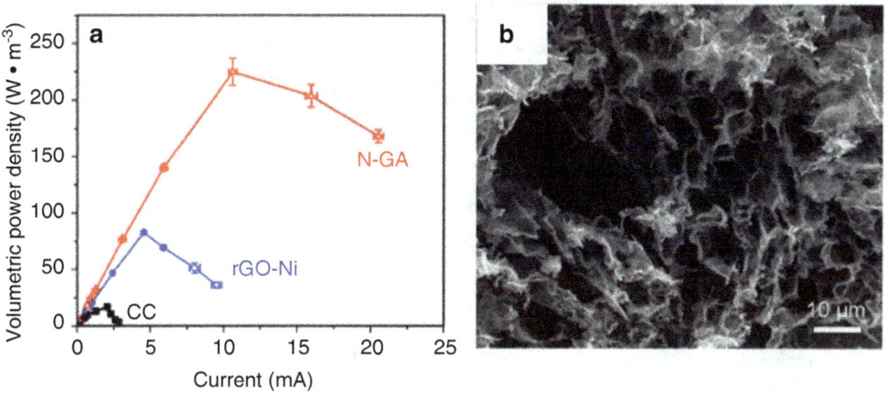

Fig. 13.11 (**a**) Volumetric power density, (**b**) SEM image of N-GA. (Credit: Yang et al. 2016b)

The graphene oxide doped with nitrogen (N-GA) has been successfully adopted as the anodic material for the development of microfluidic MFC (Fig. 13.11). The microfluidic MFCs are useful for the design of biosensor and micropower generators. The N-GA containing MFCs generated power density of 1181.4 ± 135.6 Wm^{-3} in the continuous mode and 690.2 ± 62.3 Wm^{-3} in batch mode process (Yang et al. 2016a). The presence of nitrogen doping on a low-cost material like carbon considerably reduces the cost of the MFC construction. Further, the above report added that the nitrogen doping acts as catalytic sits that enhanced the power density by oxygen reduction rate (ORR). The nitrogen-doped graphene porous aerogel anode electrode showed a power density of 225 ± 12 W m^{-3} using the dual-chamber MFC (Yang et al. 2016b). According to the report, the presence of nitrogen doping tremendously reduced the charge transport resistance and enhanced the power density of the MFCs. The porous nature of the aerogel electrode design can enhance the adhesion capacity and facilitate the cultivation of the bacterial communities. Based on the report, the generation of the power density was found to be higher for the laboratory (chamber capacity 25 mL)-level investigation. As per report, the N-GAMicrobial fuel cells (MFCs): was obtained by the reaction of ammonium hydroxide with acid-treated graphene oxide at 180 °C under hydrothermal conditions (Yang et al. 2016b).

Hanyu et al. (2013) developed the 3D reduced graphene oxide deposited Ni foam (3D rGO-Ni foam) as an anodic material for the construction of MFC (Fig. 13.3). The anode material has been obtained by the controlled deposition of graphene oxide over the Ni foam using reduction process. The power density up to 661 W m^{-3} was achieved using 3D rGO-Ni foam as anode under batch mode process. As per report, the power density achieved using the above electrode was found to be higher as compared to the anode materials derived from carbon-based materials such as carbon felt, carbon cloth and carbon paper. The uniform porous nature of Ni foam provided effective diffusion of microorganism, and more surface area offered space for the microorganism to colonise to the large extent. The MFC has been worked

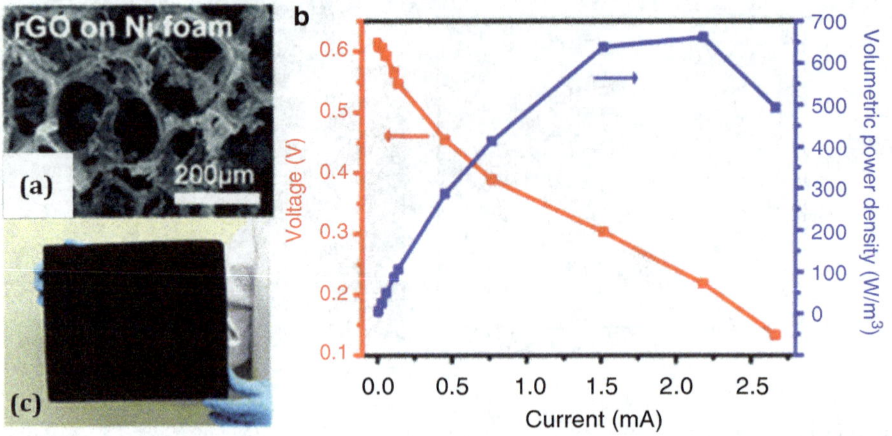

Fig. 13.12 (**a**) SEM picture of rGO on Ni foam, (**b**) polarisation and power curves collected for a MFC device with rGO-5-Ni anode. (**c**) Digital picture of a 25 cm X 20 cm rGO-Ni foam. (Credit: Hanyu et al. 2013)

more efficiently under bat-mode process using the pure strain of *Shewanella oneidensis MR-1* bacterial culture (Fig. 13.12).

The use of conducting polymer-coated carbon felt as an anode material for the evaluation of MFCs has been investigated by Chao et al. (2011). The two-chamber MFC has been constructed using carbon felt anode material coated with polyaniline (PANI) and polyaniline-co-*o*-aminophenol (PAOA). The PANI-coated anode containing MFC showed a power density of 27.4 mWm^{-2}, and the cell was found to effectively work in the presence of *Hippea maritima* bacterial culture, whereas the PAOA-coated anode material showed a power density value of 23.8 mWm^{-2} under *Clostridiales* bacterial strain. Further, the investigation stated that the power density produced by MFCs containing surface-modified anode is higher as compared to the pure counterpart. The nanographene sheets doped with nitrogen using plasma-enhanced chemical vapour deposition (CVD) have been utilised as an anode for the construction of the MFCs (Joseph Kirubakaran et al. 2015). The N-doped nanographene sheets showed porous and cross-linked framework structure as observed from electron microscope techniques. The MFC was constructed using the above as the biocompatible anode, glucose as energy source and *Escherichia coli* as redox biocatalyst. According to the report, the MFC has capacity of producing a power density of 1008 mW m^{-2}.

13.11 Future Perspectives

Based on the numerous reports, the many types of carbon electrodes have placed an ambitious focus for the development of novel low-cost electrodes for the MFC applications. The use of carbon materials would be cost-effective replacement for the expensive metal-based electrodes. Apart from the economical perspective, the carbon-based materials are biocompatible, and thus it can be suitable for wide range of microorganisms. On other hand, they can also be obtained from carbon biomass waste using suitable process. The carbon materials such as carbon cloth, carbon felt, carbon foam, graphene sheets, microporous and mesoporous carbon materials have tremendous opportunities for the fabrication of microbial fuel cells at the pilot-scale level or commercial-scale level. In addition to the electrode fabrication, the appropriate selection of the chamber design and size should be considered based on the location, type of waste to be treated, mode of operation (whether continuous or batch mode process) and source of energy.

References

Abhilasha Singh M (2016) Novel microbial fuel cell design to operate with different wastewaters simultaneously. J Environ Sci 42:105–111

Ademola A, Vijaya R, Boris T (2017) Carbon source and energy optimization in solid analyte microbial fuel cells. J Power Sources 356:324–330

Ana F, Liliana G, Joao Peixoto M, Luciana P, Antonio Brito G, Gilberto M (2017) Resource recovery in the dairy industry: bioelectricity production using a continuous microbial fuel cell. J Clean Prod 140:971–976

Anam A, Abdul Aziz AR, Wan Mohd Ashri WD (2014) Recent advances, challenges and prospects of *in situ* production hydrogen peroxide for textile wastewater treatment in microbial fuel cells. J Chem Technol Biotechnol 89:1466–1480

Atieh Z, Mostafa R, Farid T (2015) Sediment microbial fuel cells as a new source of renewable and sustainable energy: present status and future prospects. RSC Adv 5:94171–94183

Baikun L, Karl S, Micheal C (2011) Electricity generation from anaerobic wastewater treatment in microbial fuel cells. NYSEDRA Agreement 11095. Water Environment Research Foundation (WERF), USA

Booki M, Bruce Logan E (2004) Continuous electricity generation from domestic wastewater and organic substrates in a flat plate microbial fuel cell. Environ Sci Technol 38:5809–5814

Bruce Logan E (2010) Scaling up microbial fuel cells and other bioelectrochemical systems. Appl Microbiol Biotechnol 85:1665–1671

Bruce Logan E, Bert H, Rene R, Uwe S, Jurg K, Stefano F, Willy V, Koneel R (2006) Microbial fuel cells: methodology and technology. Environ Sci Technol 40:5181–5192

Byung-Min A, Yoon H, Hubdar-Ali M, Joo-Yang P (2016) Scaled-up dual anode/cathode microbial fuel cell stack for actual ethanolamine wastewater treatment. Bioresour Technol 210:68–73

Carla Jones S, Stephen MP (2016) Our energy future: introduction to renewable energy and biofuels. University of California Press, USA (ISBN: 9780520964280)

Carlo S, Catia A, Benjamin E, Ioannis I (2017) Microbial fuel cells: from fundamentals to applications. A review. J Power Sources 256:225–244

Castro C (2014) The green latrine: development of a large scale microbial fuel cell for the treatment of human waste in developing areas. Environmental & Water Resources Engineering Masters Project, 68. University of Massachusetts-Amherst, USA

Chao L, Libin Z, Lili D, Hongqiang R, Han C (2011) Effect of conductive polymers coated anode on the performance of microbial fuel cells (MFCs) and its biodiversity analysis. Biosens Bioelectron 26:4169–4176

Cohen B (1931) The bacterial culture as an electrical half-cell. J Bacteriol 21:18

Corbella C, Puigagut J (2015) Microbial fuel cells implemented in constructed wetlands: fundamentals, current research and future perspectives. Contrib Sci 11:113–120

Cotteril SE, Dolfing J, Jones C, Curstis TP, Heidrich ES (2017) Low temperature domestic wastewater treatment in a microbial electrolysis cell with 1 m^2 anodes: towards system scale-up. Fuel Cells 17:584–592

Couler J, Padgett GA, Cameron PJ, Preuss K, Titirici MM, Ieropoulos I, Di Lorenzo M (2016) Towards effective small scale microbial fuel cells for energy generation from urine. Electrochim Acta 192:89–98

Cusick RD, Bryan B, Parker DS, Merrill MD, Mehanna M, Kiely PD, Liu G, Logan BE (2011) Performance of a pilot-scale continuous flow microbial electrolysis cell fed with winery wastewater. Appl Microbiol Biotechnol 89:2053–2063

Dan W, Liuyan Y, Lin G, Qiankun C, Li L, Xu C, Xin W, Li Yun G, Aijun M (2015) Potential of novel wastewater treatment system featuring microbial fuel cell to generate electricity and remove pollutants. Ecol Eng 84:624–631

Daniele C, Daniele M, Arianna C, Andrea GC (2017) Agro-food industry wastewater treatment with microbial fuel cells: energetic recovery issues. Int J Hydrog Energy Corrected Proof 43:500. https://doi.org/10.1016/j.ijhydene.2017.07.231

Daqian J, Baikun L (2009) Granular activated carbon single chamber microbial fuel cells (GAC-SCMFC): a design suitable large-scale wastewater treatment processes. Biochem Eng J 47:31–37

Daqian J, Micheal C, Elizabeth T, Karl S, Joy M, Boxon H, Steve S, Dustin R, Baikun L (2011) A pilot scale study on utilizing multi-anode/cathode microbial fuel cells (MAC-MFCs) to enhance the power production in wastewater treatment. Int J Hydrog Energy 36(1):876–884

Dimou O, Andresen J, Feodorovich V, Goryanin I, Harper A, Simpson D (2014) Optimisation of scale-up of microbial fuel cell for sustainable wastewater treatment with positive net energy generation. New Biotechnol 31(supplement):S213

Doherty L, Zhao Y, Zhao X, Wang W (2014) Nutrient and organics removal from swine slurry with simultaneous electricity generation in an alum sludge-based constructed wetland incorporating microbial fuel cell technology. Chem Eng J 266:74–81. https://doi.org/10.1016/j.cej.2014.12.063

Elizabeth EH, Peter N, Randall P (2011) Plant biomass conversion. Wiley-Blackwell publication, USA. ISBN: 978–0–8138-1694-4

Elizabeth SH, Dolfing J, Scott K, Edwards SR, Jones C, Curtis TP (2013) Production of hydrogen from domestic wastewater in a pilot-scale microbial electrolysis cell. Appl Microbiol Biotechnol 97:6979–6989

Elizabeth HS, Stephen ER, Jan D, Cotterill SE, Curtis PT (2014) Performance of a pilot scale microbial electrolysis cell fed on domestic wastewater at ambient temperatures for a 12 month period. Bioresour Technol 173:87–95

Ellen D, Travis G, Andrew W, Chandramouli RM, Ali E (2016) Wastewater treatment with microbial fuel cells: a design and feasibility study for scale-up in microbreweries. J Bioprocess Biotech 6:267–272

Escapa A, Mateos R, Martinez EJ, Blanes J (2016) Microbial electrolysis cells: an emerging technology for wastewater treatment and energy recovery. From laboratory to pilot plant and beyond. Renew Sust Energ Rev 55:942–956

Frank D, Higson SPJ (2007) Biofuel cells-recent advances and applications. Biosens Bioelectron 22:1224–1235

Ganesh DS, Rijuta GS, Muhammad KS, Guangyin Z, Gopalakrishnan K, Han-Seung S, Young-Gyun C, Sang-Hyoun K (2017) A comprehensive overview on electro-active biofilms, role of exo-electrogens and their microbial niches in microbial fuel cells (MFCs). Chemosphere 178:534–547

Gude VG (2016a) Chapter 8: microbial fuel cells for wastewater treatment and energy generation. In: Scott K, Yu EH (eds) Microbial electrochemical and fuelcells. Fundamentals and applications, vol 2016. Elsevier/Woodhead Publishing, UK. pp 247–285

Gude VG (2016b) Wastewater treatment in microbial fuel cells – an overview. J Clean Prod 122:287–307

Hai-Liang S, Ying Z, Jie L (2015) Electron transfer mechanisms, characteristics and applications of biological cathode microbial fuel cells-a mini review. Arab J ChemIn press. https://doi.org/10.1016/j.arabjc.2015.01.008

Haishu S, Shengjun X, Guoqiang Z, Xuliang Z (2016) Performance and recent improvement in microbial fuel cells for simultaneous carbon and nitrogen removal: a review. J Environ Sci 39:242–248

Hanyu W, Gongming W, Yichuan L, Fang Q, Yong S, Shaowei C, Yexiang T, Yat L (2013) High power density microbial fuel cell with flexible 3D grphene –nickel foam as anode. Nanoscale 5:10283–10290

Haugen F, Bakke R, Lie B, Hovland J, Vasdal K (2015) Optimal design and operation of a UASB reactor for dairy cattle manure. Comput Electron Agric 111:203–213

Heinz H, Daniel H, Edith N, Manfred L, Patrick S, Karl-Georg S, Sylvia GH, Marc W (2016) An integrated 45L pilot microbial fuel cell system at a full-scale wastewater treatment plant. Bioresour Technonol 218:115–122

Heming W, Zhiyong JR (2013) A comprehensive review of microbial electrochemical systems as a platform technology. Biotechnol Adv 31:1796–1807

Henan L, Yan T, Youpeng Q, Ye Q, Jia L, Yuji F (2017) A pilot-scale benthic microbial electrochemical system (BMES) for enhanced organic removal in sediment restoration. Sci Rep 7:39802

Hina R, Nasir A, Farhat A, Iftikhar Hussain B, Abdullah Y, Shafaqat A, Tahira Y, Muhammad R (2015) Strat-up of UASB reactors treating municipal wastewater and effect of temperature/sludge age and hydraulic retention time (HRT) on its performance. Arab J Chem 8:780–786

Iwona G, John G, Chris M, Ioannis A (2016) Electricity and disinfectant production from wastewater: microbial fuel cell as a self-powered electrolyser. Sci Rep 6:25571

Jayesh Sonawane M, Samuel Adeloju B, Prkash Ghosh C (2017) Landfill leachate: a promising substrate for microbial fuel cells. Int J Hydrog Energy 42(37):23794–23798

Jeffrey JF, Miriam R, Larges Angenent T (2010) Electric power generation from municipal, food, and animal wastewaters using microbial fuel cells. Electroanalysis 22(7–8):832–843

Joseph Kirubakaran C, Santhakumar K, Gnanakumar G, Senthilkumar N, Jae-Hyung J (2015) Nitrogen doped graphene sheets as metal free anode catalysts for the high performance microbial fuel cells. Int J Hydrog Energy 40:13061–13070

Jung Rae K, Jorge R, Freda Hawks R, Richard MD, Alan Guwy J, Giuliano CP (2011) Increasing power recovery and organic removal efficiency using extended longitudinal tubular microbial fuel cell (MFC) reactors. Energy Environ Sci 4:459–465

Kelly CW, John DC (2009) Microbial fuel cells: plug-in and power-on microbiology. Microbe 4(6):281–287

Kim BH, Park DH, Shin PK, Chang IS, Kim HJ (1999a) Mediator-less biofuel cell. US Patent 5976719

Kim HJ, Hyun MS, Chang IS, Kim BH (1999b) A microbial fuel cell type lactate biosensor using a metal-reducing bacterium, Shewanella putrefaciens. J Microbiol Biotechnol 9:365–367

Kim IS, Kyu-Jung C, Mi-Jin C, Willy V (2008) Microbial fuel cells: recent advances, bacterial communities and application beyond electricity generation. Environ Eng Res 13:51–65

Kiran TM, Praneet P (2015) Performance of pilot scale microbial fuel cells treating wastewater with associated bioenergy production in the Caribbean context. Int J Energy Environ Eng 6:213–220

Krishna PK, Scott K (2010) Electricity generation from the treatment of wastewater with a hybrid up-flow microbial fuel cell. Biotechnol Bioeng 107:52–58

Marjolein H, David PBTBS, Hubertus VMH, Cees JNB (2012) The flat-plate plant microbial fuel cell: the effect of a new design on internal resistances. Biotechnol Biofuels 5:1–10

Micheal GW, Thomas AT (2013) Review of microbial fuel cells for wastewater treatment: large-scale applications, future needs and current research gaps (ES-Fuelcell 2013–18185). Proceedings of the ASME 2013, 7th international conference on energy sustainability & 11th fuel cell science, engineering and technology conference. ES Fuelcell 2013, 14–19th July 2013, Minneapolis, MN, USA

Min W, Zhenhua Y, Baoxu H, Jinsheng Z, Renmin L (2013) Electricity generation by microbial fuel cells fuelled with *Enteromorpha Prolifera* hydrolysis. Int J Electrochem Sci 8:2104–2111

Minghua Z, Hongyu W, Daniel JH, Jingyue G (2013) Recent advances in microbial fuel cells (MFCs) and microbial electrolysis cells (MECs) for wastewater treatment, bioenergy and bioproducts. J Chem Technol Biotechnol 88:508–518

Mostafa R, Arash A, Soheil D, Alizera Z, Sang-Eun O (2015) Microbial fuel cell as new technology for bioelectricity generation: a review. Alex Eng J 54:745–756

Nastro RA (2014) Microbial fuel cells in waste treatment: recent advances. Int J Perform Eng 10(4):367–376

Oliveira VB, Simeos M, Melo LF, Pinto AMFR (2013) Overview on the developments of microbial fuel cells. Biochem Eng J 73:53–64

Pablo L, Andrew S, John G, Ioannis I (2013) The first self-sustainable microbial fuel fuel cell stack. Phys Chem Chem Phys 15:2278–2281

Padma S, Dirk Hays B (2012) Future fuel technologies. National Petroleum Council (NPC) Study. Chapter 13: Microbial fuel cells, 1–19

Pant D, Van Bogaert G, Diels L, Vanbroekhoven K (2010) A review of the substrates used in microbial fuel cells (MFCs) for sustainable energy production. Bioresour Technol 101:1533–1543

Parawira W, Kudita I, Nyandoroh MG, Zvauya R (2005) A study of industrial anaerobic treatment of opaque beer brewery wastewater in a tropical climate using full-scale UASB reactor seeded with activated sludge. Process Biochem 40:593–599

Peter A, Korneel R, Hai The P, Nico B, Willy V (2006) Continuous electricity generation at high voltages and currents using stacked microbial fuel cells. Environ Sci Technol 40:3388–3394

Prashant P, Vikas NS, Rajendra LD, Sharad PK, Sunil AP, Deepak P (2016) Recent advances in the use of different substrates in microbial fuel cells toward wastewater treatment and simultaneous energy recovery. Appl Energy 168:706–723

Quanguo Z, Jiangjun H, Duu-Jong L (2016) Microbial fuel cells as pollutant treatment units: research updates. Bioresour Technol 217:121–128

Rachnarin N, Roshan R (2017) Plant microbial fuel cells: a promising biosystems engineering. Renew Sust Energ Rev 76:81–89

Ravinder K, Pradeep K (2017) Future microbial applications for bioenergy production: a perspective. Front Microbiol 8:450

Ravinder K, Lakhveer S, Zularisam AW (2016) Exoelectrogens: recent advances in molecular drivers involved in extracellular electron transfer and strategies used to improve it for microbial fuel cell applications. Renew Sust Energ Rev 56:1322–1336

Ravinder K, Lakhveer S, Zularisam AW (2017a) Microbial fuel cells: types and applications. In: Singh L, Kalia VC (eds) Waste biomass management-a holistic approach. Springer International Publishing, Cham, pp 367–384

Ravinder K, Lakhveer S, Zularisam AW, Faisal Hai I (2017b) Microbial fuel cell is emerging as a versatile technology: review on its possible applications, challenges and strategies to improve the performances. Int J Energy Res 42:369. https://doi.org/10.1002/er.3780

Rene Rozendal A, Hubertus Hamelers VM, Korneal R, Jurg K, Cees Buisman JN (2008) Towards practical implementation of bioelectrochemical wastewater treatment. Trends Biotechnol 26(8):450–459

Ritu S, Sanjeev K (2017) Green technologies and environmental sustainability. Springer International Publishing, Switzerland. ISBN: 978-3-319-50654-8

Salargarcia MJ, Ortiz-Martinez VM, Baicha Z, De Los R, Hernandez-Fernandez FJ (2016) Scaled-up continuous up-flow microbial fuel cell based on novel-embedded ionic-liquid type membrane cathode assembly. Energy 101:113–120

Shijia W, Hui L, Xuechen Z, Peng L, Xiaoyuan Z, Yong J, Xia H (2016) A novel pilot-scale stacked microbial fuel cell for efficient electricity generation and wastewater treatment. Water Res 98:396–403

Sona K, Madjid M, Khalid F (2015) Passive air breathing flat-plat microbial fuel cell operation. J Chem Technol Biotechnol 90:468–475

Sona K, Madjid M, Khalid F (2016) A systematic study of separators in air-breathing flat-plate microbial fuel cells-part 1: structure, properties, and performance correlations. Energies 9(78):1–10. https://doi.org/10.3390/en9020078

Syed Zaghum A, Mohd R, Norli I, Muhammad IS (2017) A review on sediment microbial fuel cells as a new source of sustainable energy and heavy metal remediation: mechanisms and future perspectives. Int J Energy Res 41:1242–1264

Takahasi S, Miyahara M, Kouzuma A, Watanabe K (2016) Electricity generation from rice bran in microbial fuel cells. Bioresour Bioprocess 3:50

Tonia T, Giorgia L (2017) Energy sustainability of microbial fuel cell (MFC): a case study. J Power Sources 356:438–447

Valeria A, Daniyal A, Adriano S, Valentina M, Caterina A, Marzia Q (2017) Electrochemical analysis of microbial fuel cells based on enriched biofilm communities from fresh water sediment. Electrochim Acta 237:133–143

Valesquez-Orta SB, Werner D, Varia JC, Mgana S (2017) Microbial fuel cells for inexpensive continuous in-situ monitoring of ground water quality. Water Res 117:9–17

Venkata Mohan S, Velvizhi G, Annie JM, Srikanth S (2014) Microbial fuel cells: critical factors regulating bio-catalysed electrochemical processes and recent advancements. Renew Sust Energ Rev 40:779–797

Villasenor J, Capilla P, Rodrigo MA, Canizares P, Fernandez FJ (2013) Operation of a horizontal sub-surface flow constructed wetland-microbial fuel cell treating wastewater under different organic loading rate. Water Res 47:6731–6738

Vinay S, Kundu PP (2010) Biocatalysis in microbial fuel cells. Enzyme Microb Technol 47(5):179–188

Wei Y, Jun L, Dingding Y, Liang Z, Xun Z, Qiang L (2016) A hybrid microbial fuel cell stack based on single and double chamber microbial fuel cells for self-sustaining pH control. J Power Sources 306:685–691

Wen-Wei L, Guo-Ping S, Han-Qing Y (2013) Chapter 14 - Electricity generation from food industry wastewater using microbial fuel cell technology. In: Kosseva MR, Webb C (eds) Food industry wastes. Academic Press, San Diego, pp 249–261 ISBN 9780123919212

Wu S, Kuschk P, Brix H, Vymazal J, Dong R (2014) Development of constructed wetlands in performance intensifications for wastewater treatment: a nitrogen and organic matter targeted review. Water Res 57:40–55

Xiaonan Z, Laura P, John MA (2011) Sustainable energy from dairy form waste using a microbial fuel cell (MFC). Bioenergy Technology (BE), World Renewable Energy Congress 2011, 8–13 May 2011, Linkoping, Sweden

Xuyun W, Xuehai Y, Qinglie G (2015) Production of electricity during wastewater treatment using fluidized bed microbial fuel cells. Appl Mech Mater 713-715:2807–2810

Yang Y, Tianyu L, Qiang L, Dingding Y, Xun Z, Jun L, Pengging Z, Yi P, Chaowei C, Yat L (2016a) A three dimensional nitrogen-doped graphene aerogel-activated carbon composite catalyst that enables low-cost microfluidic microbial fuel cells with superior performance. J Mater Chem A 41:15913–15919

Yang Y, Tianyu L, Xun Z, Feng Z, Dingding Y, Qiang L, Yat L (2016b) Boosting power density of microbial fuel cells with 3D nitrogen-doped graphene aerogel electrode. Adv Sci 3:1600097. https://doi.org/10.1002/advs.201600097

Yi-Chi Z, Zhao-Hong J, Ying L (2015) Application of electrochemically active bacteria as anodic biocatalyst in microbial fuel cells. Chin J Anal Chem 43(1):155–163

Yongwon J, Kyoungmin K, Hyung JK, Sunghyun K (2013) Construction and operation of a scaled-up microbial fuel cell. Bull Kor Chem Soc 34(1):317–320

Yu J, Seon J, Park Y, Cho S, Lee T (2012) Electricity generation and microbial community in a submerged-exchangeable microbial fuel cell system for low-strength domestic wastewater treatment. Bioresour Technol 117:172–179

Yue D, Youpeng Q, Weihua H, Yue D, Jia L, Xiaoyu H, Yujie F (2015) A 90-litre stackable baffled microbial fuel cell for brewery wastewater treatment based on energy self-sufficient mode. Bioresour Technol 195:66–72

Yujie F, Xin W, Bruce Logan E, He L (2008) Brewery wastewater treatment using air-cathode microbial fuel cells. Appl Microbiol Biotechnol 78:873–880

Yujie F, Weihua H, Jia L, Xin W, Youpeng Q, Nanqui R (2014) A horizontal plug flow and stackable pilot microbial fuel cell for municipal wastewater treatment. Bioresour Technol 156:132–138

Zhao Y, Collum S, Phelan M, Goodbody J, Doherty L, Hu Y (2013) Preliminary investigation of constructed wetland incorporating microbial fuel cell: batch and continuous flow trials. Chem Eng J 229:364–370

Zheng G, Liao W, Fei Z, Zhen H (2015) Energy extraction from a large-scale microbial fuel cell system treating municipal wastewater. J Power Sources 297:260–264

Zhuwei D, Haoran L, Tingyue G (2007) A state of the art review on microbial fuel cells: a promising technology for wastewater treatment and bioenergy. Biotechnol Adv 25:464–482

Chapter 14
Future Perspectives on Cost-Effective Microbial Fuel Cells in Rural Areas

C. Nagendranatha Reddy, M. P. Sudhakar, Booki Min, and P. Shanmugam

14.1 Introduction

Microbial fuel cells (MFC) are bioelectrochemical devices that generate bioelectricity by utilizing organic substrates during the microorganism's metabolic process (Rabaey and Verstraete, 2005; Rahimnejad et al., 2015; Nagendranatha Reddy et al., 2015). The numbers of researchers working on MFC for various objectives are increasing worldwide. This depicts the ever-increasing focus on MFC technology due to its advancement in producing the green and sustainable energy. The need of every individual for renewable, clean, sustainable, carbon neutral, and affordable energy by utilizing minimal resources is a desirable goal in developing countries. This process of utilizing organic carbon by bacteria has revolutionized the way of producing bioenergy, which acts as a sustainable alternative to depleting fossil fuels (Santoro et al., 2017). The anodic bacteria (anode respiring bacteria, ARB or exoelectrogens or electrochemically active bacteria, EAB) generates redox equivalents (H^+ and e^-) along with acid intermediates via., substrate oxidation under anaerobic conditions while the reduction process occurs in cathode chamber under aerobic conditions (Schroder, 2007; Logan, 2009). The MFCs can unconventionally treat waste/wastewater or can be integrated to other wastewater treatment processes depending upon the requirement. Hence, MFCs have emerged as a promising and

The original version of this chapter was revised. A correction to this chapter can be found at https://doi.org/10.1007/978-3-319-92904-0_15

C. N. Reddy · B. Min
Department of Environmental Science and Engineering, Kyung Hee University, Yongin-si, 1732, Gyeonggi-do, Republic of Korea
e-mail: bkmin@khu.ac.kr

M. P. Sudhakar (✉) · P. Shanmugam
Department of Environmental Science and Engineering, CSIR-Central Leather Research Institute, Chennai 600020, India

V. Sivasankar et al. (eds.), *Microbial Fuel Cell Technology for Bioelectricity*,
https://doi.org/10.1007/978-3-319-92904-0_14

energy efficient technology for effluent treatment while recovering electrical energy from organic waste/wastewater (Venkata Mohan et al., 2014; Kondaveeti et al., 2018). Sustainable treatment and utilization of waste/wastewater are receiving intensive consideration due to the decrease in availability of freshwater, depletion of carbonaceous fuels with associated environmental pollution. The MFC technology stands unique as most of the conventional wastewater treatment processes cause environmental pollution (sludge disposal, generation of greenhouse gases, etc.) along with high energy consumption (Li et al., 2014). For this reason, there is a great interest in valorizing waste for generating value addition to the process, thereby encouraging the bioeconomy.

The energy policies of all countries have set up sustainable development goals in the energy field. The policies are targeted to diminish the consumption of fossil fuels and increase the renewable energy utilization in all the sectors (REN21 Report). But, rural areas in all parts of the world lack electricity to meet the daily requirements. Hence, in order to make it happen in rural areas, renewable energy generation technologies such as MFCs play a significant role. MFCs can act as one of alternate sources to achieve energy independence by utilizing the excess biomass for transportation and electricity needs. MFCs are also utilized on real field application as an integration to other treatment technologies. Due to low efficiency of MFC performance utilizing waste/wastewaters, various challenges are observed in order to overcome the limitations (Liu and Cheng, 2014; Ramadan and Purwono, 2017).

14.2 MFC and its Types (at Pilot Scale)

The biological process of generating bioelectricity by converting the biodegradable organic matter in the wastewater is called bioelectrogenesis. MFC has the ability to overcome the problems associated with conventional biological reactions by converting organic (chemical) energy to electrical energy through a series of oxidation and reduction reactions with concurrent waste remediation (Velvizhi and Venkata Mohan, 2017; Butti et al., 2016; Logan, 2009). The redox reactions in MFC are determined by the half-cell potentials of anode and cathode chambers. The difference between the cathodic (positive) and anodic (negative) potentials is measured as cell voltage/ electron motive force that drives the flow of electrons. The idea of generating potential difference by utilizing microorganisms was studied by Potter in 1911. Later, Cohen used bacterial half cells in series to produce electricity which led to the development of various concepts for further practical developments (Santoro et al., 2017). The general MFC consists of two chambers viz., anode and cathode. The anodic chamber consists of various mixed microorganisms to perform the substrate oxidation and generate redox equivalents, intermediates (VFAs) and CO_2 during the metabolic process. The electrons generated in the anode chamber are transferred to electrode (anode; electron acceptor) through various electron transfer mechanisms (Direct-Cytochrome proteins and conductive pili and mediated-Mediators) (Schroder, 2007). The electrode assembly acts as a solid electron acceptor enabling bacterial respiration on its surface. The efficient movement of electrons

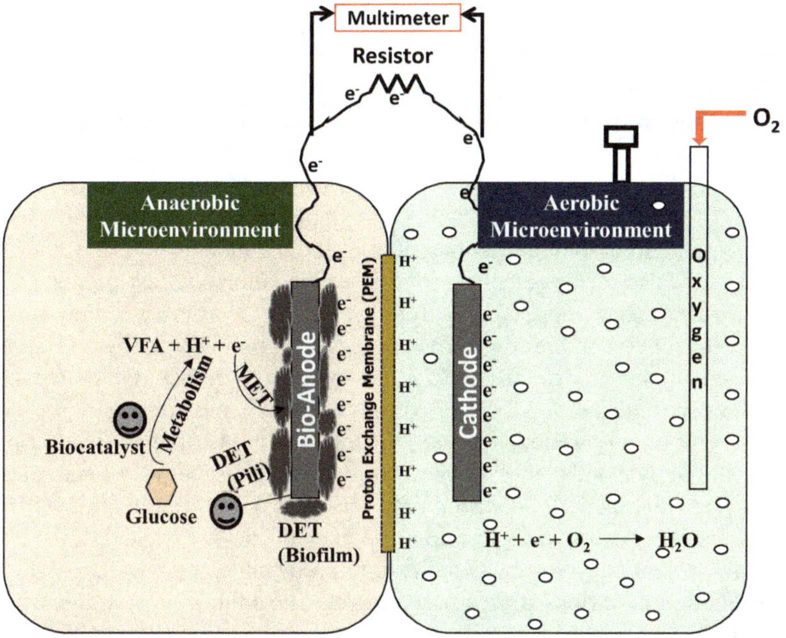

Fig. 14.1 Schematic depiction of MFC operation

on to the electrode surface without any losses is significant in the study of MFC towards commercialization. The H^+ pass through the membrane (either proton or cationic exchange membrane) to the cathode chamber. The movement of protons to cathode chamber creates a potential gradient that combines with electrons when the circuit is closed and terminal electron acceptor (TEA) to form water, which is the final and non-polluting product in the cathode chamber (Guo et al., 2012). This occurs when the oxygen is used as a TEA in the cathode chamber. Various TEAs have been studied during MFC operation to increase the overall efficiency of the MFC system (Ucar et al., 2017). The schematic of the general mechanism of bio-electrogenesis in MFC is illustrated in Fig 14.1.

The possible reactions occurring in the anode chamber during glucose fermentation shows that around one third of the electrons can be theoretically used to generate current while the remaining two thirds remain in the byproducts such as VFAs. This is because the enzyme hydrogenases that generally use electrons to produce H_2 gas are often situated in the outer surfaces of membrane which are easily accessible to electron mediators or to the electrode directly. Different substrates ranging from simple (acetate, glucose, etc.) to complex and real field wastewaters (pollutants, etc.) have been evaluated with MFC (Pant et al., 2010). The redox reactions that occurs on the working and counter electrode chambers respectively are represented as:

$$C_6H_{12}O_6 + 6H_2O \rightarrow 6CO_2 + 24H^+ + 24e^- \quad (1)$$

$$4e^- + 4H^+ + O_2 \rightarrow 2H_2O \qquad (2)$$

The other operational parameters that determine the performance of MFC are the organic feed conversion rate, membrane activity, losses and overpotentials at the anode and cathode, etc. The acid intermediates generated in the anode chamber during the metabolism acts as a rate limiting step making the process cease at lower pH conditions. The anode potentials will have significant influence on the bacterial metabolism and govern the redox potential of the mediator. The electron transport system in MFC use various electron shuttles or mediators viz., ubiquinone, NADH dehydrogenase, coenzyme Q, Fe/S proteins, and cytochrome etc. that aid in the transport of electrons and protons. The processes involving oxidative phosphorylation showed higher energy efficiencies up to 65% in MFCs. Other factors that impact the performance of MFC are wastewater (type and composition), biocatalyst, bioreactor (configuration, size etc.), membrane, mode of operation (suspended or attached), anolyte and catholyte, mediators used, electrodes (size, materials, pretreatment, porosity, activity, distance between electrodes, etc.), microenvironment (aerobic, anaerobic or anoxic) etc. (Kondaveeti et al., 2017; Park and Zeikus, 2003).

The MFC technology has the benefits of minimal adverse environmental impact, easy installation at various places based on the substrate source, stable treatment performance, high energy efficiency, neutral-energy operation, generation of value added products based on the requirement, little resource consumption, enhanced quality of effluent for water reuse requirements, low operational costs and minimal maintenance, good social equity etc. making it the potential candidate for realizing the sustainability in wastewater treatment. Various bioproducts viz., H_2, CH_4, alcohols, platform chemicals etc. are also generated in the MFC when the relevant biocatalyst and little potentials are provided in the working electrode chamber depending upon the objective of the study (Badwal et al., 2014).

Based on the application and objective of the study, many types and configurations of MFCs have been proposed and developed through the years to amplify the power generation, treat particular wastewaters, source of waste etc. The conventional MFC design and structure were modified according to the present state requirement (Logan et al., 2015; Zhang et al., 2017). Broadly, MFCs can be classified into single and double chamber fuel cells. The single chambered MFC may be operated in either membrane or membrane less conditions. Various configurations of single chambered MFC are up flow, air cathode, concentric tubular etc. while dual chambered consists of aerated cathode (simple H shaped), up flow, cuboid etc. (Zhang et al., 2017). The configurations of MFC are detailed elsewhere. Now, this chapter focusses on applications of MFCs in rural areas for practical applicability.

Few applications of MFC are benthic or sediment, submersible, plant, stacked and multielectrode, photosynthetic MFCs and other hybrid and integrated MFCs (Fig 14.2).

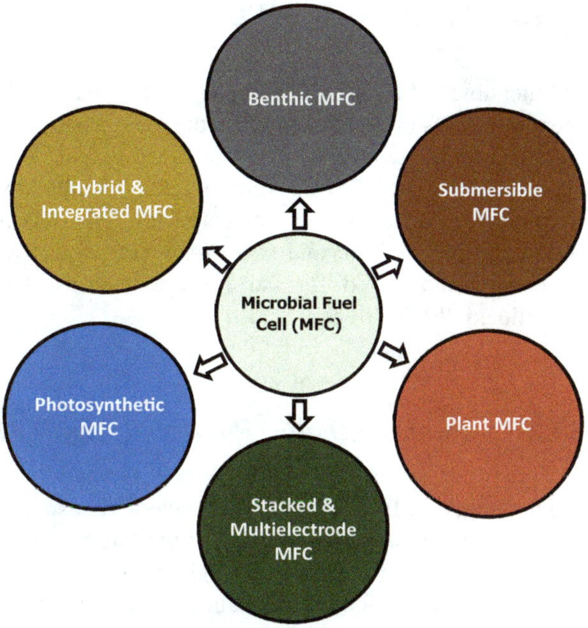

Fig. 14.2 Different applications of MFC types

14.2.1 Benthic MFC

Benthic or Sediment MFCs are the bioelectrochemical systems that follow the same principle of MFC utilizing the naturally occurring potential difference amid the anoxic sediments and oxic seawater to generate bioelectricity. The sediments possessing the intricate community of microorganisms including EAB required for MFC and nutrient rich media and other nutrients that would have accumulated over millions of year's acts as inoculum and substrate sources respectively. This abundant availability of potential biocatalyst, substrates and nutrients makes benthic MFC a promising and sustainable power source providing consistent power supply for longer periods. Though the idea of benthic MFC was presented by Reimers et al., (2001), it was first demonstrated by Tender et al., (2008), where a meteorological buoy was powered by two prototype benthic MFCs. Gong et al., (2011) powered an acoustic modem interfaced with an oceanographic sensor by benthic MFC with an average power density (PD) of 44 mW/m^2 for over 50 days. Another chamber based benthic MFC with incorporated semi closed and suspended anode, showed power output of 3.8 W/m^3 with reduced system foot print (Nielsen et al., 2007). A pilot scale benthic microbial electrochemical system (BMES) using three-dimensional anode with honeycomb structure showed maximum PD of 81 mW/m^2 (Li et al., 2017).

14.2.2 Submersible MFC

Submersible MFCs are the compact MFCs that can be integrated with other treatment processes in order to enhance the overall activity viz., long term operation, and stability, complete utilization of substrate, enhanced product generation and recovery etc. Submersible MFCs were first demonstrated by Min and Angelidaki in 2008 to combine with an anaerobic bioreactor. The maximum PD and current density (CD) was 204 mW/m^2 and 595 mA/m^2 respectively. Several configurations were developed and tested for various applications (Cha et al., 2010; Zhang and Angelidaki, 2012; Xu et al., 2014).

14.2.3 Photosynthetic (Plant and Algal) MFC

Photosynthetic MFCs are the renewable and sustainable systems that convert sunlight energy into electrical energy within the metabolic reaction of MFCs. The process of obtaining energy from sunlight by biological means is called photosynthesis. This process is carried out by plants, algae and photosynthetic bacteria. This eliminates the extensive use of fossil fuels and environmental pollution by utilizing the CO_2 emissions which is the major cause of global warming effect. A plant MFC is a unique modification of MFC technology that exercises the plant-microbe relationship at the rhizosphere region of the plants. There is no need of supplying external substrates for plant MFC when compared to conventional MFCs. In the past decade, focus is mostly on effects of plant and microbes interactions along with suitable universal configuration for effective functioning of photosynthetic MFC. Plant MFC can be categorized into biocontrol and bioprocess structures. The biocontrol (plants) obtains external energy (sunlight) to generate voltage whereas the bioprocess (microbial population) structure utilize material resources at root exudates to produce voltage output. A wide range of microbial communities at the rhizodeposits plays a significant role in bioelectricity generation. Different plants such as *O. sativa*, *I.aquatica*, *G. maxima*, *C. indica*, *A. anomola*, *L. perenne*, *S. anglica* etc., are studied as option to function as plant MFC. An estimated potential electricity generation of 5800 kW/h ha^{-1} $year^{-1}$ with Reed Manna grass plant MFC in Europe was reported by Strik et al. (2011). Various combinations of photosynthetic bacteria at anode and/or cathode chambers were explained in detail by Rosenbaum et al. (2010). However, plant MFCs can generate bioelectricity that can be used for operating small scale electrical appliances. Plant MFCs have the inherent limitations of non-versatile and slow growth, food security, competition for arable land, requirement of chemicals like pesticides etc. But the photosynthetic microorganisms like bacteria overcome the limitations to certain extent but whereas algae has the inherent advantages of fast growth (doubling periods as short as 3.5 h), food security point of view (no competition with arable land and all-round the year harvesting), aqueous media growth, CO_2 fixation, efficient converters of solar power, uptake of nutrients (NPK), generation of O_2 (to act as TEA), circumventing the utilization of exogenous and unstable mediators and precious catalysts, high lipid content, no requirement for pesticides or herbicides, and production of value addition (biomass, proteins, biodiesel, biofertilizers, pigments, nutraceuticals etc.). Also, microalgae can aid in efficient removal of pollutants, nutrients uptake, etc., as the MFC alone cannot act as stand-alone process in terms of economic viability. Photosynthetic microbes can be used as a biocatalyst in either anode or cathode or both chambers with sludge or chemical cathodic catalyst in other chamber depending on the objective of the study (Rodrigo et al., 2009; Raman and Lan, 2012). Venkata Mohan et al., (2014), studied the synergistic association between bacterial fermentation and the oxygenic photosynthesis of mixed microalgae at anode and cathode chambers respectively in dual-chambered MFC. The experiments were conducted to evaluate the power generation along with wastewater (domestic wastewater) treatment in two seasons, spring and summer. The results depicted higher bioelectrogenic activity (57 mW/m^2) in spring in comparison to summer (1.1 mW/m^2) due to the higher dissolved oxygen (DO) levels by oxygenic photosynthetic activity of microalgae. Whereas, Hu et al., (2015) evaluated the air-lift-type microbial carbon capture cell (ALMCC) by using an air-lift type photo bioreactor as the cathode chamber for the first time and the anodic effluent was integrated to cathode chamber for further treatment by microalgae (Hu et al., 2015). With the inherent advantages, ALMCC system produced a maximum PD of 972.5 mW/m^3 and removed 69% of phosphorus, 71% of ammonium nitrogen and 87% of COD. Besides this, ALMCC also demonstrated higher lipid productivity and CO_2 fixation rate suggesting the overall energy of the ALMCC was comparatively higher to control systems. And in extension of this work, Hu et al., (2016) evaluated the ALMCC for different cathodic microorganisms (*C. vulgaris* and *Chlorella sp.*) under different light intensities. But the ALMCC system with *C. vulgaris* depicted higher lipid productivity of 21.75 mg/L/d, CO_2 fixation rate of 223.68 mg/L/d and PD of 558.22 mW/m^3 at 8.9 W/m^2 optimal light intensity depicting the influence of light intensity on the operation of microalgae in biocathode. Some studies showed the transfer of CO_2 produced in the anode to cathode chamber in order to be utilized by microalgae for generation of O_2 (Wang et al., 2010). This process of transferring the inorganic carbon to maintain carbon neutrality is called as microbial carbon capture (MCC) cells.

14.2.4 Stacked and Multi-electrode MFC

A stacked MFC is the number of individual cells connected either in series or parallel. In series and parallel connections, voltages and currents are the sum of all the individual cells respectively. In comparison, parallel connected stack MFCs depicts higher currents than series connected ones. In order to enhance the substrate utilization (COD removal) and obtain maximum bioelectrogenesis, parallel connection is preferred to series connection (Sleutels et al., 2012). To obtain higher power generation in a stacked MFC, numerous small MFC modules can be combined to form a larger stack rather than enhancing the volume of individual MFC. This is because of inactive reactor volume and increased volumetric ohmic resistance in the MFC having larger volume (Ieropoulos et al., 2008). Four different types of configurations of hydraulic flow and electrodes are (1) parallel electrode connections in series flow mode; (2) parallel electrode connections in parallel flow mode, (3) series electrode connections in parallel flow mode, and (4) series electrode connections in series flow mode (Choi and Ahn, 2013). The effective strategy of stacked MFC is still in its infancy due to lower PD and energy recovery when compared to conventional anaerobic processes (Rabaey and Verstraete, 2005). Stacked MFC fed with synthetic wastewaters achieved very low PDs of 11 W/m^3 and 5.6 W/m^3 with normal (20 L) and tubular MFC stack (1 L) respectively while plug flow stacked MFC (250 L) treating municipal wastewater showed PD of 0.47 W/m^3.The low performance of stacked MFC might be due to voltage reversal in individual cells, limited biomass content on the anode, inactive surface area on the cathode, increased ohmic, kinetic and transportation resistances etc. (Dekker et al., 2009; Feng et al., 2014; Feng et al., 2017). The concept of scaling up of MFC using multiple electrodes in a single and large MFC presents enhancement in the overall performance of the system (Venkata Mohan et al., 2014). A study with multiple anodes and single air cathode with separator electrode assembly showed maximum PD of 975 and 880 mW/m^2 respectively with fed batch and continuous mode operation (Ahn and Logan, 2012). Therefore, stacked MFC and multiple electrode operation should be made efficient and feasible by reducing the scalable costs and increased power generation and recovery. Various stack configurations, electrode materials and sizes to lower the resistances associated, operate in continuous mode with different wastewaters should be evaluated in detail.

14.2.5 Other Hybrid MFCs

Several other hybrid configurations have been developed to upsurge the efficiency and sustainability of MFC technology. The other advantages of hybrid MFCs are lower production and operating costs, integration with other processes etc. An upflow MFC system in which the influent is fed from the bottom was developed by merging the benefits of the UASB system with the requirements of a dual-chamber MFC. The continuous operation treating wastewater showed maximum PD of 170 mW/m2. The novel MFC configuration, M2FC was developed by combining ferric-based MFC with a ferrous-based fuel cell (FC). In this, the catholyte in the MFC is regenerated by FC system along with bioelectricity generation (2 W/m^2) (Eom et al., 2011). Malaeb et al., (2013) developed a new hybrid, air-biocathode microbial fuel cell-membrane bioreactor (MFC-MBR) system for wastewater treatment with simultaneous ultrafiltration for direct reclamation of produced water and generated maximum PD of 0.38 W/m^2. Other configurations include tubular MFCs (Rabaey et al., 2005), baffled air cathode MFC (Feng et al., 2010), etc.

Table 14.1 Energy recovery from various renewable substrates in MFCs

S. No	Renewable substrates	Electricity produced	Reference
1.	Microalgae (*Chlorella vulgaris*) + wastewater	0.20 mA/cm^2	Velasquez-Orta et al. (2009)
2.	Seaweeds (*Ulva lactuca*) + wastewater	0.25 mA/cm^2	Velasquez-Orta et al. (2009)
3.	*Chlorella pyrenoidosa*	6030 mWm^{-2}	Xu et al. (2014)
4.	*Chlorella sp.*	0.99 V	Das (2015)
5.	Microalgae + bacteria	1.7 W m^{-2}	Strik et al. (2008)
6.	Wastewater	6.0Wm^{-3}	Wang et al. (2012)
7.	Corn stover biomass	0.15 mA/cm^2	Zuo et al. (2006)
8.	Farm manure	0.004 mA/cm^2	Scott and Murano (2007)
9.	Landfill leachate	0.0004 mA/cm^2	Greenman et al. (2009)
10.	*Chlorella vulgaris*, *Scenedesmus* powder	8.67 ± 0.10 W/m^3	Cui et al. (2014)
11.	Urine	0.053 to 0.580 W m^3	Chouler et al. (2016)

Table 14.2 Substrates available in rural areas at large scale

Substrate (wastewater) in rural area	Current density (mA/cm^2) at maximum power	Reference
Brewery	0.2	Feng et al. (2008)
Domestic	0.06	Wang et al. (2009a)
Meat processing	0.115	Heilmann and Logan (2006)
Swine	0.015	Min et al. (2005)
Starch processing	0.09	Lu et al. (2009)
Sewage sludge	73 (mA/m^2)	Yuan et al. 2012

14.3 Cost-Effective Resources for MFC Technology

Global resources are enormous but proper utilization in cost-effective ways is merely very less. Especially in agricultural countries like India, China, Australia, Sri Lanka, Malaysia, Thailand, Indonesia, United States, Sweden, etc., the renewable substrates available and wastewater generated in rural areas in these countries is enormous, but very few countries are utilizing it for renewable energy generation.

Resources such as municipal waste/wastewater, agricultural residues, algae grown in natural waters, industrial effluents are available free of cost (Tables 14.1 and 14.2). Developing and high population countries like India and China can utilize these resources and produce renewable and green energy efficiently throughout the year to meet the energy demand (Chouler et al. 2017). A natural polymer (eggshell membrane), and a synthetic polymer (polydimethylsiloxane, PDMS) can be used as membrane to reduce the cost of MFC. The use of cheap and efficient membranes in MFCs can guide the researchers on type and configuration, electrode size etc. based on the material used.

14.4 Scaling Up for Commercialization

Based on low energy consumption, utilization of biodegradable organic waste as substrate and value added product (bioelectricity) generation, MFCs are considered as theoretical energy profitable technology. Some advantages available with MFC technology are no aeration required, no potential poised, no temperature maintenance, low sludge generation, zero discharge etc. when compared to other conventional wastewater technologies. MFC consumes only 0.076 kWh/kg-COD for reactor stirring and feeding when compared to 0.6 kWh/kg-COD for the activated sludge-based aerobic process (McCarty et al., 2011). Direct conversion of organic matter to bioelectricity with high conversion efficiency in MFC makes the process an energy saving as the biogas conversion to electricity has significant energy loss of >60% (Rittmann, 2008). The sludge production is low (about 0.1 g-VSS/g-COD) in comparison to activated sludge process (0.4–0.8 g-VSS/g-COD) thereby reducing the need for secondary treatment (Foley et al., 2010). MFCs also have good operational stability (pH and temperature fluctuations) and are capable of efficiently removing a large variety of contaminants from wastewaters.

In spite of many advancements with MFC operation, the commercialization and scale up in integration with real field wastewater treatment is currently not economically feasible due to various limitations. The major limitations associated with large-scale MFC operation are relatively low power generation, instability due to longer operational times, managing power output and high input costs.

14.4.1 Enhanced Power Generation

To obtain higher power from MFC, the size of the bioreactor should be enlarged or small units should be stacked for scale up. The increase in size of MFC leads to decrease in volumetric PD by 2-4 times when compared to lab scale MFCs (Liu et al., 2008; Clauwaert and Verstraete, 2009; Logan, 2010; Cheng and Logan, 2011). This lower power generation is due to upsurge in internal resistance and losses associated, non-maintenance of homogeneity in large reactors etc. Internal resistance of a large scale MFC can be reduced by modifying the electrode configuration, buffering solutions, reducing the overpotentials by increasing the surface area of electrodes, enhance the size of air cathode and achieve homogenous distribution by increasing the hydraulic retention time and placing a low cost separator to avoid oxygen diffusion to anode. With the limitations associated with large scale MFC reactors, the small MFC units are stacked as an alternative to MFC scaling up. In order to practically apply MFCs as an energy source, one can connect MFC units in parallel and series mode to generate higher currents and voltage respectively. But the cell voltage reversal and ionic short circuits can be overcome by using air cathodes of high parallelism in performance, maintaining similar catalytic activity of anode biofilms, and increasing the homogeneity of substrate distribution in different unit cells and separating the anolyte of the unit cells to prevent ionic short circuits.

14.4.2 Low Input Costs

The large scale MFC application needs high capital costs as the materials (outer structure, membrane and separator, large electrodes, buffer agents etc.) required for constructing large scale MFC are expensive. So, reducing the costs by using cheap and efficient materials may overcome the high costs associated. Electrodes made of graphite fiber brush and activated carbon granules with high specific surface area can function as effective anodes. But, the cathode materials, which accounts for highest capital cost for constructing a MFC, can be developed by low-cost materials such as stainless steel mesh and nickel foam, oxygen diffusion layer (tetrafluoroethylene; PTFE) and catalyst binders (polydimethylsiloxane; PDMS), low cost catalysts (Co/Fe/N/CNT, MnOx, Co-OMS-2, β-MnO_2, MnPc and CoTMPP) can be used (Zhou et al., 2011). The separator in large scale MFCs is a must for enhancing the power output and so, non-woven cloth, can be used as a low cost separator (Wang et al., 2013).

14.4.3 Long-term Stability

The large scale MFCs have the limitations of anodic biofilm and cathodic activity diminution, membrane fouling, clogging due to excessive biomass and solids in effluents (Min and Logan, 2004). The electroactivity of anodic biofilm can be maintained stably by understanding various factors like extracellular electron transfer behavior between electrode and biocatalyst, increasing biofilm conductivity, decreasing internal resistance and understanding the complex interactions with symbiotic association of microorganisms. The deterioration of cathode is dependent on oxygen reduction catalyst deactivation, biofilm growth on the surface, fouling and deformation of separator materials, salt precipitation and electrode (current collector) corrosion.

14.4.4 Power Output Management

Harvesting the electricity efficiently is one of the critical issues being faced during the large scale MFC operation. It is difficult to obtain a practical power load due to low levels of current and voltage generation. Hence, power management system (PMS) is required to be incorporated for making energy feasible to power electrical devices such as wireless biosensors etc. For efficient harvesting and usage of MFC energy, various electric-storage capacitors and a DC/DC voltage boost converters are used to excerpt energy from MFCs by a high frequency switching action and boost output voltage respectively. Optimization of converter electronic circuit and developing a maximum power point tracking technique could be promising strategies for this purpose (Tommasi and Lombardelli, 2017).

14.5 Integrated Centralized MFC System

The most immediate requirement for an MFC-based technology should be scalable technology in a cost-effective manner. The MFC development related issues must be addressed to advance technical understanding of the biological basis. Integrated centralized MFC operation in the rural community helps to develop stand-alone power generation throughout the year in a self-sustainable way. The following points aid in sustainable power generation through integrated centralized MFC system (Fig 14.3).

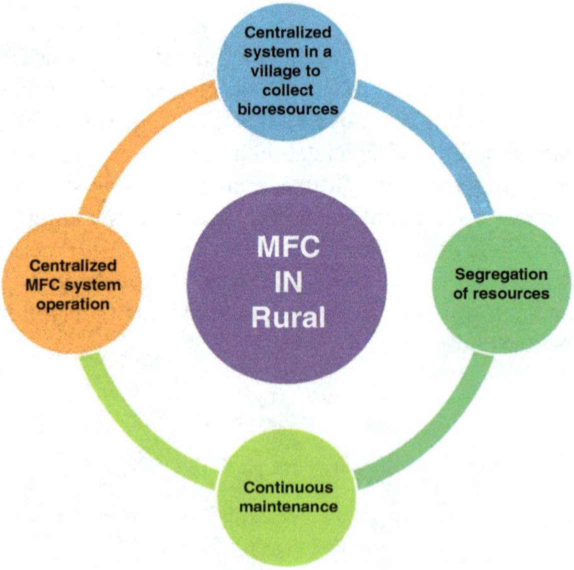

Fig. 14.3 Integrated bioresource collection for bioelectricity generation through MFC

(a) Centralized system in a village to collect bioresources

The primary activity of rural people to gain income in villages is agricultural work. Apart from this, bioresources (waste) collection would be an alternative for earning additional income. This activity not only benefits the individuals but also keeps the village hygienic and obtain bioresources for MFC operation.

(b) Segregation of resources

Resources (wastes) such as household food wastes, toilet flushing water, kitchen flushing water, vessels washing water of every street can be collected through proper pipe line which is further connected to main pipe line of centralized treatment plant. The waste collected from rural areas mainly consists of organic rather than few pesticide bottles etc. that should be segregated before sending the resources to centralized treatment system. The big solid and non-biodegradable waste leads to clogging and non-functioning of the treatment system. Therefore, segregating waste before feeding into treatment system is mandatory for sustainable operation.

(c) Centralized MFC system

The treatment should be centralized in order to achieve full operation and benefit the village in overall. The Bristol bioenergy center (BBiC) of Uganda developed a centralized treatment plant that powers the remote rural village "Kisoro" using wastewater to generate bioelectricity. The toilet block of Sesame girls' school present in the kisoro is also powered by bioelectricity generated from toilet waste. Urine and other types of wastewater are used to generate electricity by the centralized MFC developed by the BBiC.

(d) Continuous maintenance

In order for smooth operation of centralized MFC, continuous maintenance in terms of funding, man power, availability of substrate etc. is mandatory. The centralized MFC was continuously maintained by Pee Power in the 'Urine-tricity' project (funded by the Bill and Melinda Gates Foundation). With the phase III funds form the society, BBiCis developing a product from peepower and testing in the field trials at different locations of developing countries (http://www.brl.ac.uk/research-themes/bioenergyself-sustaining.aspx; https://wlvdigital.wordpress.com/2017/10/12/bioenergy-central-pee-power-lights-up-rural-uganda-as-microbial-fuel-cells-generate-electricity-from-waste).

14.6 Implementation in Rural Areas

In lab scale, MFCs that utilize waste/wastewaters have been successfully operated with adequate power generation. But, the technology has to be transferred on to practical implementation for its application in large scale. In the modern world there is huge waste generation day by day and are not reused and properly treated. Hence, to utilize those waste/wastewaters effectively for power generation in MFCs, the costs has to be shared by individual family or private sector involvement and further research on pilot scale operation in rural areas. Significant factors that contribute for continuous operation of pilot scale MFC in rural areas include optimization of operational parameters and best architectural design and selection of robust and efficient microbial inoculum which can degrade/metabolize the target pollutants present in the composite wastewaters. The varying composition of the heterogeneous wastewater that is fed as substrate will control the MFC operation (http://www.brl.ac.uk/research/researchthemes/bioenergyselfsustainable/scaleupmicrobialfuelcells.aspx).

Naturally selected electrochemically active bacteria and microalgae in anode and cathode chambers respectively, low cost and sustainable electrodes, minimizing the losses, etc. will increase the yields thereby making a way for practical implementation (Strik et al. 2008). Hence, more technological advancements are required to meet energy demand. Various industrial wastewaters/effluents that are produced in large volumes and contain high organic content will become the potential substrate for pilot scale MFC operation in a commercial level. The existing technology of treatment, conversion, separation and integration with other processes can be feasible when operated in a biorefinery approach. The syntrophic relationship of algae and bacteria in algae based MFCs will lead to postulation of mimicking natural systems for enhancing the overall power generation with simultaneous wastewater treatment (He et al., 2009). The integrated studies of anodic effluent as substrate to cathodic microalgae will have further advantages of zero liquid discharge (Zhang et al., 2011; Gude et al., 2013). The algal based MFCs can be implemented in the present infrastructure used to treat wastewater for additional advantage of biomass production for value addition (nutrient recovery, biofertilizer, biomass, feed or fodder, platform chemicals, etc.). The bioelectricity generation from the energy present in the organic content of wastewater will serve as a sustainable alternative to energy recovery balance. The integration of energy recovered from anode chamber and biomass generated at cathode chamber will have net gain of energy (Gonzalez, 2008; Gajda et al. 2015). As mentioned in the previous sections, the stack MFC can be a cost effective and efficient model to recover the energy. The miniature model MFC is feasible for minimizing the losses associated (Chouler et al. 2016). The utilization of eggshell membrane devices will help in reducing the internal resistance of the system and enhancing the cross sectional area and membrane spacing (Chouler et al. 2017). The MFC at pilot scale are utilized for generation of bioelectricity and biohydrogen, wastewater treatment and its application as biosensor. The bioreactor design and type should meet the rural area requirements in a sustainable way (Logan and Regan, 2006). To implement the MFCs in rural areas, the following schemes of loans from banks and government subsidies are required.

Fig. 14.4 Implementation in rural areas – through local community people

14.6.1 Loan from Banks and Easy Return Agreement

The rural local bodies (RLBs) can obtain loans from a private or public banks for treating wastewater and generate income out of it at a low interest rate. Some of such financial institutions are Housing and Urban Development Corporation Limited (HUDCO), Infrastructure Development Finance Company Limited (IDFC), Infrastructure Leasing & Financial Services Limited (IL&FS), National Bank for Agriculture and Rural Development (NABARD), Indian Renewable Energy Development Agency Limited (IREDA), Industrial Development Bank (TERI, 2015).

14.6.2 Government Schemes and Subsidies

The Government provides financial incentives and subsidies for "waste to energy" projects viz., biogas composters, treatment plants etc. The Ministry of New and Renewable Energy (MNRE, India) and TERI, India encourages both private and public sector companies to take part in this projects. The urban local bodies (ULBs) and state nodal agencies are provided with incentives for providing garbage free of cost and promotion and coordination of projects respectively. On an average, a subsidy of INR 15 million to INR 30 million per MW is given. Ministry of Environment Forest and Climate Change and Ministry of Agriculture provides a subsidy of up to 50% of capital cost of compost plants (TERI, 2015) (Fig.14.4).

14.7 Conclusion

Due to depletion of fossil fuels and energy crisis, researchers are encouraged to look for an alternate energy production from resources available in the world. Investigating the sustainable technologies for its successful implementation in rural areas is mandatory to meet the energy demand. The renewable source of substrates should be utilized as a perspective plan in rural areas for future.

Acknowledgement We thank the Department of Environmental Science and Engineering, CSIR-Central Leather Research Institute, Chennai, India and Department of Environmental Science and Engineering, Kyung Hee University, The Republic of Korea (Research grants from National Research Foundation of Korea (2015R1D1A1A09059935; 2018R1A2B6001507) and Korea-India S & T Co-operation Program (2016K1A3A1A19945953)) for constant support and encouragement for reviewing this chapter for the benefit of the society.

References

Ahn Y, Logan BE, (2012) A multi-electrode continuous flow microbial fuel cell with separator electrode assembly design. Applied Microbiology and Biotechnology 93 (5):2241–2248

Badwal SPS, Giddey SS, Munnings C, Bhatt AI, Hollenkamp AF (2014) Emerging electrochemical energy conversion and storage technologies. Frontiers in Chemistry. 2: 79.

Butti SK, Velvizhi G, Sulonen MLK, Haavisto JM, Koroglu EO, Cetinkaya AY, Singh S, Arya D, Annie Modestra J, Vamsi Krishna K, Verma A, Ozkaya B, Lakaniemi A-M, Puhakka JA, Venkata Mohan S (2016) Microbial electrochemical technologies with the perspective of harnessing bioenergy: maneuvering towards upscaling. Renew Sust Energ Rev 53:462–476

Cha J, Choi S, Yu H, Kim H, Kim C (2010) Directly applicable microbial fuel cells in aeration tank for wastewater treatment. Bioelectrochemistry 78:72–79

Cheng S, Logan BE (2011) Increasing power generation for scaling up single-chamber air cathode microbial fuel cells. Bioresour Technol 102:4468–4473

Choi J, Ahn Y (2013) Continuous electricity generation in stacked air cathode microbial fuel cell treating domestic wastewater. J Environ Manag 130:146–152

Chouler J, Bentley I, Vaz F, O'Fee A, Cameron PJ, Lorenzo MD (2017) Exploring the use of cost-effective membrane materials for microbial fuel cell based sensors. Electrochim Acta 231:319–326

Chouler J, Padgett GA, Cameron PJ, Preuss K, Titirici M-M, Ieropoulos I, Lorenzo MD (2016) Towards effective small scale microbial fuel cells for energy generation from urine. Electrochim Acta 192:89–98

Clauwaert P, Verstraete W (2009) Methanogenesis in membraneless microbial electrolysis cells. Appl Microbiol Biotechnol 82:829–836

Cui Y, Rashid N, Hu N, Rehman MSU, Han J-I (2014) Electricity generation and microalgae cultivation in microbial fuel cell using microalgae-enriched anode and bio-cathode. Energy Convers Manag 79:674–680

Das MP (2015) Bioelectricity production using algae in microbial fuel cell. Der Pharma Chemica 7(11):8–10

Dekker A, TerHeijne A, Saakes M, Hamelers HVM, Buisman CJN (2009) Analysis and improvement of a scaled-up and stacked microbial fuel cell. Environ Sci Technol 43:9038–9042

Electric Power Research Institute (2002) U.S. water consumption for power production—the next half century; water and sustainability, Vol. 3. www.epri.com

Eom H, Chung K, Kim I, Han JI (2011) Development of a hybrid microbial fuel cell (MFC) and fuel cell (FC) system for improved cathodic efficiency and sustainability: the M2FC reactor. Chemosphere 85:672–676

Feng C, Tsai C-C, Ma C-Y, Yu C-P, Hou C-H (2017) Integrating cost-effective microbial fuel cells and energy-efficient capacitive deionization for advanced domestic wastewater treatment. Chem Eng J 330:1–10

Feng Y, Lee H, Wang X, Liu Y, He W (2010) Continuous electricity generation by a graphite granule baffled air–cathode microbial fuel cell. Bioresour Technol 101:632–638

Feng Y, Wang X, Logan BE, Lee H (2008) Brewery wastewater treatment using air-cathode microbial fuel cells. Appl Microbiol Biotechnol 78:873–880

Feng YJ, He WH, Liu J, Wang X, Qu YP, Ren NQ (2014) A horizontal plug flow and stackable pilot microbial fuel cell for municipal wastewater treatment. Bioresour Technol 156:132–138

Foley JM, Rozendal RA, Hertle CK et al (2010) Life cycle assessment of high-rate anaerobic treatment, microbial fuel cells, and microbial electrolysis cells. Environ Sci Technol 44:3629–3637

Gajda I, Greenman J, Melhuish C, Ieropoulos I (2015) Self-sustainable electricity production from algae grown in a microbial fuel cell system. Biomass Bioenergy 82:87–93

Gong Y, Radachowsky SE, Wolf M, Nielsen ME, Girguis PR, Reimers CE (2011) Benthic microbial fuel cell as direct power source for an acoustic modem and seawater oxygen/temperature sensor system. Environ Sci Technol 45(11):5047–5053

Gonzalez C, Marciniak J, Villaverde S, Leon C, Garcia PA, Munoz R (2008) Efficient nutrient removal from swine manure in a tubular biofilm photo-bioreactor using algae-bacteria consortia. Water Sci Technol 58:95–102

Greenman J, Gálvez A, Giusti L, Ieropoulos I (2009) Electricity from landfill leachate using microbial fuel cells: comparison with a biological aerated filter. Enzym Microb Technol 44:112–119

Gude VG, Kokabian B, Gadhamshetty V (2013) Beneficial bioelectrochemical systems for energy, water, and biomass production. J Microb Biochem Technol 6:2

Guo K, Hassett DJ, Gu T (2012) Microbial fuel cells: Electricity generation from organic wastes by microbes. In: Chapter 9 in Microbial Biotechnology: Energy and Environment. edited by R. Arora, CAB International, Oxon, United Kingdom. ISBN 978-1845939564, pp 162–189

He L, Du P, Chen Y, Lu H, Cheng X, Chang B, Wang Z (2017) Advances in microbial fuel cells for wastewater treatment. Renew Sust Energ Rev 71:388–403

He Z, Kan J, Mansfeld F, Angenent LT, Nealson KH (2009) Self-sustained phototrophic microbial fuel cells based on the synergistic cooperation between photosynthetic microorganisms and heterotrophic bacteria. Environ Sci Technol 43:1648–1654

Heilmann J, Logan BE (2006) Production of electricity from proteins using a microbial fuel cell. Water Environ Res 78:531–537

http://www.brl.ac.uk/research/researchthemes/bioenergyselfsustainable/scaleupmicrobialfuelcells.aspx

http://www.brl.ac.uk/researchthemes/bioenergyself-sustaining.aspx

https://wlvdigital.wordpress.com/2017/10/12/bioenergy-central-pee-power-lights-up-rural-uganda-as-microbial-fuel-cells-generate-electricity-from-waste

Hu X, Liu B, Zhou J, Jin R, Qiao S, Liu G (2015) CO2 fixation, lipid production, and power generation by a novel air lift-type microbial carbon capturefbioelectricity: photosynthetic microbial e cell system. Environ Sci Technol 49:10710–10717

Hu X, Zhou J, Liu B (2016) Effect of algal species and light intensity on the performance of an air-lift-type microbial carbon capture cell with an algae-assisted cathode. RSC Adv 6:25094–25100

Ieropoulos I, Greenman J, Melhuish C (2008) Microbial fuel cells based on carbon veil electrodes: stack configuration and scalability. Int J Energy Res 32:1228–1240

Kondaveeti S, Kakarla R, Kim HS, Kim B, Min B (2018) The performance and long-term stability of low cost separators in single chamber bottle type microbial fuel cells. Environ Technol 39(3), 288–297

Kondaveeti S, Moon MM, Min B (2017) Optimum spacing between electrodes in an air-cathode single chamber microbial fuel cell with a low-cost polypropylene separator. Bioprocess Biosyst Eng 40(12), 1851–1858

Li H, Tian Y, Qu Y, Qiu Y, Liu J, Feng Y (2017) A pilot-scale benthic microbial electrochemical system (BMES) for enhanced organic removal in sediment restoration. Sci Rep 7:1–9

Li WW, Yu HQ, He Z (2014) Towards sustainable wastewater treatment by using microbial fuel cells-centered technologies. Energy Environ Sci 7:911–924

Liu H, Cheng S, Huang L et al (2008) Scale-up of membrane-free single-chamber microbial fuel cells. J Power Sources 179:274–279

Liu W, Cheng S (2014) Microbial fuel cells for energy production from wastewaters: the way toward practical application. Appl Phys Eng 15:841–861

Logan BE (2009) Exoelectrogenic bacteria that power microbial fuel cells. Nat Rev Microbiol 7:375–381

Logan BE (2010) Scaling up microbial fuel cells and other bioelectrochemical systems. Appl Microbiol Biotechnol 85(6):1665–1671

Logan BE, Call D, Cheng S, Hamelers HVM, Sleutels T, Jeremiasse AW, Rozendal RA (2008) Microbial electrolysis cells for high yield hydrogen gas production from organic matter. Environ Sci Technol 42:8630–8640

Logan BE, Regan JM (2006) Pennsylvania state university. Envir Sci Technol 40:5172–5180 American chemical society

Logan BE, Wallack MJ, Kim KY, He W, Feng Y, Saikaly PE (2015) Assessment of microbial fuel cell configurations and power densities. Environ Sci Technol Lett 2:206–214

Lu N, Zhou S-G, Zhuang L, Zhnag J-T, Ni J-R (2009) Electricity generation from starch processing wastewater using microbial fuel cell technology. Biochem Eng J 43:246–251

Malaeb L, Katuri KP, Logan BE, Maab H, Nunes SP, Saikaly PE (2013) A hybrid microbial fuel cell membrane bioreactor with a conductive ultrafiltration membrane biocathode for wastewater treatment. Environ Sci Technol 47:11821–11828

McCarty PL, Bae J, Kim J (2011) Domestic wastewater treatment as a net energy producer—can this be achieved? Environ Sci Technol 45:7100–7106

Min B, Angelidaki I (2008) Innovative microbial fuel cell for electricity production from anaerobic reactors. J Power Sources 180:641–647

Min B, Kim JR, Oh S, Regan JM, Logan BE (2005) Electricity generation from swine wastewater using microbial fuel cells. Water Res 39:4961–4968

Min B, Logan BE (2004) Continuous electricity generation from domestic wastewater and organic substrates in a flat plate microbial fuel cell. Environ Sci Technol 38:5809–5814

Nagendranatha Reddy C, Annie Modestra J, Kumar AN, Venkata Mohan S (2015) Waste remediation integrating with value addition: biorefinery approach towards sustainable bio-based technologies. In: Kalia VC (ed) Microbial factories, biofuels, waste treatment, vol 1. Springer India, New Delhi, pp 231–256

Ng F-L, Phang S-M, Periasamy V, Yunus K, Fisher AC (2017) Enhancement of power output by using alginate immobilized algae in biophotovoltaic devices. Sci Rep 7:16237. https://doi.org/10.1038/s41598-017-16530-y

Nielsen ME, Reimers CE, Stecher HA (2007) Enhanced power from chambered benthic microbial fuel cells. Environ Sci Technol 41:7895–7900

Pant D, Bogaert GV, Diels L, Vanbroekhoven K (2010a) A review of the substrates used in microbial fuel cells (MFCs) for sustainable energy production. Bioresour Technol 101:1533–1543

Pant D, Bogaert GV, Diels L, Vanbroekhoven K (2010b) A review of the substrates used in microbial fuel cells (MFCs) for sustainable energy production. Bioresour Technol 101(6):1533–1543

Park DH, Zeikus JG (2003) Improved fuel cell and electrode designs for producing electricity from microbial degradation. Biotechnol Bioeng 81:348–355

Rabaey K, Clauwaert P, Aelterman P, Verstraete W (2005) Tubular microbial fuel cells for efficient electricity generation. Environ Sci Technol 39:8077–8082

Rabaey K, Verstraete W (2005) Microbial fuel cells: novel biotechnology for energy generation. Trends Biotechnol 23:291–298

Rahimnejad M, Adhami A, Darvari S, Zirepour A, Oh SE (2015) Microbial fuel cell as new technology for bioelectricity generation: a review. Alex Eng J 54:745–756

Ramadan BS, Purwono (2017) Challenges and opportunities of microbial fuel cells (MFCs) technology development in Indonesia. MATEC web of conferences. 101, 02018

Raman K, Lan JC-W (2012) Performance and kinetic study of photo microbial fuel cells (PMFCs) with different electrode distances. Appl Energy 100:100–105

Reimers CE, Tender LM, Fertig S, Wang W (2001) Harvesting energy from the marine sediment-water interface. Environ Sci Technol 35:192–195

Renewable energy policy network for 21st century (REN21) Report. http://www.ren21.net/Portals/0/documents/activities/Topical%20Reports/REN21_10yr.pdf

Rittmann BE (2008) Opportunities for renewable bioenergy using microorganisms. Biotechnol Bioeng 100:203–212

Rodrigo MA, Cañizares P, García H, Linares JJ, Lobato J (2009) Study of the acclimation stage and of the effect of the biodegradability on the performance of a microbial fuel cell. Bioresour Technol 100:4704–4710

Rosenbaum M, He Z, Angenent LT (2010) Light energy to bioelectricity: photosynthetic microbial fuel cells. Curr Opin Biotechnol 21:259–264

Santoro C, Arbizzani C, Erable B, Ieropoulos I (2017) Microbial fuel cells: from fundamentals to applications. J Power Sources 356:225–244

Saratale GD, Saratale RG, Shahid MK, Zhen G, Kumar G, Shin H-S, Choi Y-G, Kim S-H (2017) A comprehensive overview on electro-active biofilms, role of exoelectrogens and their microbial niches in microbial fuel cells (MFCs). Chemosphere 178:534–547

Schroder U (2007) Anodic electron transfer mechanisms in microbial fuel cells and their energy efficiency. Phys Chem Chem Phys 7:2619–2629

Scott K, Murano C (2007) A study of a microbial fuel cell battery using manure sludge waste. J Chem Technol Biotechnol 82:809–817

Shizas I, Bagley DM (2004) Experimental determination of energy content of unknown organics in municipal wastewater streams. J Energ Eng 130:45–53

Sleutels THJA, TerHeijne A, Buisman CJN, Hamelers HVM (2012) Bioelectrochemical systems: an outlook for practical applications. Chem Sus Chem 5:1012–1019

Strik DPBTB, Terlouw H, Hamelers HVM, Buisman CJN (2008) Renewable sustainable biocatalyzed electricity production in a photosynthetic algal microbial fuel cell (PAMFC). Appl Microbiol Biotechnol 81:659–668

Strik D, Timmers R, Helder M, Steinbusch JJK, Hameters H, Buisman JN (2011) Microbial solar cells: applying photosynthetic and electrochemically active organisms. Trends Biotechnol 29:41–49

T E R I (2015) Industrial and urban waste management in India. The Energy and Resources Institute, New Delhi 34 pp

Tender LM, Gray SA, Groveman E, Lowy DA, Kauffman P et al (2008) The first demonstration of a microbial fuel cell as a viable power supply: powering a meteorological buoy. J Power Sources 179:571–575

Tommasi T, Lombardelli G (2017) Energy sustainability of Microbial Fuel Cell (MFC): a case study. J Power Sources 356:438–447

Ucar D, Zhang Y, Angelidaki I (2017) An overview of electron acceptors in microbial fuel cells. Front Microbiol 8:643

Velvizhi G, Venkata Mohan S (2017) Multi-electrode bioelectrochemical system for the treatment of high total dissolved solids bearing chemical based wastewater. Bioresour Technol 242:77–86

Velasquez-Orta SB, Curtis TP, Logan BE (2009) Energy from algae using microbial fuel cells. Biotechnol Bioeng 103(6):1068–1076

Venkata Mohan S, Srikanth S, Chiranjeevi P, Arora S, Chandra C (2014) Algal biocathode for in situ terminal electron acceptor (TEA) production: synergetic association of bacteria-microalgae metabolism for the functioning of biofuel cell. Bioresour Technol 166:566–574

Wang X, Feng Y, Liu J, Lee H, Li C, Li N, Ren N (2010) Sequestration of CO2 discharged from anode by algal cathode in microbial carbon capture cells (MCCs). Biosens Bioelectron 25:2639–2643

Wang X, Feng Y, Ren N, Wang H, Lee H, Li N, Zhao Q (2009) Accelerated start-up of two-chambered microbial fuel cells: effect of positive poised potential. Electrochem Acta 54:1109–1114

Wang YK, Sheng GP, Shi BJ et al (2013) A novel electrochemical membrane bioreactor as a potential net energy producer for sustainable wastewater treatment. Sci Rep 3:1864–1870

Wang Y-P, Liu X-W, Li W-W, Li F, Wang Y-K, Sheng G-P, Zeng RJ, Yu H-Q (2012) A microbial fuel cell–membrane bioreactor integrated system for cost-effective wastewater treatment. Appl Energy 98:230–235

Water Infrastructure Network (2001) Water infrastructure now: recommendations for clean and safe water in the 21st century. www.win-water.org/win_reports/pub2/winow.pdf

Xu GH, Wang YK, Sheng GP, Mu Y, Yu HQ (2014) An MFC-based online monitoring and alert system for activated sludge process. Sci Rep 4:6779

Yagishita T, Sawayama S, Tsukahara K, Ogi T (1997) Effects of intensity of incident light and concentrations of Synechococcus sp. and 2-hydroxy-1, 4-naphthoquinone on the current output of photosynthetic electrochemical cell. Sol Energy 61:347–353

Yuan Y, Chen Q, Zhou S, Zhuang L, Hu P (2012) Improved electricity production from sewage sludge under alkaline conditions in an insert-type air-cathode microbial fuel cell. J Chem Technol Biotechnol 87(1):80–86

Zhang Y, Angelidaki I (2012) A simple and rapid method for monitoring dissolved oxygen in water with a submersible microbial fuel cell (SBMFC). Biosens Bioelectron 38:189–194

Zhang Y, Liu L, Bruggen BV, Yang F (2017) Nanocarbon based composite electrodes and their application in microbial fuel cells. J Mater Chem A 5:12673–12698

Zhang Y, Noori JS, Angelidaki I (2011) Simultaneous organic carbon, nutrients removal and energy production in a photomicrobial fuel cell (PFC). Energy Environ Sci 4:4340–4346

Zhou M, Chi M, Luo J et al (2011) An overview of electrode materials in microbial fuel cells. J Power Sources 196:4427–4435

Zuo Y, Maness P-C, Logan BE (2006) Electricity production from steam-exploded corn stover biomass. Energy Fuel 20(4):1716–1721

Correction to: Future Perspectives on Cost-Effective Microbial Fuel Cells in Rural Areas

C. Nagendranatha Reddy, M. P. Sudhakar, Booki Min, and P. Shanmugam

Correction to Chapter 14 in: V. Sivasankar et al. (eds.), *Microbial Fuel Cell Technology for Bioelectricity*, https://doi.org/10.1007/978-3-319-92904-0_14

The correct affiliation of Dr. Sudhakar M. P, is Department of Environmental Science and Engineering, CSIR-Central Leather Research Institute, Chennai, 600020, India and not Department of Environmental Science and Engineering, Kyung Hee University, Yongin-si, 1732, Gyeonggi-do, Republic of Korea.

The correct affiliation of Dr. C. Nagendranatha Reddy, is Department of Environmental Science and Engineering, Kyung Hee University, Yongin-si, Gyeonggi-do, Republic of Korea and not Department of Environmental Science and Engineering, CSIR-Central Leather Research Institute, Chennai, 600020, India.

The updated online version of this chapter can be found at https://doi.org/10.1007/978-3-319-92904-0_14

V. Sivasankar et al. (eds.), *Microbial Fuel Cell Technology for Bioelectricity*,
https://doi.org/10.1007/978-3-319-92904-0_15

Index

V. Sivasankar et al. (eds.), *Microbial Fuel Cell Technology for Bioelectricity*,
https://doi.org/10.1007/978-3-319-92904-0

R

CPSIA information can be obtained
at www.ICGtesting.com
Printed in the USA
LVHW080050030919
629740LV00003B/36/P

9 783030 065423